METHODS IN MOLECULAR BIOLOGY

Series Editor
John M. Walker
School of Life and Medical Sciences
University of Hertfordshire
Hatfield, Hertfordshire, UK

For further volumes:
http://www.springer.com/series/7651

For over 35 years, biological scientists have come to rely on the research protocols and methodologies in the critically acclaimed *Methods in Molecular Biology* series. The series was the first to introduce the step-by-step protocols approach that has become the standard in all biomedical protocol publishing. Each protocol is provided in readily-reproducible step-by-step fashion, opening with an introductory overview, a list of the materials and reagents needed to complete the experiment, and followed by a detailed procedure that is supported with a helpful notes section offering tips and tricks of the trade as well as troubleshooting advice. These hallmark features were introduced by series editor Dr. John Walker and constitute the key ingredient in each and every volume of the *Methods in Molecular Biology* series. Tested and trusted, comprehensive and reliable, all protocols from the series are indexed in PubMed.

Capturing Chromosome Conformation

Methods and Protocols

Edited by

Beatrice Bodega

Istituto Nazionale di Genetica Molecolare "Romeo ed Enrica Invernizzi", Milan, Italy

Chiara Lanzuolo

Istituto Nazionale di Genetica Molecolare, Milan, Italy; IRCCS Fondazione Santa Lucia, Rome, Italy; Istituto di Tecnologie Biomediche – Consiglio Nazionale delle Ricerche, Milan, Italy

Humana Press

Editors
Beatrice Bodega
Istituto Nazionale di Genetica Molecolare "Romeo ed Enrica Invernizzi"
Milan, Italy

Chiara Lanzuolo
Istituto Nazionale di Genetica Molecolare
Milan, Italy

IRCCS Fondazione Santa Lucia
Rome, Italy

Istituto di Tecnologie Biomediche – Consiglio Nazionale delle Ricerche
Milan, Italy

ISSN 1064-3745 ISSN 1940-6029 (electronic)
Methods in Molecular Biology
ISBN 978-1-0716-0663-6 ISBN 978-1-0716-0664-3 (eBook)
https://doi.org/10.1007/978-1-0716-0664-3

This Humana imprint is published by the registered company Springer Science+Business Media, LLC, part of Springer Nature.
The registered company address is: 1 New York Plaza, New York, NY 10004, U.S.A.

Preface

Tridimensional chromatin conformation has emerged nowadays as a fundamental layer of epigenetic regulation of cell identity and plasticity. Indeed, the genome is spatially ordered in compartments and topologically associated domains, and the multiple levels of DNA folding generate contacts between different genomic regions that may involve chromatin loops formation. The field is evolving rapidly, together with a plethora of sophisticated novel technologies based on next generation sequencing and advanced microscopy to capture the higher order chromatin structure and its functional changes in cellular processes and diseases.

Many approaches have been developed to study 3D genome organization. Chromosome conformation-based technologies (3C, 4C, 5C, Hi-C, and derivatives) have been developed to study genome organization in fixed cells. Such approaches are based on the capability to capture the contact frequencies between genomic loci in physical proximity. Nevertheless, 3D interactions are dynamics in time and space and highly variable between individual cells, consisting in multiple and extensively heterogeneous interactions.

Therefore, this collection represents our updated comprehension of methods based on the evolution of chromosome conformation capture (3C) technique and other complementary approaches to dissect chromatin conformation with an emphasis on dissection of nuclear compartmentalization and visualization in imaging.

Milan, Italy

Beatrice Bodega
Chiara Lanzuolo

Contents

Contributors

LAURA ANTONELLI • *Institute for High Performance Computing and Networking, ICAR-CNR, Naples, Italy*
MONICA BALLARINO • *Department of Biology and Biotechnology Charles Darwin, Sapienza University of Rome, Rome, Italy*
FRÉDÉRIC BANTIGNIES • *Institute of Human Genetics, CNRS and University of Montpellier, Montpellier Cedex 5, France*
ANDREA BIANCHI • *IRCCS Fondazione Santa Lucia, Rome, Italy*
SILVIO BICCIATO • *Department of Life Sciences, University of Modena and Reggio Emilia, Modena, Italy*
MATHIEU BLANCHETTE • *School of Computer Science, McGill University, Montréal, QC, Canada*
BEATRICE BODEGA • *Istituto Nazionale di Genetica Molecolare "Romeo ed Enrica Invernizzi", Milan, Italy*
KERSTIN BYSTRICKY • *Laboratoire de Biologie Moléculaire Eucaryote, Centre de Biologie Inté grative (CBI), University of Toulouse, CNRS, UPS, Toulouse, France*
CHRISTOPHER J. F. CAMERON • *Department of Biochemistry and Rosalind and Morris Goodman Cancer Research Center, McGill University, Montréal, QC, Canada; School of Computer Science, McGill University, Montréal, QC, Canada*
DAVID CASTILLO • *CNAG-CRG, Centre for Genomic Regulation (CRG), Barcelona Institute of Science and Technology (BIST), Barcelona, Spain*
GIACOMO CAVALLI • *Institute of Human Genetics, CNRS and University of Montpellier, Montpellier Cedex 5, France*
ALICE CORTESI • *Istituto Nazionale di Genetica Molecolare "Romeo ed Enrica Invernizzi", INGM, Milan, Italy; IEO, Istituto Europeo di Oncologia, Milan, Italy*
KIM L. DE LUCA • *Hubrecht Institute, Royal Netherlands Academy of Arts and Sciences (KNAW), University Medical Center Utrecht and Oncode Institute, Utrecht, The Netherlands*
MARCO DI STEFANO • *CNAG-CRG, Centre for Genomic Regulation (CRG), Barcelona Institute of Science and Technology (BIST), Barcelona, Spain*
JOSÉE DOSTIE • *Department of Biochemistry and Rosalind and Morris Goodman Cancer Research Center, McGill University, Montréal, QC, Canada*
ZHIJUN DUAN • *Institute for Stem Cell and Regenerative Medicine, University of Washington, Seattle, WA, USA; Division of Hematology, Department of Medicine, University of Washington, Seattle, WA, USA*
STEPHANIE FANUCCHI • *Division of Chemical, Systems and Synthetic Biology, Faculty of Health Sciences, University of Cape Town, Cape Town, South Africa; Gene Expression and Biophysics Group, BTRI, CSIR Biosciences, Pretoria, South Africa*
IRENE FARABELLA • *CNAG-CRG, Centre for Genomic Regulation (CRG), Barcelona Institute of Science and Technology (BIST), Barcelona, Spain*
EZIO T. FOK • *Division of Chemical, Systems and Synthetic Biology, Faculty of Health Sciences, University of Cape Town, Cape Town, South Africa; Gene Expression and Biophysics Group, BTRI, CSIR Biosciences, Pretoria, South Africa*

MATTIA FORCATO • *Department of Life Sciences, University of Modena and Reggio Emilia, Modena, Italy*
MIKE N. GOODSTADT • *CNAG-CRG, Centre for Genomic Regulation (CRG), Barcelona Institute of Science and Technology (BIST), Barcelona, Spain*
FRANCESCO GREGORETTI • *Institute for High Performance Computing and Networking, ICAR-CNR, Naples, Italy*
KATHARINA LAURENCE JOST • *Department of Infectious Diseases, CIID, Integrative Virology, Heidelberg University Hospital and German Center for Infection Research, Heidelberg, Germany; German Center for Infection Research, Heidelberg, Germany*
NEZIH KARASU • *Institute of Genetics and Molecular and Cellular Biology (IGBMC), Illkirch, France; CNRS UMR7104, Illkirch, France; INSERM U1258, Illkirch, France; University of Strasbourg, Illkirch, France*
JOP KIND • *Hubrecht Institute, Royal Netherlands Academy of Arts and Sciences (KNAW), University Medical Center Utrecht and Oncode Institute, Utrecht, The Netherlands*
CHIARA LANZUOLO • *Istituto Nazionale di Genetica Molecolare, Milan, Italy; IRCCS Fondazione Santa Lucia, Rome, Italy; Istituto di Tecnologie Biomediche – Consiglio Nazionale delle Ricerche, Milan, Italy*
FEDERICA LO SARDO • *IRCCS Regina Elena National Cancer Institute, Rome, Italy*
BOJANA LUCIC • *Department of Infectious Diseases, CIID, Integrative Virology, Heidelberg University Hospital and German Center for Infection Research, Heidelberg, Germany; German Center for Infection Research, Heidelberg, Germany*
FEDERICA LUCINI • *Istituto Nazionale di Genetica Molecolare, Milan, Italy; Università degli Studi di Milano-Bicocca, Milan, Italy*
MARINA LUSIC • *Department of Infectious Diseases, CIID, Integrative Virology, Heidelberg University Hospital and German Center for Infection Research, Heidelberg, Germany; German Center for Infection Research, Heidelberg, Germany*
PAOLO MAIURI • *IFOM, the FIRC Institute of Molecular Oncology, Milan, Italy*
FEDERICA MARASCA • *Istituto Nazionale di Genetica Molecolare "Romeo ed Enrica Invernizzi", INGM, Milan, Italy*
MARC A. MARTI-RENOM • *CNAG-CRG, Centre for Genomic Regulation (CRG), Barcelona Institute of Science and Technology (BIST), Barcelona, Spain; Gene Regulation, Stem Cells and Cancer Program, Centre for Genomic Regulation (CRG), Barcelona, Spain; Universitat Pompeu Fabra (UPF), Barcelona, Spain; ICREA, Pg. Lluís Companys 23, Barcelona, Spain*
JULIE MARTONE • *Department of Biology and Biotechnology Charles Darwin, Sapienza University of Rome, Rome, Italy*
MUSA M. MHLANGA • *Division of Chemical, Systems and Synthetic Biology, Faculty of Health Sciences, University of Cape Town, Cape Town, South Africa; Gene Expression and Biophysics Unit, Instituto de Medicina Molecular, Faculdade de Medicina Universidade de Lisboa, Lisbon, Portugal*
PAULINA NASTAŁY • *IFOM, the FIRC Institute of Molecular Oncology, Milan, Italy*
GENNARO OLIVA • *Institute for High Performance Computing and Networking, ICAR-CNR, Naples, Italy*
DENIS PAQUETTE • *Department of Biochemistry and Rosalind and Morris Goodman Cancer Research Center, McGill University, Montréal, QC, Canada*
MICHEL PUCÉAT • *Aix-Marseille University, INSERM U-1251, MMG, Marseille, France*
JÉRÔME D. ROBIN • *Aix Marseille Univ, Marseille Medical Genetics MMG, INSERM U1251, Marseille, France*

Tiziana Santini • *Center for Life Nano Science@Sapienza, Istituto Italiano di Tecnologia, Rome, Italy*
Dana Segal • *Department of Biochemistry and Rosalind and Morris Goodman Cancer Research Center, McGill University, Montréal, QC, Canada*
François Serra • *CNAG-CRG, Centre for Genomic Regulation (CRG), Barcelona Institute of Science and Technology (BIST), Barcelona, Spain*
Tom Sexton • *Institute of Genetics and Molecular and Cellular Biology (IGBMC), Illkirch, France; CNRS UMR7104, Illkirch, France; INSERM U1258, Illkirch, France; University of Strasbourg, Illkirch, France*
Mia Stanic • *Department of Infectious Diseases, CIID, Integrative Virology, Heidelberg University Hospital and German Center for Infection Research, Heidelberg, Germany; German Center for Infection Research, Heidelberg, Germany; Department of Laboratory Medicine and Pathobiology, University of Toronto, MaRS Centre, University Ave., Toronto, ON, Canada*
Quentin Szabo • *Institute of Human Genetics, CNRS and University of Montpellier, Montpellier Cedex 5, France*
Xue Qing David Wang • *Department of Biochemistry and Rosalind and Morris Goodman Cancer Research Center, McGill University, Montréal, QC, Canada*
Julia Wegner • *Department of Infectious Diseases, CIID, Integrative Virology, Heidelberg University Hospital and German Center for Infection Research, Heidelberg, Germany; Institute for Clinical Chemistry and Clinical Pharmacology, Universitätsklinikum Bonn, Bonn, Germany*

Chapter 1

Capturing Chromosome Conformation

Michel Pucéat

Abstract

The genome is organized in 3D topology-associated domains to ensure proper gene transcriptional processes. The chromosome conformation capture (3C) is an affordable method to investigate local chromatin structure and dynamics in cells and tissue. Herein I describe an easy to design and a cost-effective protocol.

Key words Chromosome, Chromatin, Topology-associated domain

1 Introduction

Our genome is encoded in the nucleus by DNA, wrapped by proteins or chromatin including histones within a scaffold. The chromatin within the nucleus is packaged in a 3D space surrounded by the nuclear envelope in a very organized manner. Such a 3D configuration is built by the cohesin complex and the mediator [1] and allows for specific and dynamic intra- and inter-chromosomal interactions. Such an organization is instrumental for the regulation of gene transcription and more specifically for a proper function of an enhancer located far away from the gene promoter. This is thus a key feature of the nuclear organization which leads to a fine tuning of gene transcription [2] and cell fate decision during embryogenesis [3].

Among the first technologies available to look at 3D chromatin configuration, chromosome conformation capture was engineered [4]. The aim of this technology is to reveal the frequency of DNA–DNA contacts within the 3D nuclear space. Proteins and DNA of cells or tissues are cross-linked by formaldehyde in order to freeze the configuration of interacting chromatin regions. DNA is then cut using a frequent cutter restriction enzyme. Intramolecular ligation of cross-linked fragments is then carried out. Ligation products are subsequently analyzed by real-time PCR using primers

Beatrice Bodega and Chiara Lanzuolo (eds.), *Capturing Chromosome Conformation: Methods and Protocols*, Methods in Molecular Biology, vol. 2157, https://doi.org/10.1007/978-1-0716-0664-3_1,

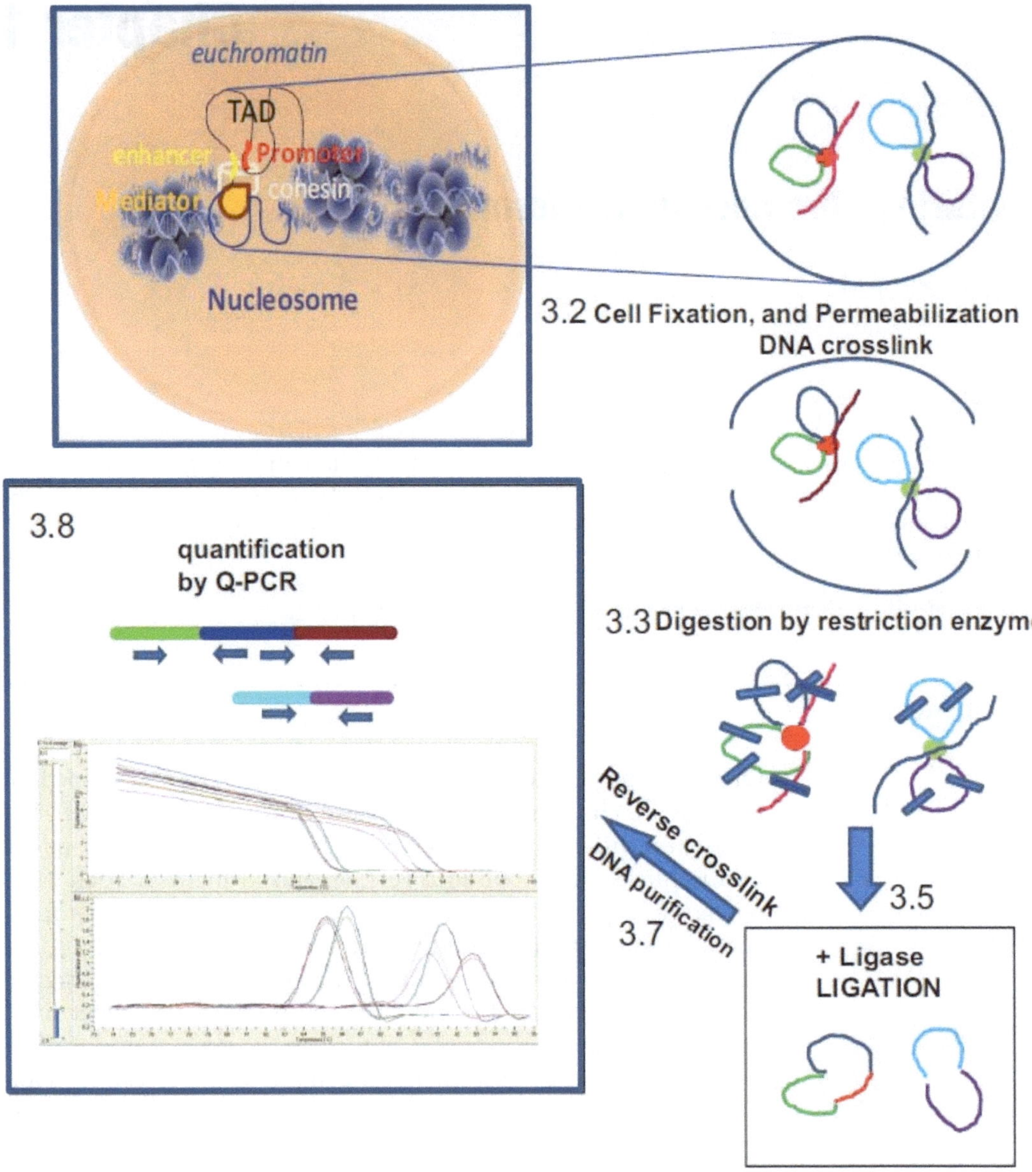

Fig. 1 Principle of the chromosome configuration capture. The numbers of each step refer to chapter sections

specific for the restriction fragments of interest (Fig. 1). The use of real-time quantitative PCR is required to ensure that the amplified interacting restriction fragments forming a 3D DNA loop are more abundant than just neighboring DNA fragments. Such a quantitative approach has been recently described [5].

The technology can be combined to chromatin precipitation assay (ChIP-3C) or Run and Cut technology [6] to specifically look at DNA loops at specific loci bound by a transcription factor as well as to high-throughput sequencing (4-C, 5-C, or Hi-C) [7].

The 3C-technology provides several advantages over other genome-wide technology approaches such as 4-C or 5-C or

Hi-C. It does face DNA sequencing resolution limitations, and it is more affordable and allows to map the DNA/DNA interactions within a locus of interest in a quantitative manner and with high resolution. The amount of material required is reasonable (i.e., a few hundreds of μg of chromatin) and much less than the amount required for 4-C or 5-C or Hi-C. Single cells or single-nucleus Hi-C has been recently reported [8, 9] but this remains a challenging technology with still poor resolution.

A guide has been recently published to help the newcomers in this field of research to better choose the technology that would better address their biological questions [10].

2 Materials

2.1 Solutions

1. 1 M glycine stock.
2. Permeabilization buffer: 15 mM HEPES pH 7.6, 15 mM NaCl, 4 mM $MgCl_2$, 60 mM KCl, 0.5% Triton X-100.
3. Lysis buffer: 50 mM Tris–HCl pH 8, 1% SDS; 10 mM EDTA. Add 2 μg/ml leupeptin, 1 μg/ml aprotinin, and 0.1 mM PMSF when ready to use.
4. 10× ligation buffer: 660 mM Tris HCl pH 7.5, 50 mM DTT, 50 mM $MgCl_2$, 10 mM ATP.
5. 10× universal restriction enzyme buffer.
6. 20% SDS w/v.
7. 10 mg/ml proteinase K (PKK).
8. 1 mg/ml aprotinin.
9. 1 mg/ml leupeptin.
10. 10 mM PMSF (phenylmethanesulfonyl fluoride).
11. 10 mg/ml RNase.

2.2 Reagents

1. 37% formaldehyde: always open in a hood.
2. SYBR Green.
3. Qubit™ 1X dsDNA HS Assay Kit (Invitrogen).

3 Methods

Carry all procedures at 4 °C unless otherwise specified.

3.1 Preparation of the Experiment

The first and important step is to choose the restriction enzyme that cuts within the locus of interest such that it allows for the analysis of expected regulatory regions (promoters, enhancers, intronic regions of the gene, etc.). The choice of non-sensitive methylation enzyme that generates cohesive ends to facilitate ligation depends upon the size of the locus that needs to be investigated. A small locus (up to 30 kb) requires the use of frequently cutting restriction enzymes such as DpnII (4-base cutters). A larger locus will require a 6-base cutter such as usual EcoRI, BglII, or HindIII. In any case, check that the restriction enzyme used will not cut the PCR fragments that you will amplify later.

3.2 Cell Harvesting and Cross-Linking of DNA

Harvest the cells or enzymatically dissociated tissues, wash twice with PBS at 4 °C, and cross-link in PBS added with 1% formaldehyde for 10 min at room temperature while shaking. Ten volumes of cross-linking buffer are required for one volume of biological material.

Stop the cross-linking process by adding 125 mM Glycine from 1 M stock, and incubate while shaking for 5 more minutes. Wash twice the cells with cold (4 °C) PBS by spinning them down for 5 min at 500 × *g*.

3.3 Cell and Nuclei Permeabilization

1. Permeabilize the cells by resuspending them for 10 min at 4 °C in permeabilization buffer.
2. Spin down the cells for 10 min at 500 × *g* at 4 °C.
3. Resuspend the cell pellet in lysis buffer and incubate on a rotating wheel at 4 °C for 10 min.
4. Spin down nuclei at 8000 × *g* for 10 min.

3.4 DNA Digestion

1. Resuspend the nuclei in 0.5 ml 1.2× RE buffer and add 7.5 μl SDS (20%).
2. Incubate for 1 h at 37 °C while shaking at 900 rpm.
3. Take 5 μl aliquot at −20 °C and label as undigested DNA (UND) to check digestion efficiency. Store at −20 °C.
4. Add 200 U of selected enzyme and incubate for 2 h at 37 °C and add again 200 U for an overnight digestion at 37 °C at 300 rpm. At this step also digest a few μg BAC (*see* **Note 1**) DNA of your region of interest to be used later for the standard curve of Q-PCR.

3.5 Second Day DNA Ligation

1. Take 5 μl aliquot and labeled digested genomic DNA (DIG).
2. Add 40 μl SDS 20% to the remaining sample (final concentration 1.6%).
3. Incubate 20 min at 65 °C (shake every 5 min) to stop the digestion.

3.6 DNA Ligation

1. Transfer to a 50 ml tube and add 6.125 ml 10× ligation buffer and 375 μl 20% Triton X-100 (final concentration 1%), and incubate 1 h at 37 °C with gentle shaking.
2. Add 5 μl ligase and incubate 4 h at 16 °C and overnight at 10 °C; at this step also ligate the digested BAC DNA of your region of interest.

3.7 Reverse Cross-Link

Add 30 μl 10 mg/ml proteinase K and incubate at 65 °C overnight.

3.8 Third Day DNA Extraction

1. Add 30 μl 10 mg/ml RNase for 45 min at 37 °C.
2. Add 7 ml phenol-chloroform and mix. Spin down 15 min at RT.
3. Transfer supernatant in new 50 ml tube and add 7 ml H_2O, 1 ml 3 M Na acetate pH 5.6.
4. Add 35 ml 100% ethanol (stored at −20 °C).
5. Place the tube at −80 °C for 1 h.
6. Spin 15 min at 1500 × *g* at 4 °C.
7. Take out the supernatant and dry the DNA pellet.
8. Dissolve DNA in Tris 10 mM pH 7.5 (up to 100 μl).
9. Quantify the DNA using SYBR Green. Use a 10,000× dilution of SYBR Green in TE buffer. Adjust the concentration to 25 ng/ml.

3.9 Real-Time QPCR

1. DNA quantification: We use a LightCycler 1.5 (Roche). Make dilution of pure DNA standard down to 0.1 ng/0.3 ng/1 ng/3 ng/10 ng per μl H_2O; use 10 μl diluted SYBR Green and 1 μl DNA standard or 1 μl of your sample ligated DNA. Run the LightCycler using the fluorimeter mode.

   Alternatively, use a Qubit™ 1× dsDNA HS Assay Kit if a Qubit™ fluorimeter is available.
2. Process to Real-Time PCR Using SYBR Green (*see* **Notes 2** and **3**).

   Use the digested (DIG) BAC DNA of your region of interest to make a standard curve by diluting $1/10^{-2}$ down to $1/10^{-10}$ into the digested DNA at 25 ng/ml PCR the chimerical product with the anchor primer (site of DNA–DNA interaction) and more distant primers. This will allow you to calculate the efficiency of each primer set. Use GAPDH primers as a loading DNA control.

   $$\text{Q Ligated} = 10^{((\text{standard curve intercept ligated} - \text{Ct ligated sample}/-\text{slope ligated}))}$$

   $$\text{QC} = 10^{((\text{standard curve intercept } C - \text{Ct Csample}/-\text{slope } C))}$$

   $$\text{Normalized quantification} = \text{QL}/\text{QC}.$$
3. Standard Curve for Digestion Efficiency

   Use from Subheading 3.8 UND and from 4.1 digested (DIG) DNA. Make dilutions of digested DNA from 20 ng/μl down to 1.25 ng/ μl.

Then PCR the standard curve and 2 ng of UND and digested DNA for all couple of primers and for loading control C primers.

Digestion efficiency = $(1 - (qR_{DIG}/qC_{DIG})/(qR_{UND}/qC_{UND}) \times 100$

$qR = 10^{((\text{standard curve intercept R} - \text{CtR sample})/-\text{slope R})\ \text{primer R}}$

$qC = 10^{((\text{standard curve intercept } C - \text{CtC sample})/-\text{slope } C)\ \text{primer C}}$

To further learn about alternative protocol: *see* **Notes 4** and **5**.

4 Notes

1. To choose the BAC DNA, go to the genome browser (http://genome.ucsc.edu); select the species genome (mouse, human, etc.), and then go to the locus of interest by entering the name of a gene.

   In the mapping and sequencing module below the image of the locus on the viewer, select clone ends and refresh the screen. To configure clone ends, right-click on the left border of the image. This will display all the clones available from different BAC libraries.
2. Careful analysis of QPCR using appropriate controls must be designed to discriminate background interactions from physiologically relevant interactions [4]. Background interactions can be checked by correlating interaction frequencies to linear distance.
3. The efficiency of primers in QPCR could bias the contact frequency. Thus the efficiency should be carefully evaluated for each primer set by calculating the slope of a standard curve [11].
4. Validation of some interactions revealed by 3C can be validated using FISH [12].
5. Novel high-resolution technologies based on FISH and super-resolution imaging are emerging to investigate the 3D scaffold of the genome [13]. To address the emerging question of cell heterogeneity in any tissue or cell populations in culture, besides the challenging single cells or nucleus Hi-C, new technologies based on microfluidics and microscopy are proposed such as multiplex chromatin interaction visualization in single cells [14].

References

1. Merkenschlager M, Nora EP (2016) CTCF and Cohesin in genome folding and transcriptional gene regulation. Annu Rev Genomics Hum Genet 17:17–43
2. Rowley MJ, Corces VG (2018) Organizational principles of 3D genome architecture. Nat Rev Genet 19(12):789–800

3. Moore-Morris T et al (2018) Role of epigenetics in cardiac development and congenital diseases. Physiol Rev 98(4):2453–2475
4. Dekker J (2006) The three 'C' s of chromosome conformation capture: controls, controls, controls. Nat Methods 3(1):17–21
5. Ea V, Court F, Forne T (2017) Quantitative analysis of intra-chromosomal contacts: the 3C-qPCR method. Methods Mol Biol 1589:75–88
6. Skene PJ, Henikoff S (2017) An efficient targeted nuclease strategy for high-resolution mapping of DNA binding sites. Elife 6:e21856
7. van de Werken HJ et al (2012) Robust 4C-seq data analysis to screen for regulatory DNA interactions. Nat Methods 9(10):969–972
8. Flyamer IM et al (2017) Single-nucleus hi-C reveals unique chromatin reorganization at oocyte-to-zygote transition. Nature 544 (7648):110–114
9. Nagano T et al (2017) Cell-cycle dynamics of chromosomal organization at single-cell resolution. Nature 547(7661):61–67
10. Grob S, Cavalli G (2018) Technical review: a Hitchhiker's guide to chromosome conformation capture. Methods Mol Biol 1675:233–246
11. Pfaffl MW (2001) A new mathematical model for relative quantification in real-time RT-PCR. Nucleic Acids Res 29:e45
12. Abboud N et al (2015) A cohesin-OCT4 complex mediates Sox enhancers to prime an early embryonic lineage. Nat Commun 6:6749
12. Bintu B et al (2018) Super-resolution chromatin tracing reveals domains and cooperative interactions in single cells. Science 362:6413
13. Zheng M et al (2019) Multiplex chromatin interactions with single-molecule precision. Nature 566(7745):558–562

Chapter 2

The Chromosome Conformation Capture (3C) in *Drosophila melanogaster*

Federica Lo Sardo

Abstract

The discovery of the DNA double helix by Watson and Crick in 1953 was the first report showing that the genomic information is not contained in a stretched linear molecule. After that, a huge advance in the knowledge of the structure of the eukaryotic genome in the nuclear space has been made over the last decades, bringing us to the widely accepted concept that the genome is packaged into hierarchical levels of higher-order three-dimensional structures. The spatial organization of the eukaryotic genome has direct influence on fundamental nuclear processes that include transcription, replication, and DNA repair. The idea that structural alterations of chromosomes may cause disease goes back to the early nineteenth century. Big effort has been devoted to the study of the three-dimensional architecture of the genome and its functional implications. In this chapter, I will describe the chromosome conformation capture (3C), one of the first techniques used to detect and measure the frequency of interactions between genomic sequences that are kept in spatial proximity in the nucleus.

Key words Chromatin architecture, DNA loops, Chromosome conformation capture, *Drosophila melanogaster*

1 Introduction

In higher eukaryotes, including *Drosophila melanogaster* and mammals, the genome is composed of thousands of megabases and can occupy a space of some meters if completely stretched out. In the cell, the genome is packaged into hierarchical levels of higher-order three-dimensional structures, for structural and functional reasons. Structurally, the genome needs to be contained into a nuclear space bearing a 6–10 μm diameter.

Functionally, the three-dimensional organization of the genome is precisely regulated in time and space and contributes to specific transcriptional programs during the cell life. In the same cell type, the genomic structure needs to be maintained through several cell divisions to preserve cell identity and tissue homeostasis. Moreover, in cycling cells, the strict control of the

Beatrice Bodega and Chiara Lanzuolo (eds.), *Capturing Chromosome Conformation: Methods and Protocols*, Methods in Molecular Biology, vol. 2157, https://doi.org/10.1007/978-1-0716-0664-3_2,

three-dimensional organization of genomic DNA is necessary for proper chromosome segregation and maintenance of genome integrity through several cell divisions. During cell commitment or reprogramming, as well as upon environmental stress, the three-dimensional structure of the genome changes following a precise and reproducible program [1–3]. The dysregulation of the mechanisms regulating chromatin architecture is involved in numerous diseases [4, 5]. In this scenario, advancements in the study of the genome architecture provided an important tool for better understanding the complexity of genome function in space and time as well as in various diseases.

The first level of genomic three-dimensional organization is the nucleosome-based chromatin fiber, a mixture of DNA, histones, and non-histone proteins (Fig. 1a). Additional folding of the

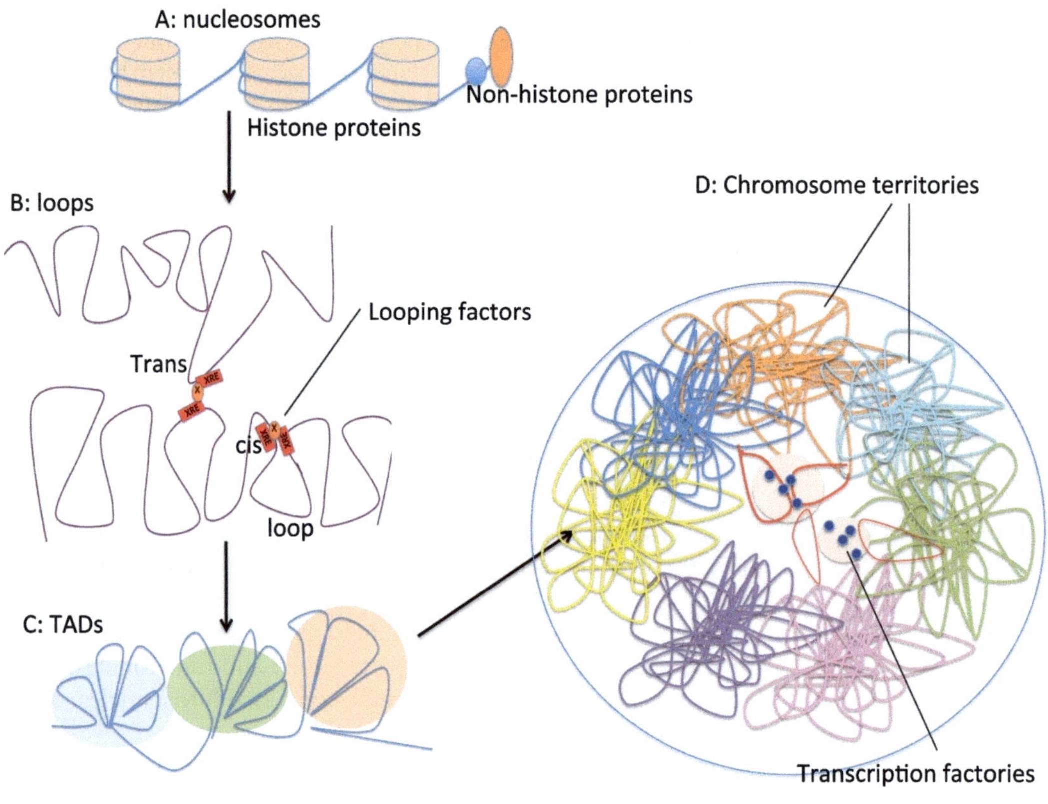

Fig. 1 Schematic representation of the hierarchical organization of chromatin into high-order structures. (**a**) Nucleosome-based chromatin fiber, with histone and non-histone proteins. (**b**) Loops, interactions between distant genomic sequences on the same chromosome (in cis) or on different chromosomes (in trans). (**c**) TADs (topologically associated domains). (**d**) Representation of the interphase chromatin which is thought to be organized into chromosome territories (multicolored areas). Active genes (in red) loop out from chromosome territories and are transcribed in transcription factories (light orange) regions of active transcription where there is a high concentration of ribonucleoproteic complexes of the transcription machinery. It has been shown that co-regulated genes share the same transcription factory

chromatin fibers into chromatin loops can bring into proximity genomic sites that are distant in the linear DNA, both on the same chromosome (in cis) and in different chromosomes (in trans) (Fig. 1b). Groups of DNA sequences that physically interact with each other can form topologically associated domains (TADs) [6] (Fig. 1c). The formation of chromatin loops and TADs is facilitated by looping factors (polycomb complexes, CTCF, cohesins) and lncRNA molecules, recruited at specific regulative genomic sites (enhancers, repressors/insulators) and the process is evolutionarily conserved between insect and mammalian cells [7–9]. Chromosomes are localized nonrandomly in the nuclear three-dimensional space and form chromosomal compartments and chromosome territories. The relative position of chromosomal compartments and of chromosome territories is cell type and tissue-specific and contributes to the specification of cell-specific transcriptional program [10]. At present, the mechanisms that maintain chromosome territories are unknown, while what is known is that these mechanisms are evolutionarily conserved [11]. The localization of specific gene promoters in proximity of specific regulative elements and into specific chromatin domains results in their expression state (Fig. 1d).

Drosophila melanogaster is a model organism used for studying the organization and functional relevance of 3D genome structure, owing to its relatively small genome and the availability of many genetic tools [12, 13].

The present chapter is aimed to describe the chromosome conformation capture (3C), a technique proposed for the first time in 2002 by Job Dekker [14] and widely used to:

1. Detect sequences that are engaged in physical interaction or are in spatial proximity in the nucleus.
2. Provide a measure of the frequency of their association.

With a resolution in the range of several Kb, the 3C technique is usually focused on the detection of loops in relatively small regions (up to 1 Mb). In some situations, these regions are screened for the presence of interactions with specific sequences considered as "baits." In other cases, 3C is aimed to detect and measure changes in the frequency of interaction between sequences already shown to interact in different cell types or experimental conditions.

Briefly, the procedure can be divided into five steps (Fig. 2):

1. Chemical cross-linking of chromosomes to covalently link chromatin segments that are in spatial proximity (Fig. 2a).
2. Digestion or fragmentation of the genomic DNA into small pieces (Fig. 2b).
3. Ligation of linked DNA fragments under diluted conditions in order to favor intramolecular ligation over intermolecular ligation (Fig. 2c).

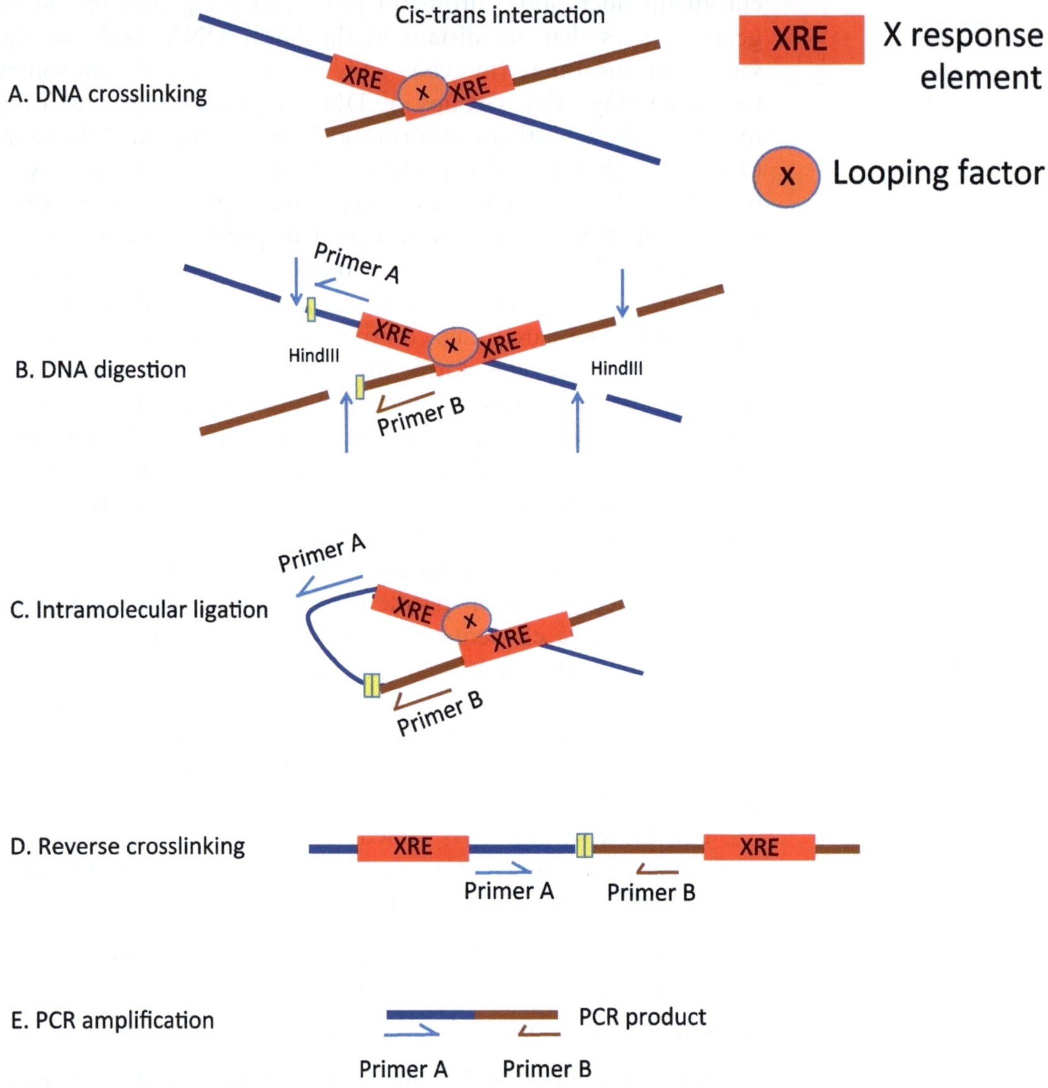

Fig. 2 Schematic representation of the main steps of chromosome conformation capture. For simplicity, XRE (general X response elements) are shown, with X factor (general looping factor) binding to XRE and mediating the formation of loops

4. Reverse of cross-linking (Fig. 2d).
5. Detection and quantification of ligated products (Fig. 2e).

2 Materials

2.1 Reagents

1. ATP.
2. Bovine serum albumin.
3. 10% SDS.

4. 20 mg/ml proteinase K: dissolve proteinase K in deionized water. Make aliquots and store at −20 °C.
5. 10 mg/ml RNase.
6. Protease inhibitors.
7. 37% formaldehyde.
8. 2.5 M glycine: dissolve in deionized water and filter. Use fresh within few weeks.
9. 10% Triton X-100 (dissolve in water, protect from light).
10. Restriction enzyme (*see* **Note 3**).
11. T4 DNA ligase.
12. Phenol-chloroform-isoamyl alcohol 25:24:1.
13. 100% ethanol.
14. 70% ethanol.
15. 3 M NaAc pH 5.2.

2.2 Buffers

1. 1× lysis buffer: 10 mM Tris pH 8.0, 10 mM NaCl, 0.2% NP40, protease inhibitors. Make fresh and store at room temperature.
2. Restriction enzyme buffer (specific for each enzyme, *see* **Note 3**).
3. 10× ligation buffer: 500 mM Tris–HCl pH 7.5, 100 mM $MgCl_2$, 100 mM dithiothreitol, autoclaved water. Store 20 ml and 1 ml aliquots at −80 °C, and dilute to 1× just before use with sterile water.

2.3 Equipments

1. Water bath.
2. Thermomixer.
3. Centrifuge.
4. Microcentrifuge.

3 Methods

Carry out all procedures at room temperature unless otherwise specified.

3.1 Cell Fixation and Lysis

1. Fix 10^7 cells with 1% formaldehyde added in the cell culture medium for 10 min at room temperature shaking (*see* **Note 1**).
2. Use also one tube with10^7 cells to be treated without cross-linking as a control (*see* **Note 2**).
3. Quench cross-linking by the addition of glycine to a final concentration of 0.125 M, and shake 5 min at room temperature.

4. Store on ice for at least 15 min.
5. Centrifuge cross-linked cells at 2000 rpm for 5–10 min.
6. Resuspend in 1 ml of 4 °C cold 1× cell lysis buffer and incubate on ice for 15 min.
7. Complete the lysis with ten strokes using a Dounce homogenizer (pestle A) on ice.
8. Wash nuclei with 0.5 ml of restriction enzyme buffer.
9. Pellet nuclei at 2000 rpm for 10 min.

3.2 Digestion of Cross-Linked DNA

1. Resuspend nuclei in 362 μl of restriction enzyme buffer (*see* **Note 3**).
2. Add SDS to a final concentration of 0.1%.
3. Incubate nuclei at 37 °C for 15 min.
4. Add Triton X-100 to the final concentration of 1%. Mix by pipetting up and down, avoiding the formation of bubbles (*see* **Note 4**).
5. Digest DNA with 400U of restriction enzyme at 37 °C for 1.5 h.
6. Inactivate the restriction enzyme by adding SDS to 2% and incubating at 65 °C for 30 min.

3.3 Ligation and Reverse Cross-Linking

1. Transfer reaction to 15 ml plastic tubes.
2. Dilute digested chromatin into 8 ml of 1× ligation reaction buffer with the addition of 1% Triton X-100, 0.1 mg ml^{-1} bovine serum albumin, 1 mM ATP, and 4000U of T4 DNA ligase.
3. Incubate at 16 °C for 2 h.
4. Add EDTA to a concentration of 10 mM to stop the reaction.
5. Add 500 μg of proteinase K and incubate for 5 h at 50 °C.
6. Incubate overnight at 65 °C to reverse the formaldehyde cross-links.

3.4 DNA Purification

1. Transfer solutions to sterile polypropylene tubes and extract once with 4 ml of p alcohol 25:24:1 (vortex and then spin 10 min at 10,000 rpm).
2. Transfer the aqueous phase to fresh tubes.
3. Precipitate DNA by addition of 0.8 ml 3 M NaAc pH 5.2 and 20 ml 100% ethanol.
4. Incubate at −80 °C for 20 min.
5. Centrifuge at 4000 rpm for 60 min at 10,000 rpm for 20 min to pellet DNA.
6. Wash with 70% ethanol.

7. Air-dry.
8. Resuspend pellets in 50–100 μl water with1 μl of 20 mg/ml RNase.

3.5 Detection of Ligation Products

Perform quantitative PCR with the designed primers using as a template the cross-linked genome and the control library (*see* **Notes 2** and **5–7**).

4 Notes

1. Fixation conditions need to be optimized in different cell types and tissues and standardized for experiment reproducibility. An appropriate fixation will correctly capture looping interactions without false positive or negative.
2. Control library: different primer pairs can amplify the target template with different efficiencies. In order to avoid misinterpretation of interaction frequencies because of different amplification efficiencies, use the same primer pairs to amplify a control library (prepared by digesting and randomly ligating non-cross-linked purified DNA, *see* Subheading 3.1, **step 2**). In this way, you can obtain information on differences in amplification efficiencies and consider this variable in the evaluation of interaction frequencies measured in the cross-linked genome. The relative interaction frequency of a pair of loci is calculated by dividing the amount of PCR product obtained with the 3C ligation product library by the amount of PCR product obtained with the control library.
3. Choice of restriction enzyme.

   The choice of restriction enzyme determines the resolution of the experiments. In general, enzymes that recognize and cut 6 bp sites are recommended such as EcoRI, HindIII, BglII, and NcoI that were successfully used in previous works [15–17]. Such enzymes will cut the genome approximately once every 4 kb. If desired, after a first analysis, a higher resolution analysis can be achieved with a restriction enzyme that recognizes 4 bp sites, which cuts on average every 256 base pairs. Enzymes that make sticky ends are recommended, as these ends are more efficiently ligated with respect to blunt ends.
4. SDS is a ionic denaturing detergent and can be disruptive to many protein-protein interactions and inhibit enzymatic reactions. Therefore, it is common to dilute out SDS as well as "sequester" it with non-ionic detergents, such as Triton X-100.
5. Primer design: in order to avoid false positive, design unidirectional primers annealing on the same strand so that all pairs of primers will amplify ligation products that are the result of

head-to-head ligation. Design primers ~80–150 bp away from the restriction cut site (Fig. 2b–d, see the orientation of primer A and primer B).

6. Signal-to-noise ratio: 3C signals typically decay with genomic distance. This is the reason why 3C analyses are limited to regions up to 1 Mb.
7. Control primers: design primers amplifying one or more genomic regions where long-range interactions are assumed to be the same in the different cell types or experimental conditions (i.e., two regions of a housekeeping gene previously observed to strongly interact each other).

References

1. Bonev B, Mendelson Cohen N, Szabo Q, Fritsch L, Papadopoulos GL, Lubling Y, Xu X, Lv X, Hugnot JP, Tanay A, Cavalli G (2017) Multiscale 3D genome rewiring during mouse neural development. Cell 171 (3):557–572. e524. https://doi.org/10.1016/j.cell.2017.09.043
2. Narendra V, Bulajic M, Dekker J, Mazzoni EO, Reinberg D (2016) CTCF-mediated topological boundaries during development foster appropriate gene regulation. Genes Dev 30 (24):2657–2662. https://doi.org/10.1101/gad.288324.116
3. Di Carlo V, Mocavini I, Di Croce L (2019) Polycomb complexes in normal and malignant hematopoiesis. J Cell Biol 218(1):55–69. https://doi.org/10.1083/jcb.201808028
4. Winick-Ng W, Rylett RJ (2018) Into the fourth dimension: Dysregulation of genome architecture in aging and Alzheimer's disease. Front Mol Neurosci 11:60. https://doi.org/10.3389/fnmol.2018.00060
5. Krumm A, Duan Z (2019) Understanding the 3D genome: emerging impacts on human disease. Semin Cell Dev Biol 90:62–77. https://doi.org/10.1016/j.semcdb.2018.07.004
6. Szabo Q, Jost D, Chang JM, Cattoni DI, Papadopoulos GL, Bonev B, Sexton T, Gurgo J, Jacquier C, Nollmann M, Bantignies F, Cavalli G (2018) TADs are 3D structural units of higher-order chromosome organization in drosophila. Sci Adv 4(2):eaar8082. https://doi.org/10.1126/sciadv.aar8082
7. Lanzuolo C, Roure V, Dekker J, Bantignies F, Orlando V (2007) Polycomb response elements mediate the formation of chromosome higher-order structures in the bithorax complex. Nat Cell Biol 9(10):1167–1174. https://doi.org/10.1038/ncb1637
8. Matharu N, Ahituv N (2015) Minor loops in major folds: enhancer-promoter looping, chromatin restructuring, and their association with transcriptional regulation and disease. PLoS Genet 11(12):e1005640. https://doi.org/10.1371/journal.pgen.1005640
9. Wang Q, Sun Q, Czajkowsky DM, Shao Z (2018) Sub-kb hi-C in *D. melanogaster* reveals conserved characteristics of TADs between insect and mammalian cells. Nat Commun 9 (1):188. https://doi.org/10.1038/s41467-017-02526-9
10. Szalaj P, Plewczynski D (2018) Three-dimensional organization and dynamics of the genome. Cell Biol Toxicol 34(5):381–404. https://doi.org/10.1007/s10565-018-9428-y
11. Tanabe H, Muller S, Neusser M, von Hase J, Calcagno E, Cremer M, Solovei I, Cremer C, Cremer T (2002) Evolutionary conservation of chromosome territory arrangements in cell nuclei from higher primates. Proc Natl Acad Sci U S A 99(7):4424–4429. https://doi.org/10.1073/pnas.072618599
12. Li Q, Tjong H, Li X, Gong K, Zhou XJ, Chiolo I, Alber F (2017) The three-dimensional genome organization of *Drosophila melanogaster* through data integration. Genome Biol 18(1):145. https://doi.org/10.1186/s13059-017-1264-5
13. Fontanillas P, Hartl DL, Reuter M (2007) Genome organization and gene expression shape the transposable element distribution in the *Drosophila melanogaster* euchromatin. PLoS Genet 3(11):e210. https://doi.org/10.1371/journal.pgen.0030210
14. Dekker J, Rippe K, Dekker M, Kleckner N (2002) Capturing chromosome conformation. Science 295(5558):1306–1311. https://doi.org/10.1126/science.1067799

15. Lanzuolo C, Orlando V (2007) The function of the epigenome in cell reprogramming. Cell Mol Life Sci 64(9):1043–1062. https://doi.org/10.1007/s00018-007-6420-8
16. Lo Sardo F, Lanzuolo C, Comoglio F, De Bardi M, Paro R, Orlando V (2013) PcG-mediated higher-order chromatin structures modulate replication programs at the drosophila BX-C. PLoS Genet 9(2):e1003283. https://doi.org/10.1371/journal.pgen.1003283
17. Duan Z, Andronescu M, Schutz K, Lee C, Shendure J, Fields S, Noble WS, Anthony Blau C (2012) A genome-wide 3C-method for characterizing the three-dimensional architectures of genomes. Methods 58(3):277–288. https://doi.org/10.1016/j.ymeth.2012.06.018

Chapter 3

4C-Seq: Interrogating Chromatin Looping with Circular Chromosome Conformation Capture

Nezih Karasu and Tom Sexton

Abstract

Chromosome conformation capture and its variants have allowed chromatin topology to be interrogated at a superior resolution and throughput than by microscopic methods. Among the method derivatives, 4C-seq (circular chromosome conformation capture, coupled to high-throughput sequencing) is a versatile, cost-effective means of assessing all chromatin interactions with a specific genomic region of interest, making it particularly suitable for interrogating chromatin looping events. We present the principles and procedures for designing and implementing successful 4C-seq experiments.

Key words Circular chromosome conformation capture, Genome topology, Chromatin loops, Chromatin fixation, Restriction digestion, Ligation, Inverse PCR, High-throughput sequencing

1 Introduction

The eukaryotic genome is highly compacted to be contained within the nucleus, yet is also exquisitely spatially organized to allow tight control of DNA transcription, replication, and repair [1]. Nearly 20 years ago, our understanding of chromatin architecture was revolutionized by the development of the chromosome conformation capture (3C) method [2]. Briefly, the method involves fixation of chromatin in its native state (primarily using formaldehyde), followed by restriction digestion of the fixed chromatin and subsequent re-ligation of the cut chromatin under conditions favoring intramolecular ligation. This generates hybrid DNA restriction fragments comprised of sequences which may have been very far away from one another in linear distance, but which were physically proximal at the time of fixation and therefore "captured" as ligation events. Initially, 3C used PCR to assess specific chromatin interactions in a "one-to-one" manner, but derivatives of this technique have improved the throughput, particularly when coupled with next-generation sequencing. Chromosome conformation capture

Beatrice Bodega and Chiara Lanzuolo (eds.), *Capturing Chromosome Conformation: Methods and Protocols*,
Methods in Molecular Biology, vol. 2157, https://doi.org/10.1007/978-1-0716-0664-3_3,

versions have now been developed for "one-to-all" (4C-seq [3]), "many-to-many" (5C [4], ChIA-PET [5], Capture-HiC [6]), and "all-to-all" (Hi-C [7]) applications; *see* [8] for an overview of these different techniques. The best choice of technique to use depends on the biological question being asked. In general, the methods allowing detection of the greatest repertoire of chromatin interactions require the highest numbers of sequences to obtain comparable resolutions to lower-throughput techniques. Due to its potential to measure chromatin interactions at single restriction fragment resolution with only ~1–2 million reads [3], 4C-seq is a popular method to assess chromatin looping interactions, such as those reported between gene promoters and distal regulatory elements like enhancers [9, 10], or CTCF-mediated architectural loops [11, 12], in a relatively unbiased manner. The canonical "one-to-all" method, 4C-seq, entails the restriction digestion and re-ligation steps of a 3C, followed by redigestion of the purified DNA with a second enzyme and re-ligation under dilute conditions to obtain circular DNA fragments [13] (Fig. 1a). All of the chromatin interactions with a specific restriction fragment of interest (the "bait") can then be amplified by inverse PCR with specific primers; the addition of compatible adapters directly to these primers allows the products to be directly assessed by high-throughput sequencing. Typically, 4C-seq results are visualized as plots of read counts mapping to the non-bait fragments, interpreted as a proxy of interaction strength, against their genomic coordinates (Fig. 1b). In line with polymer physics models, the interaction frequency decays as a power law with increasing genomic distance between the bait and non-bait fragments [2, 7]; chromatin looping events are thus detected as "peaks" of signal where the interaction is stronger than with regions that are closer on the linear chromosome fiber to the bait.

In addition to being a powerful technique to detect specific looping events, dramatic losses of 4C-seq interaction signal on crossing transition points have also been used to identify chromatin domain boundaries (in particular, topologically associated domains (TADs) of highly self-interacting regions) and interrogate their changes during development or on genetic perturbation [10, 14, 15]. However, the dependence of 4C-seq on one or a few bait viewpoints can give skewed information on domains, and techniques such as 5C or Hi-C will give a more comprehensive idea of domain borders. A major limitation of 4C-seq in its classical form is that each bait-non-bait ligation product generates an identical sequence to be read, so there is no means to distinguish PCR amplification biases from true interaction frequencies. To overcome this limitation, analyses are usually performed on sliding windows of fragments, also limiting the resolution of the assay, and/or multiple replicates of 4C-seq experiments are sequenced, which also increases confidence in the called interactions [16]. Recently, the variant UMI (unique molecular identifier)-4C was also developed to completely remove PCR duplication biases [17], at the cost

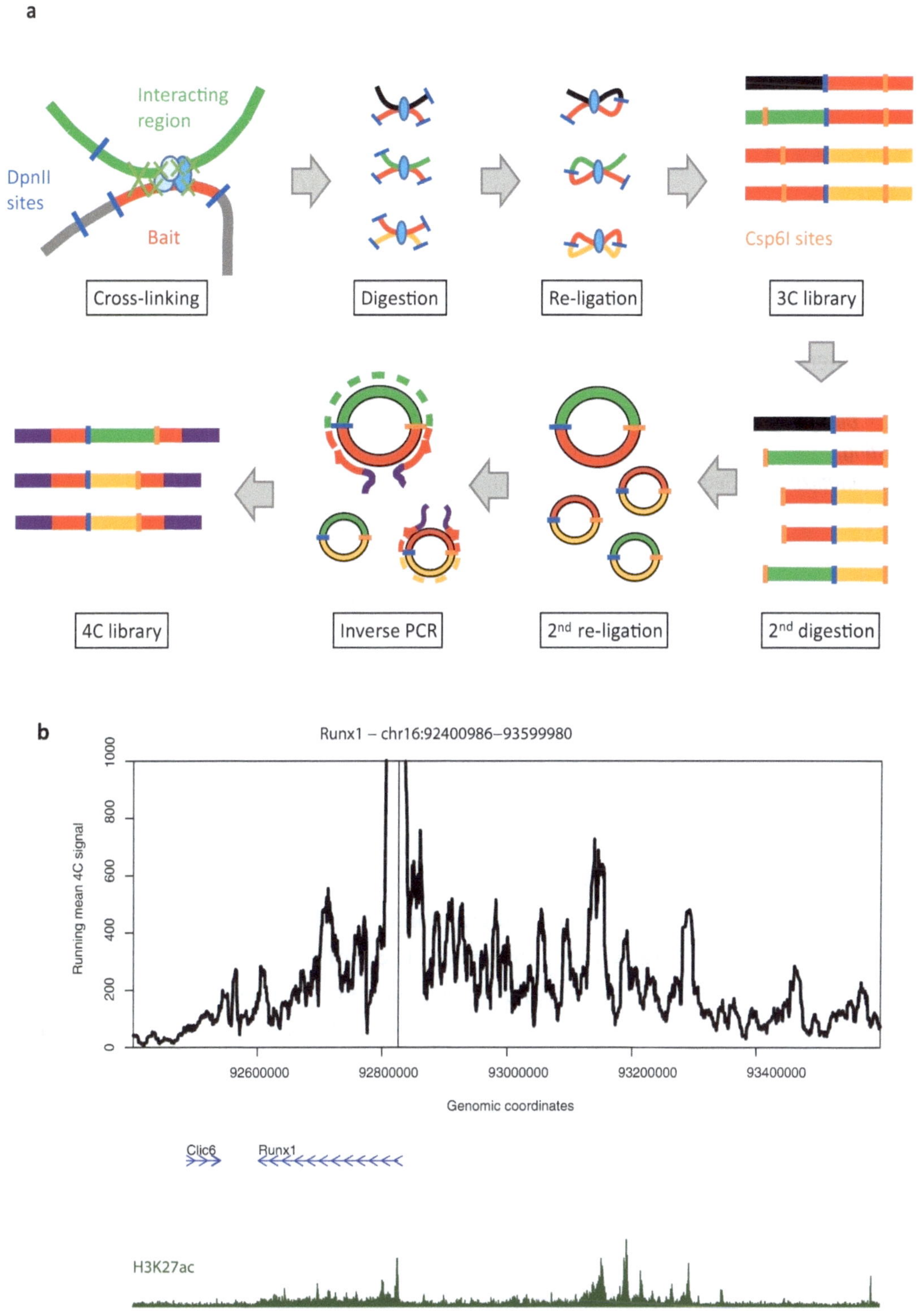

Fig. 1 Overview of the 4C-seq method. (**a**) Schematic workflow of the 4C-seq method. Chromatin interactions are fixed, including those between the bait (red) and other sequences, before digestion and re-ligation. The 3C library is redigested (secondary sites denoted in orange) and circularized, before inverse PCR with primers containing bait-complementary (red) and Illumina adapter (purple) sequences. The PCR products are directly loaded on a high-throughput sequencing machine. (**b**) Representative 4C-seq profile for the *Runx1* promoter bait, performed on the thymocyte cell line P5424, generated in the group. The running mean of sequence

of requiring more expensive paired-end sequencing and recovering fewer "analyzable" reads from the experiment. A major technical hurdle of 4C-seq, described in detail in this chapter, is the requirement and optimization of efficient and specific primers for the inverse PCR step. We present the 4C-seq protocol, applied to the mouse T cell line P5424 [18], with the promoter of the gene *Runx1* as an example, and describe how to adapt the method to other cell types and baits.

2 Materials

2.1 Designing, Testing, and Optimizing 4C-Seq Primers

1. Purified genomic DNA from cells or tissue of the species of interest.
2. 1.5 mL LoBind Eppendorf tubes.
3. LoBind guarded pipette tips.
4. Nuclease-free water.
5. Buffer B (10× stock, ThermoFisher # BB5).
6. *Csp*6I (10 U/μL, ThermoFisher #ER0211).
7. Temperature-controlled shaker, with adapters for 1.5 mL microcentrifuge tubes.
8. Molecular biology grade phenol/chloroform/isoamyl alcohol (25:24:1), pH 8.
9. Isopropanol.
10. 3 M sodium acetate, pH 5.2.
11. 70% ethanol.
12. Low TE buffer: 10 mM Tris–HCl pH 8, 0.1 mM EDTA.
13. Qubit fluorometer (ThermoFisher).
14. Qubit dsDNA BR assay kit (ThermoFisher).
15. Gel loading dye 6× (NEB #7024S).
16. Electrophoresis equipment and 1% TAE agarose gel.
17. T4 ligase buffer (10× stock, NEB # B0202S).
18. T4 DNA Ligase (2000 U/μL, NEB #M0202M).
19. Roche Expand Long Template PCR system.
20. 5 mM dNTP mix.
21. PCR tubes.
22. Thermal cycler.

Fig. 1 (continued) counts (25 fragment window) is plotted against genomic coordinate. Gene positions (blue arrows) and a ChIP-seq profile for histone H3 lysine-27 acetylation in thymocytes (dark green; data taken from [26]) are shown below the profile. The position of the bait is highlighted in red, and interactions with putative enhancers, based on histone acetylation, are highlighted in yellow

2.2 4C-Seq

1. T75 cell culture grade flask.
2. Culture medium (e.g., DMEM).
3. 15 mL Falcon tubes.
4. PBE buffer: 1× PBS, 0.5% BSA, 2 mM EDTA.
5. LoBind guarded tips.
6. 16% fresh formaldehyde.
7. Rocker.
8. 1 M glycine.
9. 1× PBS.
10. Lysis buffer: 10 mM Tris–HCl 1 M pH 8, 100 mM NaCl, 0.2% IGEPAL CA-630 (*aka* NP-40, Merck # I3021), 1× protease inhibitor cocktail (Roche).
11. 1.5 mL LoBind Eppendorf tubes.
12. 20 mg/mL BSA.
13. NEB DpnII buffer (10× stock, NEB #B0543S).
14. Nuclease-free water.
15. 20% SDS.
16. Temperature-controlled shaker, with adapters for 1.5 mL microcentrifuge tubes.
17. 20% Triton X-100.
18. *Dpn*II (50 U/μL, NEB #R0543M).
19. 100 mM Tris–HCl, pH 8.
20. 20 mg/mL proteinase K.
21. T4 ligase buffer (10× stock, NEB # B0202S).
22. T4 DNA Ligase (2000 U/μL, NEB #M0202M).
23. 10 mg/mL RNase A.
24. Molecular biology grade phenol/chloroform/isoamyl alcohol (25:24:1), pH 8.
25. Isopropanol.
26. 3 M sodium acetate, pH 5.2.
27. 70% ethanol.
28. Low TE buffer: 10 mM Tris–HCl pH 8, 0.1 mM EDTA.
29. Qubit fluorometer (ThermoFisher).
30. Qubit dsDNA BR assay kit (ThermoFisher).
31. Gel loading dye 6× (NEB #7024S).
32. Electrophoresis equipment and 1% TAE agarose gel.
33. Buffer B (10× stock, ThermoFisher # BB5).
34. *Csp*6I (10 U/μL, ThermoFisher #ER0211).

35. Roche Expand Long Template PCR system.
36. 5 mM dNTP mix.
37. PCR tubes.
38. Thermal cycler.
39. AMPure beads.
40. DynaMag-2 magnet.
41. Agilent BioAnalyzer Tape station 2000.
42. Agilent High Sensitivity DNA kit.
43. Illumina HiSeq 2500 or 4000.

3 Methods

3.1 Designing, Testing, and Optimizing 4C-Seq Primers

The design of the inverse PCR primers is a crucial technical step of a 4C-seq experiment, since this dictates the specificity and amplification efficiency of the bait-linked ligation products that are sequenced. In turn, the exact genomic regions where primers can be designed depend on the availability and location of restriction sites within the bait locus. The resolutions of all 3C-based assays are limited to restriction fragments, so we recommend the use of frequently cutting enzymes with four base-pair recognition motifs for both the primary and secondary digestion steps. In order to use the same 4C material to assess interactions with multiple baits, it is convenient to try and use the same enzymes for all baits, although the experimental constraints described below may hamper such a design (Fig. 2a). Our group nearly always uses *Dpn*II as the primary enzyme and *Csp*6I as the secondary enzyme, as in the worked example here. Once primers are designed, they are tested on mock 4C material, comprising genomic DNA that is digested with the secondary enzyme and re-ligated to make circularized 4C templates. Inverse PCR on this material should amplify only one specific product, corresponding to the restriction fragment contiguous with the bait (Fig. 2b).

1. Using software for cloning design (e.g., SnapGene, Serial Cloner), obtain a restriction map for different combinations of four-cutter restriction enzymes, such as *Dpn*II and *Csp*6I in this example (*see* **Note 1**). Find a restriction fragment as close as possible to the bait of interest satisfying these conditions: *Dpn*II-*Dpn*II fragment between 500 bp and 1500 bp and more than 300 bp between *Dpn*II site used as bait and the closest *Csp*6I site (*see* **Note 2** and Fig. 2a).
2. Using Primer3 [19, 20] (http://primer3.ut.ee/), design inverse PCR primers with the following constraints: GC content between 30 and 70%, primers' length between 18 and

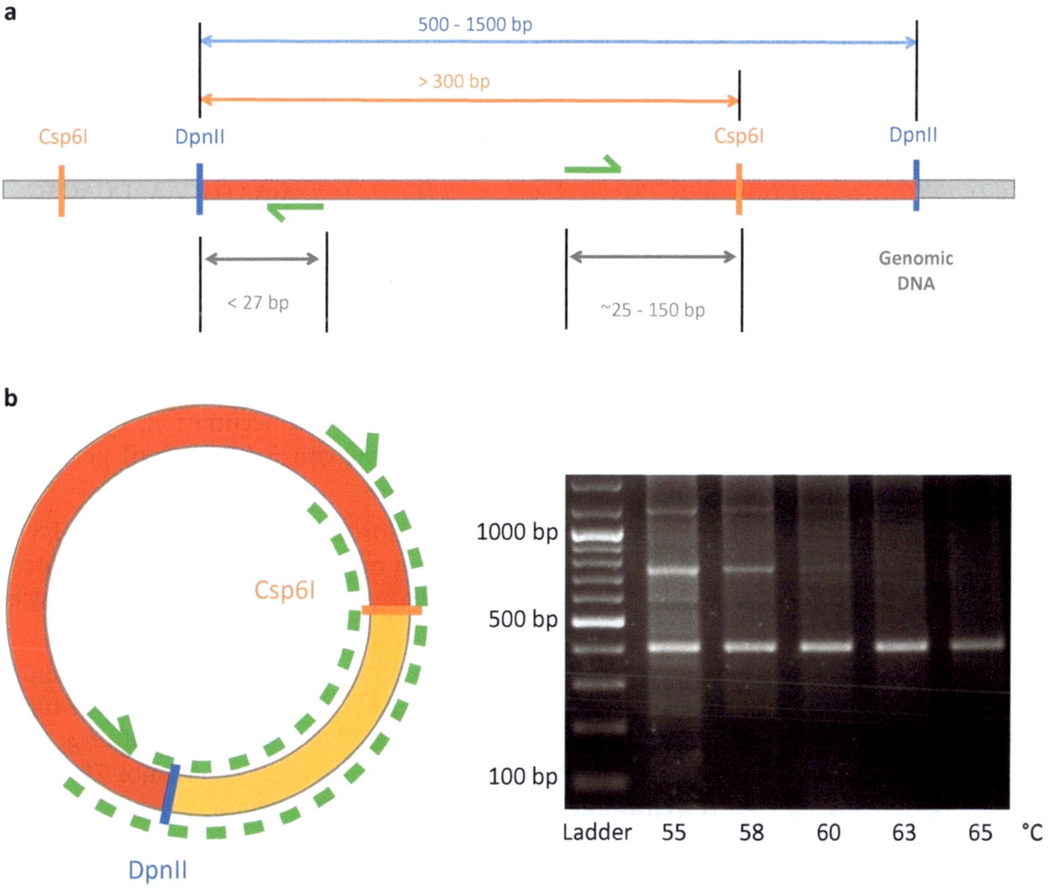

Fig. 2 Design and assessment of 4C-seq primers. (**a**) Schematic of the distance constraints for required bait restriction map and primer locations (green arrows). The bait *Dpn*II-*Csp*6I fragment is denoted in red, and the contiguous fragment amplified from circularized genomic DNA is denoted in yellow. (**b**) Schematic of amplified contiguous region (left) and TAE agarose gel (right) of products after PCR at different annealing temperatures. In this case, annealing at 65 °C generates a specific product of the expected size (*see* Table 1 for primer sequences)

23 bp, maximum melting temperature of 65 °C, and a difference of 3 °C between the two primers; the primer facing the *Dpn*II site (reading primer) is a maximum distance of 27 bp from the restriction site; the other, non-reading primer is a maximum distance of 150 bp from the *Csp*6I site (*see* **Note 3** and Fig. 2a).

3. Check the specificity of these primers in silico by mapping to the reference genome with BLAST (BLASTN with near match sensitivity, http://www.ensembl.org/Multi/Tools/Blast?db=core). Primers with extra off-target hits can be tolerated, as long as there is a minimum of a two-nucleotide mismatch at the 3′ end.

Table 1
Primers used in this example 4C-seq experiment

Reading primer	Non-reading primer
5′ -*AATGATACGGCGACCACCGAGATCTACACTCTTTCCCTACACGACGCTCTTCCGATCT*∗**GGATTTGGTGGCTTTCAGAT**—3′	5′ -*CAAGCAGAAGACGGCATACGAGCTCTTCCGATCT*∗**GAATGCAAACCACAGGCTTT**—3′

Underlined and italicized sequence denotes Illumina adapter sequence; bold sequence denotes sequence complementary to bait. ∗ denotes where unique barcode sequences can be inserted, if performing the 4C-seq on multiple cell types or conditions

4. Dilute ~50 μg of genomic DNA to 100 ng/μL using nuclease-free water and buffer B at 1× final concentration. Incubate at 37 °C and 600 rpm overnight with 5 U of *Csp*6I per μg of genomic DNA.
5. Add 1 volume of phenol/chloroform/isoamyl alcohol pH 8 to 1 volume of digestion reaction, mix by inversion, and vortex well. Centrifuge at 15,500 × *g*, 23 °C and for 5 min. Transfer the aqueous phase to a new tube.
6. Add 0.9 volumes of isopropanol and 0.1 volume of sodium acetate pH 5.2, and then mix by inversion. Incubate the tubes at −80 °C for 30 min and then centrifuge at 15,500 × *g*, 4 °C for 30 min. Wash the pellet with 500 μL ethanol 70%, and re-centrifuge at 15,500 × *g*, 4 °C for 5 min.
7. Remove supernatant and resuspend the pellet in 50 μL low TE buffer. Quantify purified digested genomic DNA with the Qubit dsDNA BR assay, following the manufacturer's instructions (*see* **Note 4**).
8. Run an aliquot (~250 ng) on 1% TAE agarose gel to confirm digestion (*see* Fig. 3). Dilute to 5 ng/μL with T4 ligase buffer (1× final) and add 200 U of ligase per μg of DNA. Incubate at 16 °C and 400 rpm overnight (*see* **Note 5**).
9. Add 1 volume of phenol/chloroform/isoamyl alcohol pH 8 for 1 volume of digestion reaction, mix by inversion, and vortex well. Centrifuge at 15,500 × *g*, 23 °C and for 5 min. Transfer the aqueous phase to a new tube.
10. Add 0.9 volumes of isopropanol and 0.1 volume of sodium acetate pH 5.2, and then mix by inversion. Incubate the tubes at −80 °C for 30 min and then centrifuge at 15,500 × *g*, 4 °C for 30 min. Wash the pellet with 500 μL ethanol 70%, and re-centrifuge at 15,500 × *g*, 4 °C for 5 min.
11. Remove supernatant and resuspend the pellet in 50 μL low TE buffer. Quantify purified 4C mock template with the Qubit dsDNA BR assay, following the manufacturer's instructions (*see* **Note 4**).

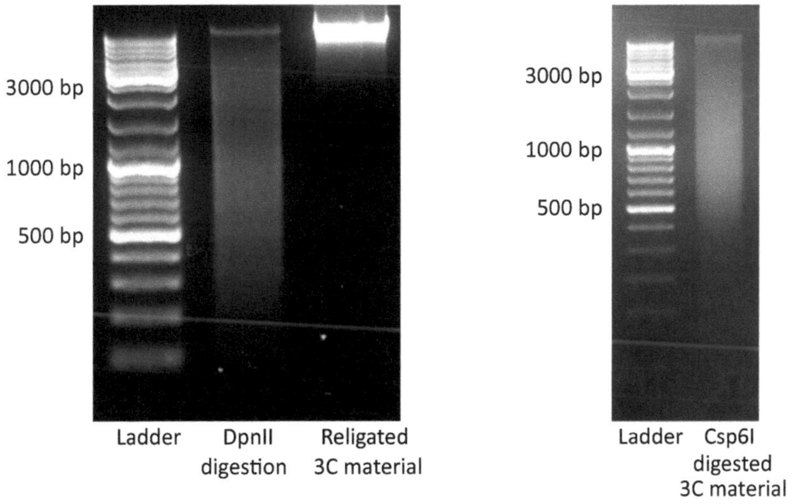

Fig. 3 4C-seq digestion and ligation tests. TAE agarose gel photos showing the profile of the processed material at different steps over the 4C-seq experiment: *Dpn*II digestion and ligation (left) and *Csp*6I redigestion (right). The digested products should form a smear, whose size depends on the restriction enzyme. Re-ligated product should have a tight band >10 kb

12. Test primers using the following composition per PCR reaction: 2.5 μL Roche Expand Long Template Buffer 1 (10×), 0.5 μL dNTP (10 mM), 0.25 μL of each 4C primer to be tested (each at 100 μM), 50 ng 4C mock template DNA, 1 μL Taq Longrange, and H_2O to a final volume of 25 μL. Perform the following thermal cycle program: 2 min at 94 °C, 30× (15 s at 94 °C, 1 min at temperature gradient (55–65 °C), 1 min at 68 °C), and 7 min at 68 °C (*see* **Note 6** and Fig. 2b).
13. Add 5 μL of 6× loading dye after amplification and load half of the reaction on a 1.5% TAE agarose gel (*see* **Note 6** and Fig. 2b).
14. Once appropriate primers and PCR conditions have been found, repeat **steps 4–14** with new primers containing Illumina sequences at their 5′ ends (*see* **Note 7**).

3.2 4C-Seq

Once optimal 4C-seq primers and PCR conditions have been found for all baits of interest, the inverse PCR can now be applied to real 4C template. This protocol describes 4C-seq in the mouse T cell line P5424.

1. Transfer material from one T75 flask of confluent P5424 cells (~20 mL at ~1 million cells/mL) evenly to two 15 mL Falcon tubes, and centrifuge at 200 × *g*, 23 °C, 5 min. Remove supernatant, resuspend in 5 mL PBE buffer, and pool to one Falcon tube (*see* **Notes 8** and **9**).

2. Centrifuge cells at 200 × *g*, 23 °C, 5 min. Carefully remove supernatant, and resuspend the pellet slowly by pipetting up and down in 3.5 mL of culture medium without serum at 23 °C. Add 500 μL 16% formaldehyde and incubate on a rocker for 10 min at 23 °C (*see* **Note 10**).
3. Add 500 μL 1 M cold glycine and put on ice for 5 min. Centrifuge at 200 × *g*, 4 °C, 3 min.
4. Remove supernatant and wash by resuspending in 10 mL cold PBS and centrifuging at 800 × *g*, 4 °C, 3 min.
5. Remove supernatant and resuspend the pellet well in 10 mL lysis buffer. Put on ice for 1 h, occasionally mixing the tubes by inverting.
6. Split nuclei at five million nuclei per LoBind Eppendorf tube and spin down at 3000 × *g*, 4 °C, 4 min. Remove supernatant.
7. Layer the pellet with 500 μL 1.25× DpnII buffer. Centrifuge at 1500 × *g*, 4 °C, 2 min and remove supernatant (*see* **Note 11**).
8. Resuspend pellet with 250 μL 1.25× DpnII buffer. Add 10 μL 20% SDS, mix well by pipetting, and incubate at 65 °C, 1000 rpm for 20 min, and then at 37 °C, 1000 rpm for 40 min (*see* **Note 12**).
9. Add 100 μL 20% Triton X-100, mix well, and incubate at 1000 rpm, 37 °C, 1 h (*see* **Note 12**).
10. Add 20 μL of 50 U/μL DpnII, mix well, and incubate overnight at 37 °C, 800 rpm.
11. Take a 50 μL aliquot for the digestion test and add 450 μL 100 mM Tris–HCl pH 8 and 5 μL 20 mg/mL proteinase K. Incubate overnight at 65 °C, 750 rpm (*see* **Note 13** and Fig. 3).
12. For the rest of the digested nuclei, centrifuge at 1500 × *g*, 4 °C, 5 min, and remove supernatant (*see* **Note 14**).
13. Resuspend in 800 μL 1× T4 ligase buffer and centrifuge at 1500 × *g*, 4 °C, 2 min.
14. Remove supernatant and resuspend pellet in 400 μL 1× T4 ligase buffer. Add 10 μL high concentrated T4 ligase. Incubate at 16 °C, 400 rpm for at least 8 h (or overnight).
15. Add 20 μL 20 mg/mL proteinase K and incubate at 65 °C, 750 rpm, and overnight.
16. Add 20 μL 10 mg/mL RNase A to the ligated sample and 5 μL to the digestion test aliquots, and incubate at 37 °C, 750 rpm for 1 h.
17. Add 1 volume of phenol/chloroform/isoamyl alcohol pH 8 for 1 volume of digestion reaction, mix by inversion, and

vortex well. Centrifuge at 15,500 × *g*, 23 °C and for 5 min. Transfer the aqueous phase to a new tube.

18. Add 0.9 volumes of isopropanol and 0.1 volume of sodium acetate pH 5.2, and then mix by inversion. Incubate the tubes at −80 °C for 30 min and then centrifuge at 15,500 × *g*, 4 °C for 30 min. Wash the pellet with 500 μL ethanol 70%, and re-centrifuge at 15,500 × *g*, 4 °C for 5 min.
19. Remove supernatant and resuspend the pellet in 50 μL low TE buffer. Quantify purified DNA with the Qubit dsDNA BR assay, following the manufacturer's instructions (*see* **Note 4**).
20. Load the digestion test aliquot (**step 11**) and an aliquot of 3C material (~250 ng) on 1% TAE agarose gel to confirm digestion and ligation (*see* **Note 13** and Fig. 3).
21. Dilute 3C DNA to 100 ng/μL in B buffer and nuclease-free water. Incubate at 37 °C, 600 rpm for 15 min.
22. Add 5 U of Csp6I per μg of DNA and incubate at 37 °C, 600 rpm for overnight.
23. Inactivate the enzyme by incubating at 65 °C, 600 rpm for 20 min.
24. Add 1 volume of phenol/chloroform/isoamyl alcohol pH 8 for 1 volume of sample to be extracted, invert the tubes up and down a couple of times, and vortex well. Centrifuge at 15,500 × *g*, 23 °C and for 5 min. Transfer the aqueous phase to a new tube.
25. Add 0.9 to 1 volume of isopropanol and 0.1 volume of sodium acetate pH 5.2, and then mix them by inverting up and down a couple of times. Incubate the tubes at −80 °C for half an hour and then spin down at 15,500 × *g*, 4 °C for 30 min. Wash the pellet with 500 μL ethanol 70%, and re-centrifuge at 15,500 × *g*, 4 °C for 5 min.
26. Quantify DNA with Qubit dsDNA HS assay, following the manufacturer's instructions (*see* **Note 4**). Run an aliquot (~250 ng) on 1% TAE agarose gel to confirm digestion (*see* **Note 13** and Fig. 3).
27. Dilute DNA to 5 ng/μL with 1× T4 ligase buffer and add 200 U of ligase per μg of DNA. Incubate at 16 °C, 400 rpm for overnight (*see* **Note 5**).
28. Add 1 volume of phenol/chloroform/isoamyl alcohol pH 8 for 1 volume of sample to be extracted, invert the tubes up and down a couple of times, and vortex well. Centrifuge at 15,500 × *g*, 23 °C and for 5 min. Transfer the aqueous phase to a new tube.
29. Add 0.9 to 1 volume of isopropanol and 0.1 volume of sodium acetate pH 5.2, and then mix them by inverting up and down a

couple of times. Incubate the tubes at −80 °C for half an hour and then spin down at 15,500 × *g*, 4 °C for 30 min. Wash the pellet with 500 μL ethanol 70%, and re-centrifuge at 15,500 × *g*, 4 °C for 5 min.

30. Quantify DNA with Qubit dsDNA HS assay, following the manufacturer's instructions (*see* **Note 4**).
31. Amplify 4C material for each bait using the primers and conditions that were previously optimized (*see* Subheading 3.1, **step 14**). Perform six PCR reactions per bait.
32. Pool the reaction mixtures to an Eppendorf tube, and perform two rounds of 1.8× AMPure bead purification following the manufacturer's instructions. Quantify with Qubit dsDNA HS.
33. Run samples on BioAnalyzer High Sensitivity chip to check the profile and calculate the quantity and molarity of the material (*see* **Note 15**). Pool equimolar quantities of 4C-seq material from at least six baits (*see* **Note 16**), which is now ready for loading on an Illumina sequencing machine.
34. Map the 4C-seq fastq files and analyze the interactions using the available tools, given in [3, 16, 21–23].

4 Notes

1. Primary enzymes need to digest fixed chromatin efficiently. When trying a new enzyme, perform **steps 1–10** of Subheading 3.2, and check digestion efficiency on 1% TAE agarose gels (*see* Fig. 3) before proceeding with full 4C-seq experiment.
2. Too long a *Dpn*II-*Dpn*II fragment reduces resolution; too short reduces 3C ligation efficiency. Too short a *Dpn*II-*Csp*6I fragment reduces circularization efficiency due to torsional effects.
3. The overall amplicon should be small to minimize biases in amplifying larger 4C ligation products (due to greater *Dpn*II-*Csp*6I distance in the interacting fragment). The reading primer should be as close as possible to the *Dpn*II site to maximize sequence which can be used in mapping the interacting regions.
4. DNA quantification based on absorbance at 260 nm (e.g., with a NanoDrop spectrophotometer) is not reliable for 3C and 4C material, presumably due to the high amounts of free ATP used in ligation reactions. Use of the Qubit system is recommended.
5. The DNA concentration and ligase amount at this step are critical, since intermolecular ligation events create false positives and/or increase background signal.

6. As shown in Fig. 2, inverse PCR on the mock 4C template is expected to produce one specific product, corresponding to amplification of the contiguous *Dpn*II-*Csp*6I fragment. Absence of product or only obtaining primer dimers suggests that the primers do not work or that the PCR conditions are too stringent. Multiple bands or a smear from non-specific products suggests that the reaction is not stringent enough. We recommend the following steps when troubleshooting 4C-seq PCR conditions: (1) alter hybridization temperatures; (2) try buffers 2 and 3 from the Roche Expand Long Template system, which provide different stringencies; (3) try new primers on the same bait fragment; (4) design new primers to a different bait fragment; and (5) design new primers to a bait fragment from a different combination of restriction enzymes.
7. Due to the possibility of primer dimers or other secondary structures, bait primers may have unexpected specificity or efficiency problems once the Illumina adapter sequences are added, so they should be tested. It is cost-effective to design multiple primer pairs without Illumina sequences and to reorder primers with Illumina sequences for a restricted number of the best-performing pairs from the first test.
8. P5424 cells grow in suspension, so the cells are readily obtained by pipetting the medium up and down the flask and then transferring to a Falcon tube. Adherent cells should first be removed by trypsinization or other means, following the procedure suitable to their cell culture. Tissue samples need to be homogenized to single-cell suspensions by the means appropriate to the sample. A minimum of five million cells is recommended for one 4C-seq experiment; up to 50 million cells can be fixed following the protocol described here. Volumes of fixation reagents can be scaled up for larger cell numbers if needed.
9. For **steps 1–16**, material loss can be avoided by pre-coating the Eppendorf tubes with PBE (fill the tubes the day before, and empty just before use) and pipette tips with 20 mg/mL BSA (pipette up and down a few times in this solution, and then eject all liquid just before use), which prevents nuclei or cells sticking to the plastic.
10. Always use fresh batches of formaldehyde that has been sealed in glass vials for efficient cross-linking and higher reproducibility.
11. Fixed nuclei at this step can be flash frozen in liquid nitrogen and stored at −80 °C for several months. When continuing the 4C-seq on this material, leave on ice for 30 min before proceeding with **step 8**.

12. Do not exceed 20 min at 65 °C or de-cross-linking will occur. Note that P5424 thymocyte-like cells have compact chromatin, so require a harsh permeabilization treatment with 0.8% final SDS concentration. For most cell types, digestion is efficient after 1 h incubation at 37 °C with 0.4% final SDS. The amount of Triton-X100 added needs to be scaled with the prior amount of SDS to ensure neutralization within the buffer.
13. The digestion and ligation efficiencies of the 3C stages are critical for a successful 4C-seq experiment. An easy way to qualitatively test them is to load aliquots of the sample at different stages of the experiment to compare the average molecular weights of the DNA (*see* Fig. 3).
14. Initially, the ligation step of 3C/4C was performed under dilute conditions, believed to favor intramolecular interactions [2]. Subsequently, it has been proposed that performing in situ ligation within the nuclei is actually more efficient [24, 25]. For the majority of samples, nuclei remain intact during this protocol, so the ligation can be performed in situ (and diluting the sample just reduces the concentration of nuclei for in situ reactions). To test whether the primary digestion step maintains intact nuclei, centrifuge the digestion reaction at 15,500 × *g*, 23 °C, 1 min. Transfer the supernatant to a new Eppendorf tube, and resuspend the pellet in an equal volume of restriction buffer. Purify the DNA from both fractions following the same principles as elsewhere in this protocol: overnight proteinase K digestion and reverse cross-linking, RNase A digestion, phenol/chloroform/isoamyl alcohol extraction, and ethanol precipitation. Compare the yields of the two fractions by quantifying with a Qubit dsDNA BR assay. More than 90% of the DNA usually originates from the pellet fraction, suggesting that the majority of the chromatin is maintained within intact nuclei. If a large proportion of DNA is found in the supernatant, consider either altering the digestion conditions (greater fixation, more gentle lysis, and/or SDS/Triton-X100 concentrations; but be sure to verify that chromatin is still digested) or performing ligation under dilute conditions (see steps after secondary digestion).
15. If the quantity of adapter dimers or uninformative material (<125 bp depending on primers' size) is too high, perform an extra round of AMPure bead purification at 0.9× concentration, following the manufacturer's instructions. Due to the different restriction maps for the regions directly surrounding a bait, BioAnalyzer profiles can be quite variable. Discernible peaks from frequently amplified products, especially the non-digested contiguous fragment, are often discernible. However, they should have little to no material <125 bp (likely

adapter dimers) and have a background "smear" from 125 to 2000 bp. Discrete peaks from frequently amplified products, especially the non-digested contiguous fragment, are often discernible. The molarity estimations are made from the BioAnalyzer software. Choose the same size range for this quantification for all 4C samples that are to be pooled before sequencing, to ensure as much as possible that equimolar quantities are mixed.

16. Illumina sequencing runs fail if there is an insufficient complexity of nucleotides present at each position of the read for the first few nucleotides. Since the first nucleotides of a 4C-seq experiment to be sequenced correspond to the bait primer, they are identical for all interactions with the same bait. As a consequence, a minimum number of baits must be multiplexed for Illumina sequencing, or else the 4C material must be spiked into an unrelated sample such as a phiX DNA library. Similarly, to distinguish 4C experiments from different conditions or tissues on the same bait, a barcode sequence needs to be intercalated between the Illumina adapter sequence and the reading primer. Since only ~1–2 million reads are required per 4C experiment, it is in any case cost-effective to multiplex 4C experiments.

References

1. Sexton T, Cavalli G (2015) The role of chromosome domains in shaping the functional genome. Cell 160:1049–1059
2. Dekker J, Rippe K, Dekker M et al (2002) Capturing chromosome conformation. Science 295:1306–1311
3. van de Werken HJG, Landan G, Holwerda SJB et al (2012) Robust 4C-seq data analysis to screen for regulatory DNA interactions. Nat Methods 9:969–972
4. Dostie J, Richmond TA, Arnaout RA et al (2006) Chromosome conformation capture carbon copy (5C): a massively parallel solution for mapping interactions between genomic elements. Genome Res 16:1299–1309
5. Fullwood MJ, Liu MH, Pan YF et al (2009) An oestrogen-receptor-alpha-bound human chromatin interactome. Nature 462:58–64
6. Hughes JR, Roberts N, McGowan S et al (2014) Analysis of hundreds of cis-regulatory landscapes at high resolution in a single, high-throughput experiment. Nat Genet 46:205–212
7. Lieberman-Aiden E, van Berkum NL, Williams L et al (2009) Comprehensive mapping of long-range interactions reveals folding principles of the human genome. Science 326:289–293
8. Denker A, de Laat W (2016) The second decade of 3C technologies: detailed insights into nuclear organization. Genes Dev 30:1357–1382
9. Ghavi-Helm Y, Klein FA, Pakozdi T et al (2014) Enhancer loops appear stable during development and are associated with paused polymerase. Nature 512:96–100
10. Lupiáñez DG, Kraft K, Heinrich V et al (2015) Disruptions of topological chromatin domains cause pathogenic rewiring of gene-enhancer interactions. Cell 161:1012–1025
11. de Wit E, Vos ESM, Holwerda SJB et al (2015) CTCF binding polarity determines chromatin looping. Mol Cell 60:676–684
12. Guo Y, Xu Q, Canzio D et al (2015) CRISPR inversion of CTCF sites alters genome topology and enhancer/promoter function. Cell 162:900–910
13. Simonis M, Klous P, Splinter E et al (2006) Nuclear organization of active and inactive chromatin domains uncovered by chromosome conformation capture-on-chip (4C). Nat Genet 38:1348–1354

14. Noordermeer D, Leleu M, Splinter E et al (2011) The dynamic architecture of Hox gene clusters. Science 334:222–225
15. Narendra V, Rocha PP, An D et al (2015) CTCF establishes discrete functional chromatin domains at the Hox clusters during differentiation. Science 347:1017–1021
16. Geeven G, Teunissen H, de Laat W et al (2018) peakC: a flexible, non-parametric peak calling package for 4C and capture-C data. Nucleic Acids Res 46:e91
17. Schwartzman O, Mukamel Z, Oded-Elkayam N et al (2016) UMI-4C for quantitative and targeted chromosomal contact profiling. Nat Methods 13:685–691
18. Mombaerts P, Terhorst C, Jacks T et al (1995) Characterization of immature thymocyte lines derived from T-cell receptor or recombination activating gene 1 and p53 double mutant mice. Proc Natl Acad Sci U S A 92:7420–7424
19. Koressaar T, Remm M (2007) Enhancements and modifications of primer design program Primer3. Bioinformatics 23:1289–1291
20. Untergasser A, Cutcutache I, Koressaar T et al (2012) Primer3--new capabilities and interfaces. Nucleic Acids Res 40:e115
21. Thongjuea S, Stadhouders R, Grosveld FG et al (2013) r3Cseq: an R/bioconductor package for the discovery of long-range genomic interactions from chromosome conformation capture and next-generation sequencing data. Nucleic Acids Res 41:e132
22. Raviram R, Rocha PP, Müller CL et al (2016) 4C-ker: a method to reproducibly identify genome-wide interactions captured by 4C-Seq experiments. PLoS Comput Biol 12:e1004780
23. Klein FA, Pakozdi T, Anders S et al (2015) FourCSeq: analysis of 4C sequencing data. Bioinformatics 31:3085–3091
24. Rao SSP, Huntley MH, Durand NC et al (2014) A 3D map of the human genome at kilobase resolution reveals principles of chromatin looping. Cell 159:1665–1680
25. Nagano T, Várnai C, Schoenfelder S et al (2015) Comparison of hi-C results using in-solution versus in-nucleus ligation. Genome Biol 16:175
26. Klein-Hessling S, Rudolf R, Muhammad K et al (2016) A threshold level of NFATc1 activity facilitates thymocyte differentiation and opposes notch-driven leukaemia development. Nat Commun 7:11841

Chapter 4

Analysis, Modeling, and Visualization of Chromosome Conformation Capture Experiments

Marco Di Stefano, David Castillo, François Serra, Irene Farabella, Mike N. Goodstadt, and Marc A. Marti-Renom

Abstract

Chromatin Conformation Capture techniques have unveiled several layers of chromosome organization such as the segregation in compartments, the folding in topologically associating domains (TADs), and site-specific looping interactions. The discovery of this genome hierarchical organization emerged from the computational analysis of chromatin capture data. With the increasing availability of such data, automatic pipelines for the robust comparison, grouping, and classification of multiple experiments are needed. Here we present a pipeline based on the TADbit framework that emphasizes reproducibility, automation, quality check, and statistical robustness. This comprehensive modular pipeline covers all the steps from the sequencing products to the visualization of reconstructed 3D models of the chromatin.

Key words Chromosome Conformation Capture data analysis, Read mapping, Interaction filtering, Matrix normalization, Detection of compartments, Detection of TADs, 3D modeling, 3D visualization

1 Introduction

Genome-wide Chromosome Conformation Capture (3C) technologies [1] fostered a huge improvement in the field of genome structural biology [2]. The outcome of these techniques is a set of DNA–DNA frequency interactions reflecting the spatial proximity between pairs of chromatin regions [3, 4].

The applications of 3C technologies to many organisms and cell lines shed light into the hierarchical genome structural organization. In particular, Hi-C confirmed the existence of chromosome territories [5] initially characterized by imaging [6–8] and unveiled the partition of chromosomes into active and inactive compartments [5], into topologically associating domains (TADs) [9, 10],

Marco Di Stefano and David Castillo contributed equally to this work.

Beatrice Bodega and Chiara Lanzuolo (eds.), *Capturing Chromosome Conformation: Methods and Protocols*, Methods in Molecular Biology, vol. 2157, https://doi.org/10.1007/978-1-0716-0664-3_4,

and in chromatin loops [11], which facilitate interactions between genes and regulatory elements [12]. However, prior to the detection of these structural features, the data from a Hi-C experiment has to be cleaned, normalized, and quantitatively characterized by several steps of bioinformatic analysis. For this purpose, we developed TADbit (https://github.com/3DGenomes/TADbit), a comprehensive Python framework, to perform the analysis of Hi-C interaction datasets covering all the steps from aligning the sequenced reads (paired-end reads) up to the inference and analysis of three-dimensional (3D) models. Specifically, TADbit can perform (1) Hi-C-specific quality control of the reads, (2) mapping to the corresponding reference genome, (3) filtering of artifactual reads, (4) normalization of experimental biases, (5) generation of binned interaction maps, (6) statistical analysis of the differences and consistencies between experiments, (7) identification and comparison of the structural features in the interaction maps, (8) generation and analysis of 3D models using restraint-based methods, and (9) visualization of the 3D models using the companion visualizer TADkit. Here, we present a hands-on protocol for the analysis of any Hi-C datasets with minimal previous knowledge of bioinformatics, including an exhaustive description of each methodological step and a detailed supplementary notebook (https://github.com/3DGenomes/MethodsMolBiol).

2 Materials

2.1 Input Experimental Data

The complete TADbit pipeline has been applied to 5C [13], dilution Hi-C [14, 15], in situ Hi-C [16], and Promoter Capture Hi-C (https://doi.org/10.1101/400291). In this chapter, we will specifically refer to Hi-C experiments, for which we also provide a set of notes and suggestions to help to design the experiment (Materials-1-Design of the HiC experiment). Additionally, other types of 3C-based datasets can be analyzed with ad hoc changes in the parameter choice, mainly after the filtering of artifactual reads. Chromatin feature tracks (such as ChIP-seq, methylation, GC content, chromHMM, or RNA-seq, among others) can be loaded and used to label the genomic structural domains. Information as the nuclear size and the chromatin compaction can be used as additional parameters for the generation of 3D models.

2.2 Hardware Requirements and Performances

The requirements and performances listed below are for a Hi-C experiment of about 200 million reads binned at 50 kilobases (kb) on a mammalian genome of about 2–3 gigabases (Gb):

- Random-access memory (RAM): it is especially important when loading matrices at high resolution. About 32 gigabytes (GB) of RAM should be sufficient.

- Disk space: the data processing occupies around 500 GB per experiment (temporary storage). Some intermediate files are compressed or erased after processing; thus, the storage of the final output files occupies around 50 GB per experiment (long-term storage).
- Calculation (CPU) time: an 8-core computer can perform the data processing in about 1 day. The time for 3D model generation and analysis depends on the size of the considered region and the specific calculations required.

2.3 Software

For data analysis:

- TADbit (https://github.com/3DGenomes/TADbit) [15].
- SciPy (https://www.scipy.org) [17].
- Matplotlib (https://matplotlib.org) [18].
- IMP (https://integrativemodeling.org) [19].
- SAMtools (http://samtools.sourceforge.net) [20].
- SRAtools (https://github.com/ncbi/sra-tools).
- GEM (http://algorithms.cnag.cat/wiki/The_GEM_library) [21].
- OneD (https://github.com/qenvio/dryhic) [22].
- DSRC (http://sun.aei.polsl.pl/dsrc) [23].

For 3D model visualization:

- TADkit (https://github.com/3DGenomes/TADkit).

For more details on hardware requirement and software installation, see the notebook Materials-2-Preparing your computer for the Hi-C data analysis.

3 Methods

Here, we provide a detailed explanation of TADbit framework for Chromosome Conformation Capture dataset analysis. The task of each processing step is described, and the core functions needed to run the corresponding TADbit commands are provided in code blocks. In these blocks, the input values to functions are in italic, strings or lists of strings in green, numbers or list of numbers in red, and Python constants in orange (Table 1). The complete code to perform all the steps of the pipeline is provided in the supplementary notebooks (https://github.com/3DGenomes/MethodsMolBiol). In this chapter, for demonstration purposes, the pipeline is applied on two experiments carried out on two mouse

Table 1
List and description of all the variable used in the code blocks

Variable name	Description
bam_file	Path to the compressed and indexed bam file containing the intersection of the pair of mapped reads file after filtering
counts_per_column	List with the sum of counts per column (bin) of the matrix
Crm	Python object containing one or more experiments
Cutoff	Distance threshold (in nm) to determine if two particles interact. The default value is twice the particle size
destination_dir	Path to store the results of the function
end_bin	Number of the last bin in the matrix for the region we want to model
Exp	Python object containing the experiment and its description
factor_eq_positions	Factor to define the percentage of equivalent positions to be considered in the clustering
file_reads1,file_reads2	Paths to the mapped reads files
gc_content_per_column	List with GC content ratio per column
hic_data, hic_data1, hic_data2	Python objects containing the Hi-C matrix and its properties
input_fastq	Path to the FASTQ file containing the read-ends
list_of_chromosomes	List of chromosomes names
list_of_cutoffs	List of cutoffs in number of particles
list_of_experiments	List of the experiment names included in the chromosome object
list_of_filters_numbers	List of numbers of filter's categories
Mappability	List with the mappability score per column (bin)
Masked	List of paths of the files containing the reads filtered
max_fragment_size	Maximum fragment size of reads considered as being too long in the filtering
max_molecule_length	Maximum distance in nucleotides of two read-ends that are coming from two different REs. It is used to define the extra-dangling-ends category
minimum_distance_to_re	Minimum distance of the reads to a RE cut site
Models	Python object containing the structural models and their properties
nbr_cpus	Number of cpu cores to use
nbr_models	Number of models to compute
nbr_models_keep	Number of best models to keep
number_ev_to_compare	Number of top eigenvectors to compare
number_of_iterations	Number of iterations for the ICE algorithm

(continued)

Table 1
(continued)

Variable name	Description
number_of_re_sites_per_column	List with the numbers of restriction sites per column (bin)
number_of_reads	Number of reads to use in the function
optimal_params	Python dictionary containing the list of best parameters after the optimization
path_to_gem_index	Path to the indexed GEM file
read_length	Read length in nucleotides
reads_file	Path to the file containing the intersection of the pairs of mapped read-ends
Resolution	Resolution in base-pairs to use for binning the Hi-C matrix
restriction_enzyme	Name of the restriction enzyme used in the Hi-C experiment
rich_in_A_marker	Path to a BED or BEDGraph file with a list of genes or active epigenetic marks
start_bin	Number of the first bin in the matrix for the region we want to model
(start_lowfreq, end_lowfreq, step_lowfreq)	Range of *lowfreq values* from *start_lowfreq* to *end_lowfreq* in steps of *step_lowfreq*
(start_maxdist, end_maxdist, step_maxdist)	Range of maxdist values from *start_maxdist* to *end_maxdist* in steps of *step_maxdist*
(start_upfreq, end_upfreq, step_upfreq)	Range of upfreq values from *start_upfreq* to *end_upfreq* in steps of *step_upfreq*
valid_reads_file	Path to the file containing the intersection of the pair of mapped reads file after filtering
win1_start,win1_end, win2_start,win2_end	Starts and ends of the windows of nucleotides to consider in the read for the iterative mapping strategy

cell types (B-cells and induced pluripotent stem cells (iPSC)) each analyzed in two replicas, which were down-sampled at 150 million (M) reads each (Accession number GSE53463) [16].

3.1 Data Management

3.1.1 Hi-C Data

FASTQ files, either input by the user or downloaded from public repositories (see Methods-1-Retrieve published Hi-C datasets), contain the sequenced paired-end reads to be mapped to a reference genome (*see* **Note 1**).

3.1.2 Other Data

A reference genome file contains the reference genome for the species of interest in a FASTA format. Several databases provide reference genomes for downloading, including the National Center for Biotechnology Information (NCBI) that we use in the online tutorial (Methods-2-Preparation of the reference genome). The following information is generated accordingly to the reference genome of interest:

- *Genome index file:* to speed up the mapping process, the FASTA file of the reference genome is converted into an indexed file used to efficiently map each read-end.
- *Positions of restriction enzyme (RE) cut-site file:* to assign each mapped read-end to a given RE fragment and filter for nonspecific products of the chromosome capture experiment.
- *GC content file:* to identify A and B compartments. It has been shown that active genomic regions tend to have higher GC content and are thus used to label type A compartments [24]. However, if available, cell-specific markers of activity (e.g., ChIP-seq for chromatin marks or RNA-seq data) should be used.
- *Mappability file*: it stores the mappability score, which is the probability that a genomic region produces uniquely mapped reads. This is one of the biases fitted by the OneD normalization procedure, and it is computed a priori for a given read length [25].

3.1.3 Hi-C-Specific FASTQ Quality Check

TADbit computes and plots several metrics from the FASTQ files to assess the quality of the Hi-C experiment and the sequencing (Fig. 1 and **Note 2**). Such metrics are:

- The PHRED score and the number of unidentified nucleotides (*Ns*) in the read sequence, which are routinely computed to address the quality of high-throughput sequencing experiments.
- The numbers of undigested and unligated RE sites per nucleotide along the read to assess the quality of the Hi-C experiment.
- The overall percentage of digested sites, which relates directly to the RE efficiency.
- The percentage of non-ligated digested (dangling-ends), which relates to the ligation efficiency.
- The percentage of read-ends with a ligation site, which is negatively correlated with the percentage of dangling-ends.

These quality measurements are performed on a subset of reads, usually *number_or_reads* = 1000000 (Methods-3-Hi-C qual ity check):

```
quality_plot(input_fastq, r_enz=restriction_enzyme,

nreads=number_of_reads)
```

3.2 Paired-End Read Mapping

The first step in the analysis of a Hi-C experiment consists on uniquely mapping all reads in the input FASTQ files to a location on the reference genome. In TADbit this is achieved using the GEM mapper [21] (Fig. 2a). Currently, three mapping strategies are implemented in TADbit:

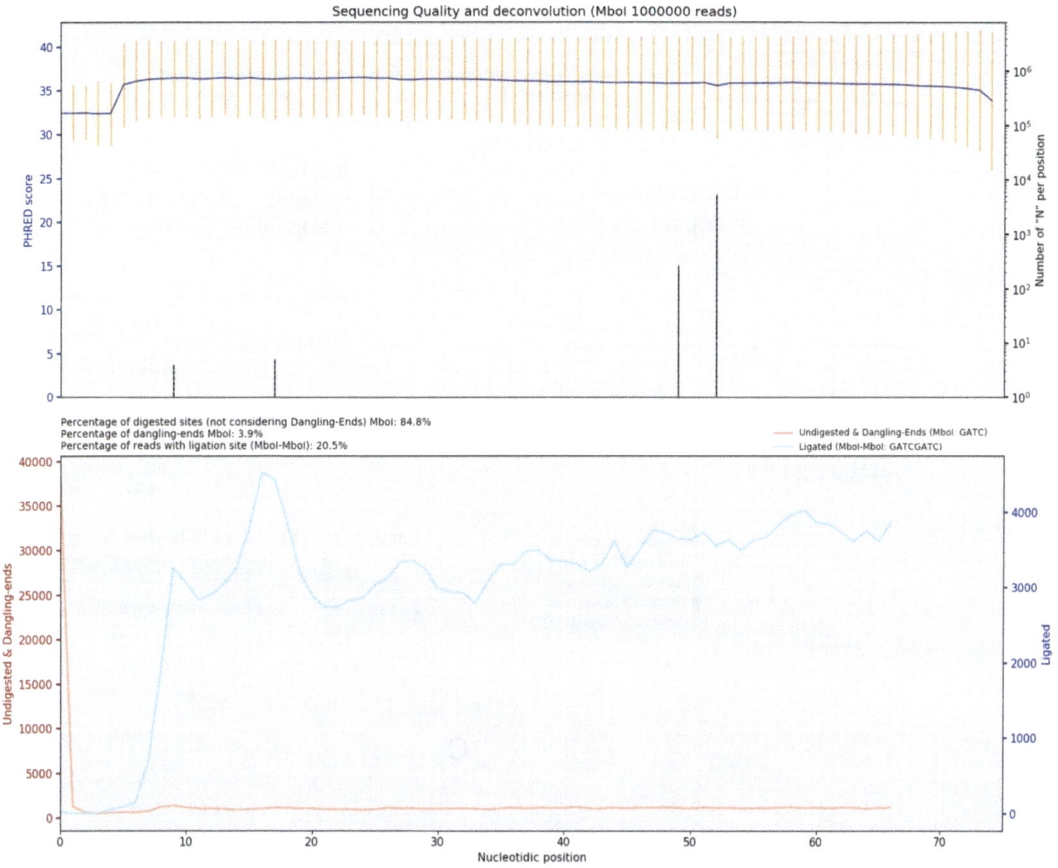

Fig. 1 Hi-C quality check plots. The upper plot shows two classic metrics of NGS experiments: the PHRED score (average in blue and standard deviation in yellow) and the proportion of *Ns* at each position of the read. The lower panel is specific to Hi-C experiments (in the example for MboI restriction enzyme). The two curves indicate the number of undigested sites (dangling-ends) in red, which is expected to be peaked at the beginning of the read, and the number of digested and ligated products in blue, which usually have a more homogeneous distribution along the read. For MboI, the patterns of undigested (GATC) and of digested and ligated (GATCGATC) sites are similar, but for other RE (e.g., HindIII), these patterns are distinct. Above the graph, three quantities are displayed: the proportions of ligated sites (84.8%), undigested sites (dangling-ends) (3.9%), and reads carrying a digestion site (20.5%). Note that TADbit can perform the same analysis in experiments where multiple REs are used at the same time

- *Full-length mapping*: The mapping of the read-end is attempted once (single iteration) taking into account the full length of the read. Unmapped read-ends are discarded.

```
full_mapping(gem_index_path=path_to_gem_index,
    out_map_dir=destination_dir, fastq_path=input_fastq,
    frag_map=False)
```

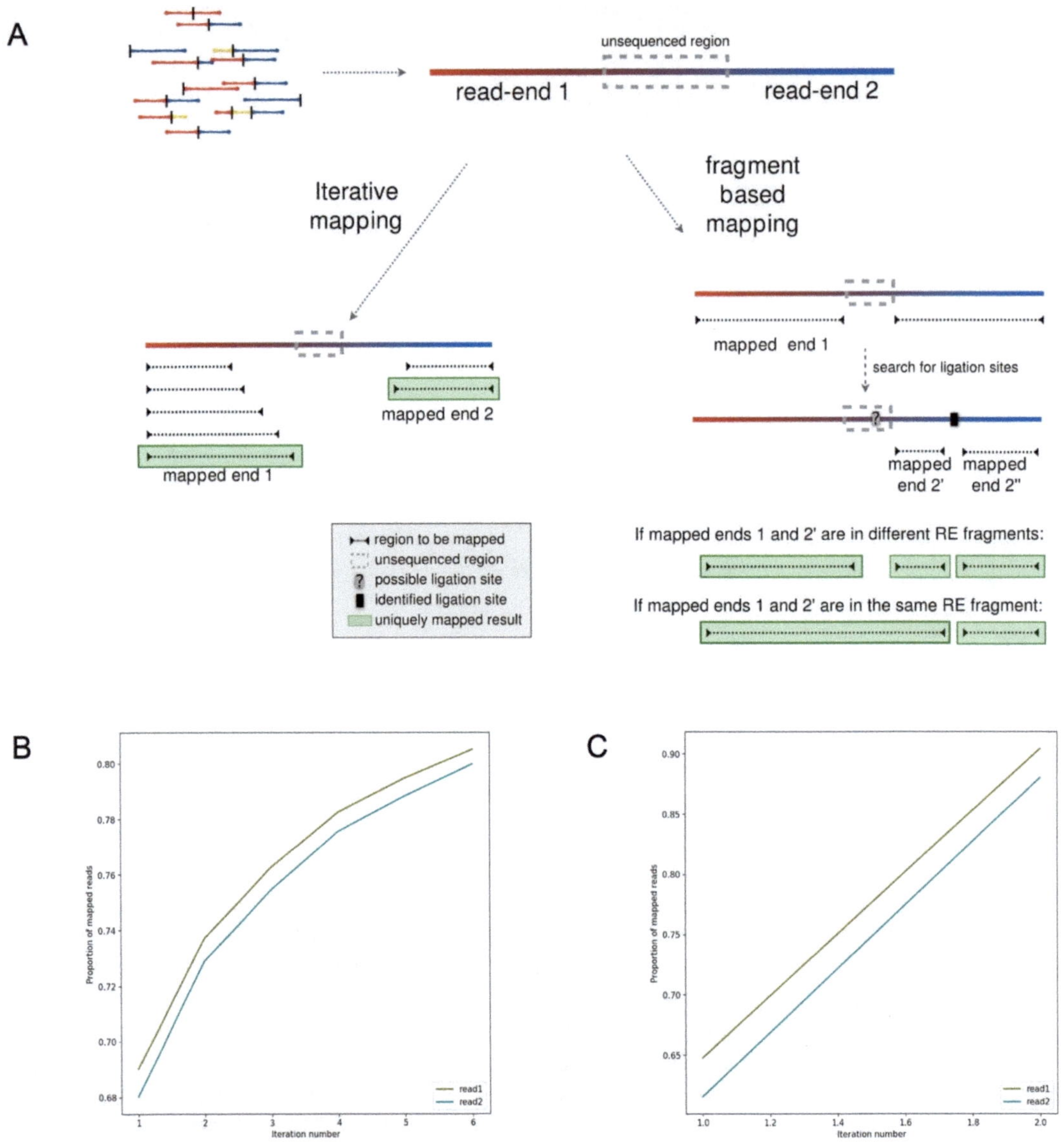

Fig. 2 Mapping of paired-end reads. (**a**) Schematic cartoons of the iterative mapping *(left)* and the fragment-based mapping *(right)* approaches. (**b** and **c**) The proportion of the read-ends mapped in each round of the iterative mapping (**b**, here 6 steps are shown) and of the two rounds of the fragment-based mapping, that is, full and fragment-based iterations (**c**)

- *Iterative mapping*: The mapping of the read-end is attempted with iteratively increasingly larger reads. Therefore, to use this option the user has to define the window parameter as a set of different ranges of nucleotide indexes all starting from 1 and of increasing sizes. Usually (in the human genome) the first section of the read-end goes from 1 to 25 bp and is followed by ranges that are incrementally extended by 5 bp ((1, 30), (1, 35), etc.).

The mapping procedure is applied iteratively until the read is uniquely mapped using the ranges defined in the set [26]. Note that TADbit accepts any combination of window ranges.

```
full_mapping(gem_index_path=path_to_gem_index,
        out_map_dir=destination_dir, fastq_path=input_fastq,
        frag_map=False,
        windows=((win1_start, win1_end), (win2_start, win2_end),
        ...))
```

- *Fragment-based mapping*: This procedure consists of two iteration steps per read-end: (1) the full-length is attempted, and (2) if the read-end is unmapped and contains ligation sites, it is split at the ligation sites, and the split sequences are separately mapped.

```
full_mapping(gem_index_path=path_to_gem_index,
        out_map_dir=destination_dir, fastq_path=input_fastq,
        r_enz=restriction_enzyme_name, frag_map=True)
```

A detailed account of the commands to perform the mapping step is provided in the notebook Methods-4-Mapping. A good metric for assessing the quality of the mapping procedure is the proportion of reads that have been uniquely mapped to the reference genome (Fig. 2b). To render the number of reads mapped at each iteration of the mapping, one can use the command:

```
plot_iterative_mapping(file_reads1, file_reads2, number_of_reads)
```

3.3 Intersection of Paired-End Reads

After mapping, a browser extensible data (BED) file is created with the reads that have been uniquely mapped on both ends. This contains the genomic coordinates of each pair of mapped read-ends as well as their position relative to the closest RE site. Note that for reliably assembled genomes, more than 80% of the reads are typically mapped for any of the two ends and more than 60% on both ends.

```
get_intersection(file_reads1, file_reads2, reads_file)
```

3.3.1 Descriptive Statistics after the Intersection of the Pair-End Reads

- *Interaction count vs. genomic separation*. Given the intrinsic polymeric structure of chromatin, the number of captured interactions is expected to decay as the genomic separation between the interacting sequences increases. In mammals, this decay typically follows a power law with an exponent around -1 in the range between hundreds of kb and tens of Mb [5] (Fig. 3a).

```
plot_distance_vs_interactions(reads_file, resolution=resolution)
```

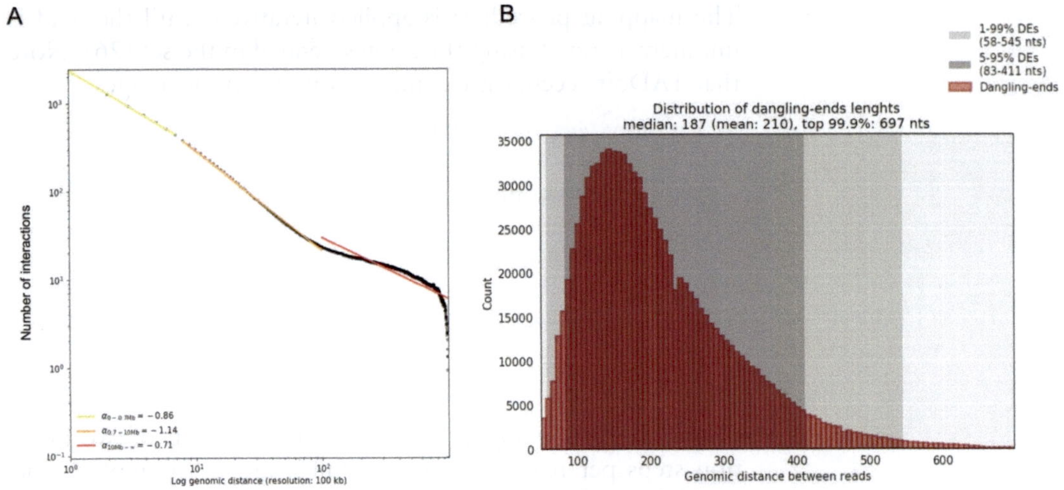

Fig. 3 Descriptive statistics of the mapped read-ends. (**a**) Decay of the number of Hi-C interactions with respect to the genomic distance between interacting *loci*. (**b**) The distribution of RE fragment lengths is estimated considering fragments mapped in a single RE fragment and in *facing* orientation (see dangling-ends in Subheading 3.4). Although the dangling-ends are expected to be shorter than most fragments in the library, we use their average length as a conservative estimate of the mapped fragments for the application of filters

- *Coverage per bin.* The genome is partitioned into regions of the same size (called "bins"), and the number of mapped reads per bin (i.e., coverage) is computed. In an ideal situation, the coverage is expected to be homogeneous across the genome. However, variations in the distribution of GC content, mappability, or the number of RE sites along the genome can cause heterogeneity. The source of such biases will be corrected during the normalization of the interaction matrices (Subheading 3.6).

```
plot_genomic_distribution(reads_file, resolution=resolution,

show=True)
```

- *Sequenced DNA fragment size.* One can measure the genomic length of the sequenced DNA strands using only the reads that were digested, but not ligated and therefore belong to the same DNA linear segment between two consecutive RE cut sites (e.g., the two paired-ends have been mapped within the same RE fragment). These reads (dangling-ends) are mapped in *facing* orientation (Fig. 3b and **Note 3**).

```
fragment_size(reads_file, show=True, nreads=number_of_reads)
```

For details see Methods-5-Parsing mapped reads.

3.4 Filtering of Mapped Reads

At this stage, the intersection file contains uniquely mapped paired-end reads of DNA sequences, including those pairs with no relevant structural information. After the filtering step, typically around

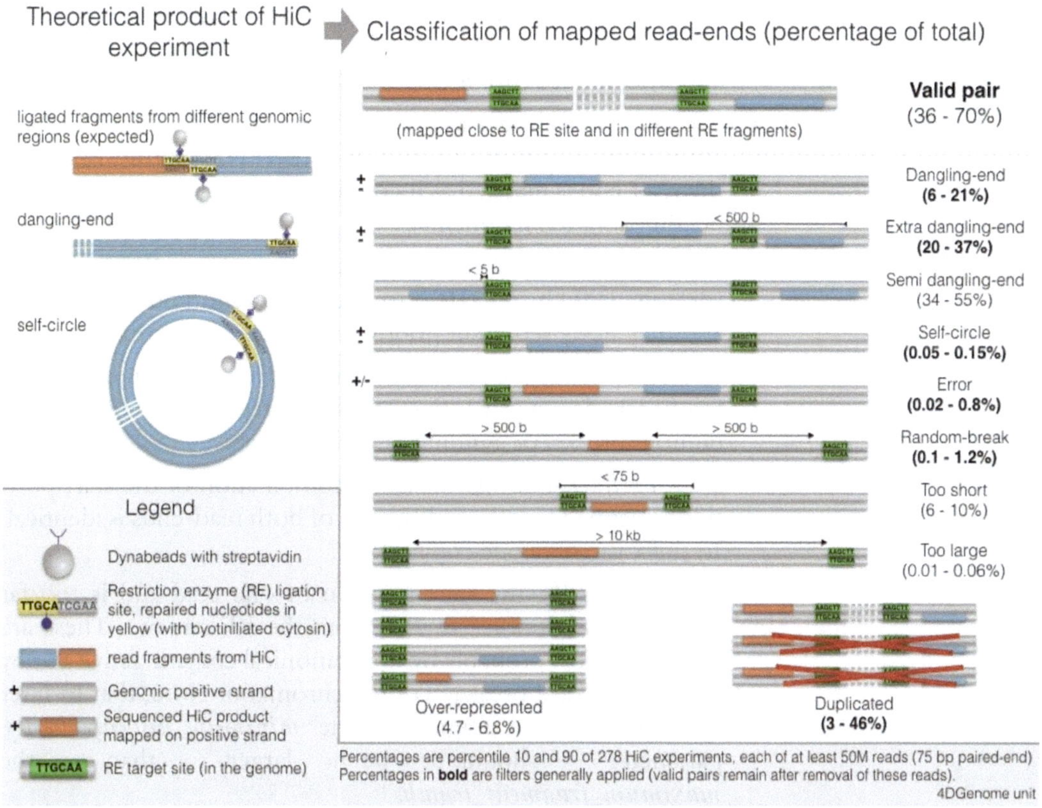

Fig. 4 Summary of the different products of a Hi-C experiment (top left) and the categories of mapped reads (right). Here, the proportions are estimated from the analysis of 278 Hi-C experiments produced by the 4DGenome unit of the CRG. All experiments were performed using 75 bp paired-end sequencing. Fair quality experiments would have proportions falling within the ranges displayed here

40–70% of the pairs are considered informative (also called "valid pairs"). TADbit will classify the remaining (or discarded) mapped pairs into one (or several) of the following categories (Fig. 4 and **Note 4**):

- *Self-circle*: both read-ends are mapped to the same RE fragment in *opposed* orientation.
- *Dangling-end*: both read-ends are mapped to the same RE fragment in *facing* orientation.
- *Error*: both read-ends are mapped to the same RE fragment in the same orientation.
- *Extra dangling-end*: the read-ends are mapped to different RE fragments in *facing* orientation, but are close enough (< *max_-molecule_length* bp) from the RE cut site to be considered part of adjacent RE fragments that were not separated by digestion. The *max_molecule_length* parameter can be inferred from the *fragment_size* function previously detailed.

- *Too close from RE sites (or semi-dangling-end)*: the start position of one of the read-end is too close (5 bp by default) from the RE cutting site (*see* **Note 5**).
- *Too short*: one of the read-ends is mapped to RE fragments of less than 75 bp. These are removed since there is ambiguity on where the read-end is mapped as it could also belong to any of the two neighboring RE fragments.
- *Too large*: the read-ends are mapped to long RE fragments (default: 100 kb, $P < 10^{-5}$ to occur in a randomized genome), and they likely represent poorly assembled or repetitive regions.
- *Overrepresented*: the read-ends coming from the top 0.5% most frequently detected RE fragments may represent PCR artifacts, random breaks, or genome assembly errors.
- *PCR artifacts or duplicated*: the combination of the start positions, mapped length, and strands of both read-ends is identical. In this case, only one copy is kept.
- *Random breaks*: the start position of one read-end is too far (>*minimum_distance_to_RE*) from the RE cut site. These are produced most probably by non-canonical enzyme activity or by random physical breakage of the chromatin. Note, that to filter all these types of fragments, the *minimum_distance_to_RE* parameter should be larger than the *maximum_fragment_length*.

Once TADbit has classified each read-end pair (*filter_reads* command below), one can define valid genomic interactions (valid pairs) by removing the paired-end reads belonging to a set of non-informative selected categories (*apply_filter* command).

```
masked = filter_reads(reads_file,

                      max_molecule_length=max_molecule_length,

                      max_frag_size=max_fragment_size,

                      min_dist_to_re=minimum_distance_to_re)


apply_filter(reads_file, valid_reads_file, masked,

                      filters=list_of_filters_numbers)
```

For the detailed list of commands to perform the filtering step, see the notebook Methods-6-Filtering mapped reads.

In all the step performed so far, the mapped read-ends were stored in tab-separated value (TSV) files. At this stage, it is convenient for faster processing to convert the TSV file to binary alignment map (BAM) format:

```
bed2D_to_BAMhic(valid_reads_file, valid=True, outbam=bam_file)
```

3.5 Generation of the Raw Interaction Matrix

The interaction matrices are generated by partitioning the genome in *loci* of equal length (bins) and assigning each end of the read to its binned genomic location. We normally refer to this process as binning, which will define the resolution of the Hi-C matrix. Therefore, it is important to choose a suitable bin size (*see* **Note 6**).

3.5.1 Plotting the Interaction Matrix

Plotting and visually inspecting the interaction matrix is a key quality control of a Hi-C experiment. In TADbit, this is achieved by the *hic_map* function:

```
hic_map(hic_data, resolution=resolution, show=True)
```

In this view, TADbit also provides some statistics on the quality of these interactions as (1) the ratio between the intra-chromosome and the total number of interactions (i.e., cis-to-trans ratio), which is expected to be at least 50% in mammals, and (2) the three first eigenvectors of the matrix that summarize the principal structural features of the matrix (Fig. 5a and Subheading 3.7).

3.5.2 Filter Bins with Low Interaction Counts

Very low coverage bins are identified and filtered out from the data analysis (Fig. 5b). For example, these removed bins can contain no RE site (or no mappable region around its RE sites) and so result in zero valid pairs. However, those bins may still contain some valid pairs as the results of mapping errors or random breakage of the DNA before the cross-linking. It is also likely that very low coverage bins present a cis-to-trans ratio typical of random ligation events [26]. Such bins could also cause technical problems, as they will introduce a strong variability in most of the normalization strategies. To visualize the number of bins to be filtered, TADbit can produce a histogram plot using the *filter_columns* function (Methods-7-Bin filtering and normalization).

```
hic_data.filter_columns(draw_hist=True, by_mean=True)
```

3.6 Normalization

Interaction matrices derived from Hi-C experiments contain different types of biases [26, 27], which need to be removed (Fig. 5c). This process is referred to as normalization, which can be done within TADbit using different normalization procedures:

3.6.1 Normalization by Visibility

- *Iterative Correction and Eigenvector decomposition (ICE)*: ICE assumes equal experimental visibility of each bin and seeks iteratively for biases that equalize the sum of counts per bin in the matrix [26]. At each iteration, a new matrix is generated by dividing each cell by the product of the sum of counts in its row times the sum of counts in its column. The process converges to a matrix in which all bins have identical sums.

```
hic_data.normalize_hic(iterations=number_of_iterations)
```

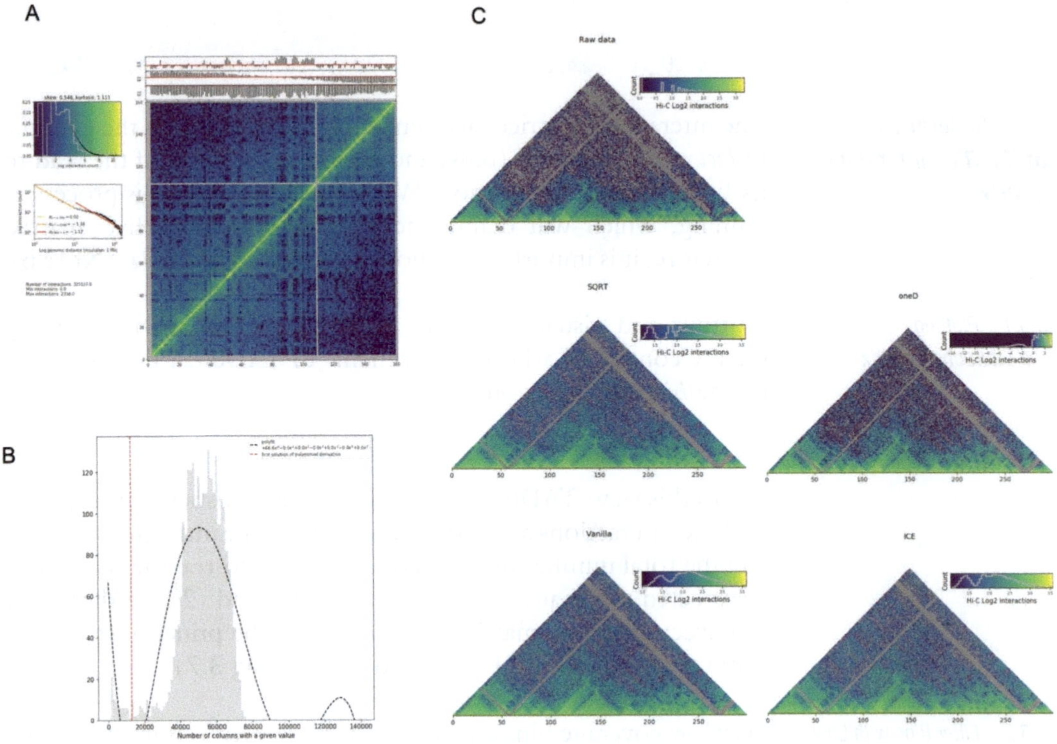

Fig. 5 Matrix binning and normalization. (**a**) The standard Hi-C map representation in TADbit also summarizes other relevant information. On the left, the histogram of the number of interactions (*top*), the plot of the interaction counts vs genomic distance (*middle*), and descriptive statistics (the sum, the minimum and the maximum of the pairwise interactions) (*bottom*) are shown. On the right, there is the binned interaction matrix represented in Log2 color scale and the first three eigenvectors of the matrix on top. (**b**) Histogram of the number of entries with non-zero values per bin. The histogram is expected to be bimodal. One peak is usually close to zero and indicates artifactual bins that are almost empty. The other peak appears at higher values and corresponds to the expected number of non-zero entries for informative bins. A polynomial function is fitted to this distribution in order to filter out bins falling into the peak with almost all the cells empty. (**c**) The Hi-C maps are shown in the normalized forms obtained with the different normalization algorithms

- *Vanilla coverage*: is a variation of the ICE where a single iteration is performed [5].

```
hic_data.normalize_hic(iterations=0)
```

- *Square root vanilla coverage*: is a variation of the vanilla coverage where each element in the matrix is divided by the square root of the product of sums of counts [11].

```
hic_data.normalize_hic(iterations=0, sqrt=True)
```

3.6.2 Normalization Via Individual Bias Estimation

- *OneD*: is based on fitting a non-linear model between the total amount of contacts per bin and the known biases, which are by default the GC content, the number of RE sites, and the mappability [22].

```
biases = oneD(tot=counts_per_column, map=mappability,

              res=number_of_re_sites_per_column,

              cg=gc_content_per_column) hic_data.bias =

              list(biases)
```

The detailed list of command to normalize a Hi-C matrix is given in the notebook Methods-7-Bin filtering and normalization.

3.7 Call for Structural Features

3.7.1 Compartments

Previous analysis of Hi-C experiments [5, 11] showed that the genome is structurally organized in two major types of compartments named A and B, which are enriched with active and inactive chromatin, respectively. Various approaches exist to determine genomic sub-compartments using Hi-C data [5, 11, 26, 27]. TADbit uses the change in sign in the components of the first eigenvector (calculated on the autocorrelation matrix of the normalized Hi-C matrix) to detect boundaries between A and B compartments [5] (Fig. 6a and **Note 7**). To define the compartment to the corresponding type, TADbit can use various type of genetic or epigenetic activity markers such as GC content (Subheading 3.1.2 and Fig. 6a).

```
hic_data.find_compartments(show_compartment_labels=True, show=True,

              crms=list_of_chromosomes,

              rich_in_A=rich_in_A_marker)
```

3.7.2 Topologically Associating Domains

Topologically Associating Domains (TADs) constitute the next level of organization of the genome structure. Several TAD callers exist based on a variety of metrics and statistics. Briefly they consist of two types: (1) tools that detect TAD borders (also known as breakpoints) assuming that the whole genome consists of a succession of TADs (e.g., methods based on the insulation score [28]) and (2) tools that identify TADs as denser interacting regions of the genome (e.g., methods using the directionality index [9]). Importantly, the choice of the suitable TAD caller is mainly dependent on the hypothesis to be tested. However, it has been shown that the results among the most used tools are overall consistent [29]. TADbit implements several TAD callers giving the user the possibility to choose the one that best fits its needs. Thus, it allows the comparison between different strategies (Fig. 6b–d). The available TAD callers in TADbit are:

- *TADbit*: is the TAD caller algorithm after which TADbit framework was named. It is a breakpoint detection algorithm that defines TADs within a chromosome (or genomic region)

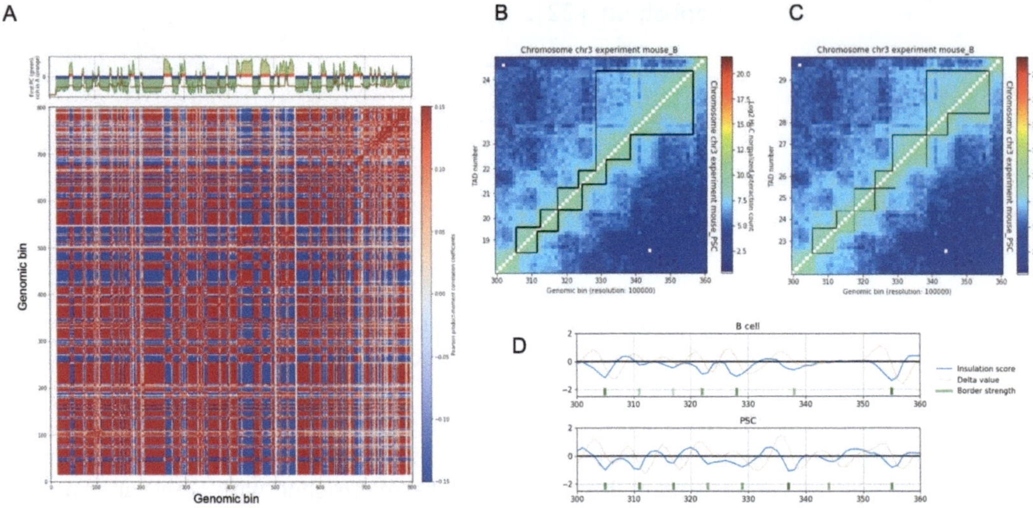

Fig. 6 Detection of the structural features in Hi-C interaction matrices. (**a**) Autocorrelation matrix of a OneD normalized matrix of mouse chromosome 3 at 200 kb resolution. The following steps were performed to obtain the autocorrelated matrix: (1) the raw Hi-C matrix was normalized with the OneD algorithm to correct for experimental and genomic biases, (2) the decay of the number of interactions with the genomic distance was corrected by dividing each cell by the average interactions in its diagonal, and (3) the obtained matrix was used to compute each element (a_{ij}) of the autocorrelation matrix as the Pearson correlation of the *i*-th row and the *j*-th column. On the top panel, we show the first eigenvector of the autocorrelation matrix, the average GC content in each compartment, and the prediction of compartment labeling (based on the average GC content). The changes in sign of the eigenvector components mark the boundaries between compartments (*green*), but the assignment of each compartment to type A or B is based on the average GC content with A compartments on top of the red bands (high CG content) and B compartments on top of blue bands (low GC content). (**b**) and (**c**) TAD border detections obtained using tadbit (**b**) and TopDom (**c**) in a region of mouse chromosome 3 for two different cell types: B cells (*top left*) and iPS cells (*bottom right*). The width of the partitioning lines is proportional to the confidence of the predicted border. (**d**) Insulation score in blue, the delta inferred from it in orange, and the prediction of TAD border in scales of green depending on the strength of the border. The two panels correspond to the same region of chromosome 3 and the same cell types of panels (**b**) and (**c**), respectively

under the BIC-penalized (Bayesian Information Criterion) likelihood [15]. The detected TAD borders are associated with a score (from 1 to 10) quantifying the accuracy of the border detection (Fig. 6b).

```
crm.find_tad(list_of_experiments)
```

- *TopDom*: identifies TAD borders based on the assumption that contact frequencies between regions upstream and downstream of a border are lower than those between two regions within a TAD. The algorithm only depends on a single parameter corresponding to the window size [30]. The algorithm provides a measure (from 0 to 10) of confidence on the accuracy of the border detection (Fig. 6c).

```
crm.find_tad(list_of_experiments, use_topdom=True)
```

- *Insulation score*: can be used to build an insulation profile of the genome and, with a simple transformation, to identify TAD borders [28, 31] (Fig. 6d).

```
insulation_values = insulation_score(input_matrix, locus_size,

delta_span)

borders = insulation_to_borders(insulation_values,

minimum_strength)
```

TAD borders called using *TADbit* or *TopDom* can be saved to text files and visualized using the *visualize* function. More details are in notebook Methods-8-Compartments and TADs detection.

```
crm.visualize(list_of_experiments, paint_tads=True)
```

3.8 Comparison between Hi-C Experiments

An important aspect of the Hi-C data analysis is the comparison of different experiments (e.g., technical or biological replicas, different time points or conditions, etc.) that can unveil fundamental biological insights. The comparison can be carried out at different levels of the data processing from the interaction matrices to the structural features or between 3D models of the chromatin (Met hods-9-Compare and merge Hi-C experiments, Subheading 3.9).

3.8.1 Comparison Between Interaction Matrices

In TADbit the user can monitor several quantities to compare Hi-C matrices:

- The Spearman rank correlation of the matrix diagonals at increasing genomic distances and the stratum-adjusted correlation coefficient (SCC) score [32]. Both quantities range from −1 (anti-) to 1 (direct correlation) (Fig. 7a).

```
spearmans, dists, scc, std = correlate_matrices(hic_data1,

                                               hic_data2, show=True,

                                               normalized=True)
```

- The correlation of the eigenvectors. Since the eigenvectors of a matrix capture its internal correlations [26], two matrices with highly correlation of eigenvectors are considered to have similar structure (Fig. 7b).

```
eig_correlate_matrices(hic_data1, hic_data2, show=True,

normalized=True)
```

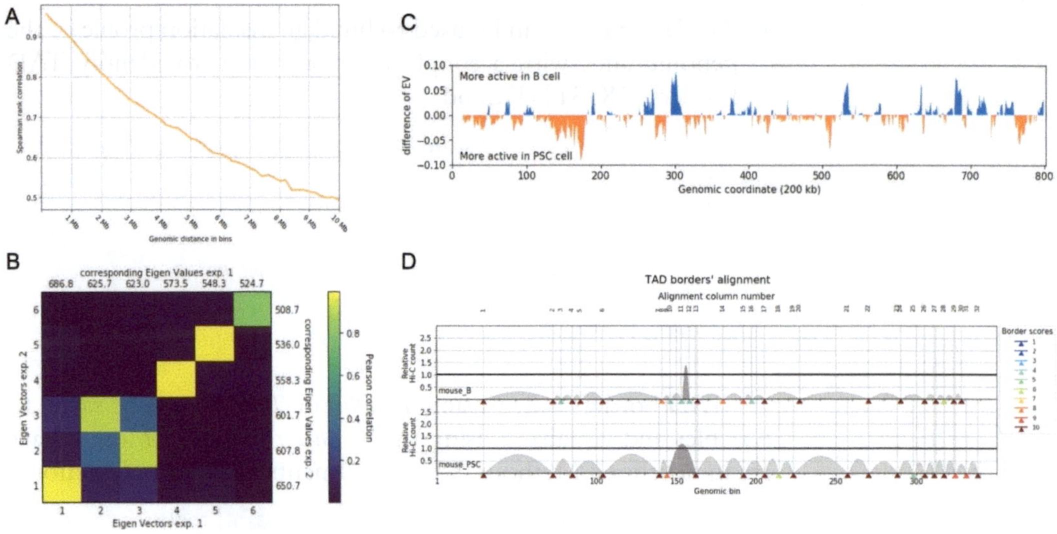

Fig. 7 Comparison between Hi-C experiments. (**a**) Spearman correlation coefficient between diagonals of the two replicates matrices in B cells. The profile in this plot is typical of highly similar matrices. (**b**) Pearson correlation of all combinations of the first six eigenvectors of these same two matrices. As in panel (**a**), these eigenvector correlations are typical of similar matrices. (**c**) The comparison of compartment partitions in B vs iPS cells is shown as the difference between the eigenvector of the autocorrelation matrix (Fig. 6a) computed in each of the two conditions. Blue peaks indicate increased activity in B cells and red peaks in iPS cells. (**d**) Alignment of TAD borders in B vs iPS cells. Each arc represents a TAD and each colored triangle a TAD border. The confidence score of each border is conveyed by the color scale from 1 (low in blue) to 10 (high confidence in red)

- The reproducibility score. Computed as in HiC-spector [33], it is also based on comparing eigenvectors. The reproducibility score ranges from 0 (low similarity) to 1 (identity).

```
reprod = get_reproducibility(hic_data1, hic_data2,

                             num_evec=number_ev_to_compare,

                             normalized=True)
```

For details see Methods-9-Compare and merge Hi-C experiments.

3.8.2 Comparison of the Structural Features

- *Compartments.* The eigenvector analysis performed to detect compartments can also be used to compare experiments. For instance, the simple difference between the first eigenvector (i.e., the one that captures the compartmentalization of the genome) of the different experiments allows detecting changes in the type of compartments between two different conditions (Fig. 7c).

- *TADs*. TAD border conservation across experiments can be used as an additional comparative measure. However, the assessment of co-occurrence of TAD borders in two different matrices is not trivial due to the border strength variability and their dynamic behavior under different conditions. For an effective TAD border comparison, TADbit allows the alignment of TAD borders using a reciprocal best hit strategy (Fig. 7d).

```
ali = crm.align_experiments(list_of_experiments)
```

For details on structural features comparison, see the notebook Methods-8-Compartments and TADs detection.

3.9 3D Modeling

In TADbit, 3D models of genomic regions can be generated via a restraint-based modeling approach [34], by transforming the normalized Hi-C interaction matrix (Subheading 6) into a set of spatial restraints that are satisfied using the Integrative Modeling Platform (IMP) [19]. Note, that TADbit provides an assessment score named Matrix Modeling Potential (MMP) score [35] to predict whether an interaction matrix can be used to produce reliable 3D models. It is recommended to assess such score before 3D models generation.

3.9.1 TADbit Modeling Approach

TADbit modeling approach is based on three main steps. These steps are briefly described [34]:

- *System representation*. Each bin of the matrix is described as a spherical particle of size proportional to its DNA content.
- *Scoring*. Each normalized Hi-C interaction count is Z-score transformed (Fig. 8a) and associated with a harmonic restraint. The simple rationale behind this transformation is that the more two particles interact, the stronger is the spring constant and the smaller is the equilibrium distance of the associated harmonics [34]. Briefly, this transformation depends on three parameters: *maxdist* (i.e., the maximal target distance between two non-interacting particles), as well as *upfreq* and *lowfreq*, which determine particle pairs that interact more than expected as well as those that do less than expected, respectively. Pairs of particles with a Z-score larger than the *upfreq* will be restrained to be close in 3D space, and particles with a Z-score smaller than *lowfreq* will be kept apart in 3D space. The sum of all the harmonic restraints is the so-called scoring function. The lower the *scoring function*, the closer the corresponding 3D structure is to the target pattern of interactions [34].
- *Sampling*. To find the 3D arrangement of the particles that better represent the input Hi-C interactions pattern, the scoring function is minimized using a Monte Carlo simulated annealing sampling protocol with the Metropolis criterion [34].

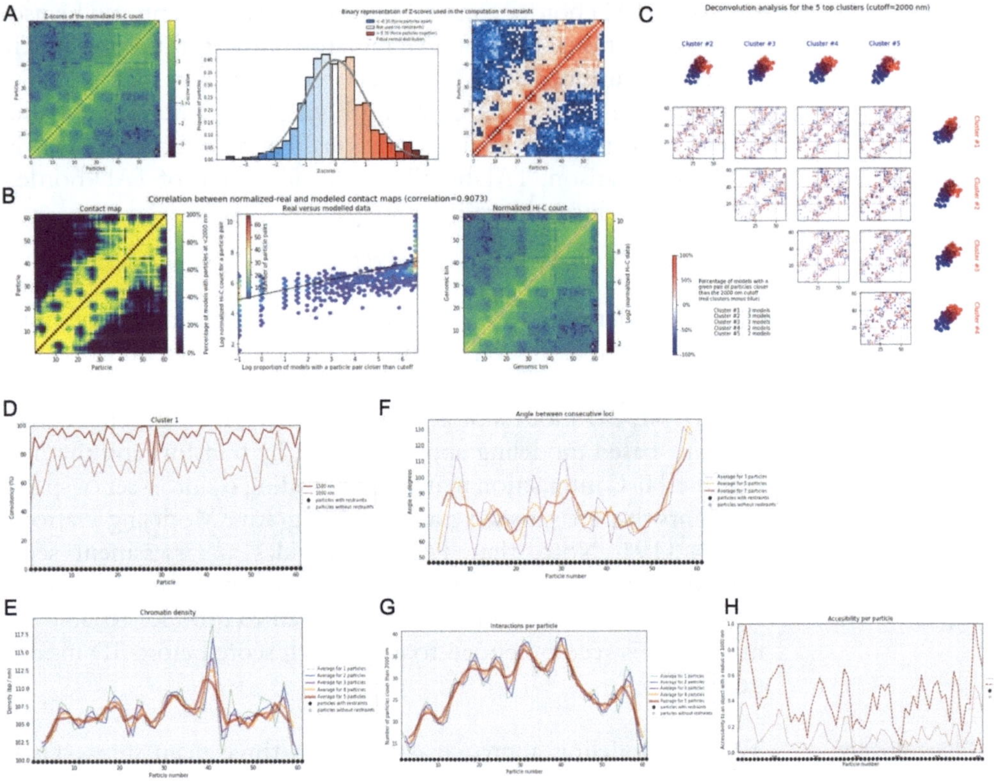

Fig. 8 3D model generation and analysis. (**a**) Interaction matrix for B cells (*left*) and its Z-score transformation (*right*). The central panel shows the distribution of the Z-score values divided into three categories: the pairs of *loci* that are restrained to be far in blue, unrestrained pairs in white, and pairs with higher interaction counts restrained to be close to each other in red. (**b**) Quantitative comparison between the contact map computed from the generated 3D models using a distance cutoff of 2000 nm at 100 kb resolution (*left*) and the original Hi-C interaction matrix of B cells (*right*). The central panel shows the scatter plot of the number of contacts observed in the simulated models (the virtual contact matrix) and the original Hi-C interaction matrix. The Spearman correlation coefficient between the two matrices is 0.91 in this specific case. (**c**) Deconvolution analysis generated ensembles of models, which are compared and clustered to identify local differences between subpopulations of 3D models. (**d**), (**e**), (**f**), (**g**), and (**h**) Structural measures from the 3D model ensemble. As discussed in the main text, they are the consistency, chromatin density, walking angles, number of interactions (*dcutoff* = 2000 nm), and accessibility, respectively

3.9.2 Parameter Optimization

In TADbit, the optimal parameters *(maxdist, lowfreq, upfreq)* are determined empirically via a grid search defined by the user. Other parameters used in the reconstruction of the model can also be optimized at the same time, like the distance cutoff that defines contact between particles. However, those can be ignored by the end-user and are less sensitive to the specifics of the Hi-C matrix experiment to be modeled. The optimization command can be run in parallel in different computers using in each case a predefined number of CPUs:

```
optimal_params = exp.optimal_imp_parameters(start=start_bin,
                                            end=end_bin, n_models=nbr_models,
                                            n_keep=nbr_models_keep,
                                            n_cpus=nbr_cpus,
                                            upfreq_range=(start_upfreq,
                                            end_upfreq, step_upfreq),
                                            lowfreq_range=(start_lowfreq,
                                            end_lowfreq, step_lowfreq),
                                            maxdist_range=(start_maxdist,
                                            end_maxdist, step_maxdist),
                                            dcutoff_range=list_of_cutoffs)
```

For each possible combination of the three parameters, TADbit will produce a set of models (*nbr_models*) from which it will keep the ones best satisfying the imposed restraints (top *nbr_models_-keep*). The ratio between the numbers of kept and generated models is typically 20%. During the search of the optimal parameters, *nbr_models_keep* = 100 and *nbr_models* = 500 are the recommended values. For assessing the best combination of parameters, a contact matrix is computed from the best models choosing an optimal distance cutoff (*dcutoff*) and compared with the input normalized Hi-C matrix using the Spearman correlation coefficient. The results of the optimization grid search can be visualized as a 2D matrix:

```
optimal_params.plot_2d(show_best=5)
```

The best set of parameters, those associated with the highest Spearman correlation coefficient, can be retrieved and used for the next modeling step. For details, see notebook Methods-10-Modeling parameters optimization:

```
optimal_params.get_best_parameters_dict()
```

3.9.3 Model Building

The optimized modeling parameters are used next to build a more exhaustive *ensemble* of 3D models usually ten times larger than the one used in the optimization step (i.e., *nbr_models* = 5000 and *nbr_models_keep* = 1000). The larger number of models is meant to represent the heterogeneity of chromatin conformations in the cell population of the Hi-C experiment; for details see the notebook Methods-11-3D Models production and analysis.

```
models = exp.model_region(start=start_bin, end=end_bin,
                          n_models=nbr_models, n_keep=nbr_models_keep,
                          n_cpus=nbr_cpus, config=optimal_params)
```

3.9.4 3D Model Assessment and Analysis

TADbit offers a set of functions for the evaluation and analysis of the generated ensemble of models. For example, one can:

- Visualize how the experimental data were converted into harmonic restraints (Fig. 8a).

```
models.zscore_plot()
```

- Verify the convergence of the Monte Carlo optimization step by checking that the IMP scoring function decays as a function of the Monte Carlo iteration and reaches a stable *plateau*.

```
models[0].objective_function(log=True)
```

- Compare a set of models with the original experimental data using Spearman correlation coefficient (Fig. 8b).

```
models.correlate_with_real_data(plot=True, cutoff=cutoff)
```

3.9.5 Subpopulations of Models

TADbit can measure similarities between the models in the generated ensemble via the structural alignment of each pair of models using a pairwise rigid-body superposition that minimizes their root-mean-square deviation (RMSD).

```
models.align_models(in_place=True)
```

Using the aligned models and clustering them based on 3D comparative metrics, the ensemble of models can be deconvolved into subpopulations of structures that might represent the variability in the subpopulations of cells in the experimental sample (Fig. 8c).

```
models.deconvolve(fact=factor_eq_positions, dcutoff=cutoff)
```

This analysis generates differential contact maps between each pair of clusters and allows the detection of shared or exclusive sets of contacts within the ensemble. Alternatively, it is possible to cluster the model ensemble using other measures as the RMSD, distance-RMSD, number of equivalent positions, or a combination of all the above measures. Each cluster will represent a given subpopulation of chromatin structures.

```
models.cluster_models(fact=factor_eq_positions, dcutoff=cutoff)
```

3.9.6 Quantitative Characterization of the 3D Models

TADbit also provides a set of measures averaged over the models (or in a given subset of models such as those in a cluster) to extract useful biological insights (Fig. 8d–h). Next, we briefly describe such measures:

- The consistency score (Fig. 8d) is the percentage of models that have a given particle superimposed within a predefined distance cutoff. The lower is the consistency value, the less deterministic (more variable) are the models in the corresponding position. The consistency score should be measured independently for each cluster.

```
models.model_consistency()
```

- The density (or local compactness, Fig. 8e) is the number of base-pairs per nanometer (nm) and is computed as the ratio of the number of base-pairs per particle and the distance between consecutive particles.

```
models.density_plot()
```

- The walking angle (Fig. 8f) is the angle between triplets of contiguous particles. The higher are these values, the straighter are the models.

```
models.walking_angle()
```

- The number of interactions (Fig. 8g) measures contacts made by each particle within a given cutoff distance.

```
models.interactions(cutoff=cutoff)
```

- The accessibility score (Fig. 8h) measures how accessible a particle is to a spherical object of a predefined radius. Note that the accessibility is measured only for the internal space of the models as the 3D neighborhood of the modeled region is normally not determined.

```
models.accessibility()
```

3.10 3D Visualization

The 3D models generated by TADbit can be saved in different formats, which can then be input to other external software for visualization. TADbit currently outputs models in three main formats:

- JSON file is the most efficient to store the results of the TADbit analysis. This file contains the particle coordinates, the experimental data structure (models, clusters, centroids, normalized interaction matrix, and restraints), and the project metadata. These can be visualized alongside genomic features and matrix datasets within the TADkit visualization tool (https://github.com/3DGenomes/TADkit).

```
models.write_json(filename=output_file)
```

- XYZ/CMM file of particle coordinates. These can be examined and rendered in the majority of the molecular visualization software, e.g., Chimera [36], Delta [37], PyMOL [38], or VMD [39].

```
models.write_cmm(directory=output_directory)

models.write_xyz(directory=output_directory)
```

4 Notes

1. To speed up the data preprocessing, we recommend using split FASTQ files, one for each read-end. We also suggest working with compressed files to save disk space. To this end, TADbit accepts FASTQ files compressed in many formats such as zip, bzip2, gzip, tar, and DSRC [23].
2. The Hi-C-specific quality plots provided by TADbit (Fig. 1) use the first million paired-ends of the input FASTQ file to assess the quality of the entire dataset. From the quality plot measures, one can infer useful indicators of the quality of the data:
   (a) The first plot, that is common in next-generation sequencing (NGS), describes the quality of the sequencing with PHRED scores and proportion of *Ns* per position. A PHRED score larger than 30 (which represents a 99.9% accuracy of the base call) and a number of *Ns* close to zero are indicative of high confidence in the sequenced product.
   (b) The percentage of digested sites (the ratio of digested over undigested sites) should be larger than 60%. Lower values indicate a problem in the digestion step of the Hi-C experiment.
   (c) The percentage of dangling-ends (the number of time a digested site is found at the beginning of a read) should be between 1% and 10%. This percentage varies also depending on the average length of the restriction fragments; the larger they are, the fewer dangling-ends should be expected. Large percentages indicate a problem in ligation efficiency.
   (d) The percentage of ligation sites (the number of times a ligation site is found in the processed reads) should be higher than 15% for read length of 75 nucleotides (nt) and

sequenced fragment size ~300 nt. Low numbers here indicate a problem in ligation efficiency. These percentages are calculated as follows. Assuming that the ligation sites are homogeneously distributed along the genome, their average number per sequenced fragment can be estimated by extrapolating it to the full length of the sequenced fragment ~300 nt. For example, in the case of Figs. 1, 20.5% of the reads present (at least) one ligation site. Given that our search space for this set of reads is 67 nt (read length is 75, minus the size of the 8 nt ligation pattern GATCGATC), and assuming a constant number of ligation sites in all the 300 nt fragment, we would expect to find a ligation site in about 92% of the reads. Usually, we expect that all the reads in the library contain biotinylated nucleotides. It means that each read has either a ligation site or a dangling-end. In this case, we have calculated that the proportions of ligation sites and dangling-ends sum up to about 100% as expected.

3. The inference of the average RE fragment size in the sequenced library is relevant for the interpretation of the Hi-C data because valid interactions consist of read-ends mapped close to a RE cut site, and this definition of proximity relies on the estimated size of the sequenced DNA fragments.
4. Some of the filters applied to the set of pairs of mapped read-ends may considerably reduce the number of valid interactions while not increasing the general quality of the resulting interaction matrix. In the TADbit implementation, none of the filters is mandatory but can be switched on or off making it easy to adapt to other 3C-based methods. According to our experience, for a Hi-C experiment, the user should define a minimum set of filters to apply including (1) any paired-end reads mapped in a single restriction fragment or very close to the diagonal (self-circle, dangling-end, error, and extra dangling-end), (2) read-ends mapped in too short or too long defined RE fragments, (3) PCR artifacts (duplicates), and (4) random breaks.
5. This filter may be too conservative for the in situ Hi-C protocol [11], and it is usually not applied.
6. In Hi-C data analysis, the choice of the binning is limited usually by the amount and the quality of data or more rarely by the available computational power. However, some of the structural features that can be detected may appear only at certain scales above or below the chosen binning. For example, in mammalian genomes, the compartments are usually detected optimally at 100 kb binning. Depending on the interaction density, it may be more difficult to determine them at

finer scales. To ensure that a given binning is adequate to the scale of the measured structure or, more simply, does not entail too much noise, a good strategy is to check for the consistency between replicates or different resolutions. In this chapter, we provide an example of the detection of TAD borders in the notebook Methods-8-Compartments and TADs detection and Fig. 6. Additionally, binning is important in the modeling step. The genomic region to model could vary in size from few kb to Mb because it can be defined as a chromosome part, an entire chromosome, or also a set of chromosomes. Hence, the resolution used in each situation should be tailored to the computational power to avoid having too many particles and restraints during modeling. Normally the number of particles to model should be maintained below 5000.

7. The sign variation of the first eigenvector of the autocorrelated Hi-C matrix usually represents the borders between compartments (Fig. 6a), but, especially in the case of low-quality Hi-C experiments or aberrant karyotypes, it could also describe other correlations in the matrix. In such cases, the A/B compartments can be explained by the second or the third eigenvector. To this end, it is essential to visually check the correspondence between the checkerboard pattern characteristic of compartments and the eigenvectors, especially when working with atypical datasets. Other more automatic metrics can be used to confirm the choice of a given eigenvector, like the level of correlation with some activity marker (GC content, RNA-seq, or epigenetic marks), or the consistency with the compartment call done at a lower resolution.

   In the case of the detection of TAD borders, a typical problem is the choice of the adequate binning. The first concern is that an increased resolution is correlated with increased noise [29]. A good strategy is to test the consistency of several resolutions between independent replicates. An example is provided using the insulation score strategy (Fig. 9, Methods-8-Compartments and TADs detection). It shows that in these datasets a resolution between 50 kb and 100 kb is optimal to call TAD border, while that at higher resolutions (<50 kb) the consistency between time points is lost.

Acknowledgments

We thank all the current and past members of the Marti-Renom lab for their continuous discussions and support to the development of TADbit. I.F. was supported by the Ministerio de Ciencia, Innovación y Universidades of Spain (IJCI-2015-23352). M.A.M-R was supported by the European Research Council under the seventh

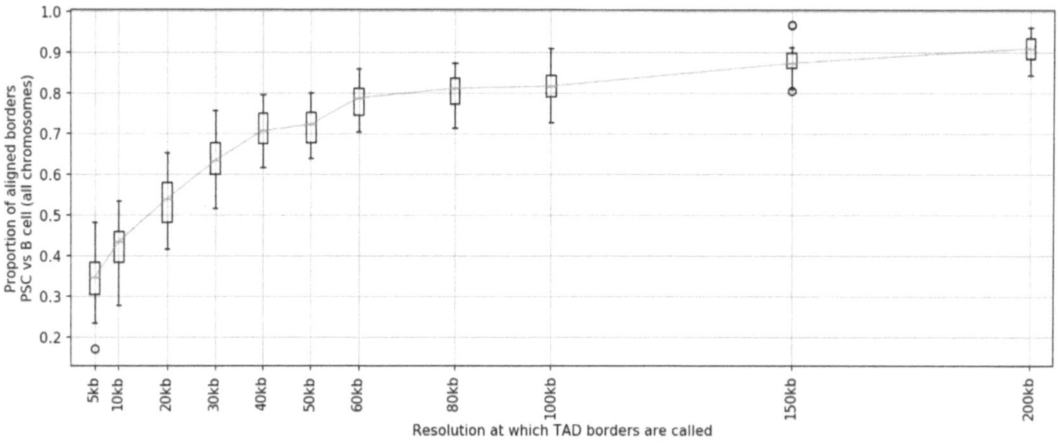

Fig. 9 TAD border consistency as a function of matrix binning. Comparison of the number of TAD borders that are shared (plus-minus one bin) between the two replicates in two different cell types (B and iPS cells) when TAD borders are called at a different matrix resolution. The TAD borders are called using the insulation score with a window of 500 kb and a delta of 100 kb

Framework Program FP7/2007-2013 (ERC grant agreement 609989), the European Union's Horizon 2020 research and innovation program (grant agreement 676556), and the Spanish Ministry of Economy and Competitiveness (BFU2013-47736-P and BFU2017-85926-P). We also acknowledge support from "Centro de Excelencia Severo Ochoa 2013-2017," SEV-2012-0208, and the CERCA Programme/Generalitat de Catalunya to the CRG. Marco Di Stefano and David Castillo contributed equally to this work.

References

1. Dekker J, Rippe K, Dekker M, Kleckner N (2002) Capturing chromosome conformation. Science 295:1306–1311
2. Dekker J, Marti-Renom MA, Mirny LA (2013) Exploring the three-dimensional organization of genomes: interpreting chromatin interaction data. Nat Rev Genet 14:390–403
3. Grob S, Cavalli G (2018) Technical review: a Hitchhiker's guide to chromosome conformation capture. Methods Mol Biol 1675:233–246
4. Kim TH, Dekker J (2018) 3C-based chromatin interaction analyses. Cold Spring Harb Protoc 2018:pdb top097832
5. Lieberman-Aiden E, van Berkum NL, Williams L, Imakaev M, Ragoczy T, Telling A, Amit I, Lajoie BR, Sabo PJ, Dorschner MO, Sandstrom R, Bernstein B, Bender MA, Groudine M, Gnirke A, Stamatoyannopoulos J, Mirny LA, Lander ES, Dekker J (2009) Comprehensive mapping of long-range interactions reveals folding principles of the human genome. Science 326:289–293
6. Zink D, Cremer T, Saffrich R, Fischer R, Trendelenburg MF, Ansorge W, Stelzer EH (1998) Structure and dynamics of human interphase chromosome territories in vivo. Hum Genet 102:241–251
7. Cremer T, Cremer C (2001) Chromosome territories, nuclear architecture and gene regulation in mammalian cells. Nat Rev Genet 2:292–301
8. Boyle S, Gilchrist S, Bridger JM, Mahy NL, Ellis JA, Bickmore WA (2001) The spatial organization of human chromosomes within the nuclei of normal and emerin-mutant cells. Hum Mol Genet 10:211–219
9. Dixon JR, Selvaraj S, Yue F, Kim A, Li Y, Shen Y, Hu M, Liu JS, Ren B (2012)

Topological domains in mammalian genomes identified by analysis of chromatin interactions. Nature 485:376–380

10. Nora EP, Lajoie BR, Schulz EG, Giorgetti L, Okamoto I, Servant N, Piolot T, van Berkum NL, Meisig J, Sedat J, Gribnau J, Barillot E, Bluthgen N, Dekker J, Heard E (2012) Spatial partitioning of the regulatory landscape of the X-inactivation Centre. Nature 485:381–385
11. Rao SS, Huntley MH, Durand NC, Stamenova EK, Bochkov ID, Robinson JT, Sanborn AL, Machol I, Omer AD, Lander ES, Aiden EL (2014) A 3D map of the human genome at kilobase resolution reveals principles of chromatin looping. Cell 159:1665–1680
12. Rowley MJ, Corces VG (2018) Organizational principles of 3D genome architecture. Nat Rev Genet 19:789–800
13. Bau D, Sanyal A, Lajoie BR, Capriotti E, Byron M, Lawrence JB, Dekker J, Marti-Renom MA (2011) The three-dimensional folding of the alpha-globin gene domain reveals formation of chromatin globules. Nat Struct Mol Biol 18:107–114
14. Le Dily F, Bau D, Pohl A, Vicent GP, Serra F, Soronellas D, Castellano G, Wright RH, Ballare C, Filion G, Marti-Renom MA, Beato M (2014) Distinct structural transitions of chromatin topological domains correlate with coordinated hormone-induced gene regulation. Genes Dev 28:2151–2162
15. Serra F, Bau D, Goodstadt M, Castillo D, Filion GJ, Marti-Renom MA (2017) Automatic analysis and 3D-modelling of hi-C data using TADbit reveals structural features of the fly chromatin colors. PLoS Comput Biol 13: e1005665
16. Stadhouders R, Vidal E, Serra F, Di Stefano B, Le Dily F, Quilez J, Gomez A, Collombet S, Berenguer C, Cuartero Y, Hecht J, Filion GJ, Beato M, Marti-Renom MA, Graf T (2018) Transcription factors orchestrate dynamic interplay between genome topology and gene regulation during cell reprogramming. Nat Genet 50:238–249
17. Virtanen P, Gommers R, Oliphant TE, et al. (2020) SciPy 1.0: fundamental algorithms for scientific computing in Python. Nat Methods 17(3):261-272
18. Hunter JD (2007) Matplotlib: a 2D graphics environment. Comput Sci Eng 9:90–95
19. Russel D, Lasker K, Webb B, Velazquez-Muriel J, Tjioe E, Schneidman-Duhovny D, Peterson B, Sali A (2012) Putting the pieces together: integrative modeling platform software for structure determination of macromolecular assemblies. PLoS Biol 10:e1001244
20. Li H, Handsaker B, Wysoker A, Fennell T, Ruan J, Homer N, Marth G, Abecasis G, Durbin R, 1000 Genome Project Data Processing Subgroup (2009) The sequence alignment/map format and SAMtools. Bioinformatics 25:2078–2079
21. Marco-Sola S, Sammeth M, Guigo R, Ribeca P (2012) The GEM mapper: fast, accurate and versatile alignment by filtration. Nat Methods 9:1185–1188
22. Vidal E, le Dily F, Quilez J, Stadhouders R, Cuartero Y, Graf T, Marti-Renom MA, Beato M, Filion GJ (2018) OneD: increasing reproducibility of hi-C samples with abnormal karyotypes. Nucleic Acids Res 46:e49
23. Roguski L, Deorowicz S (2014) DSRC 2--industry-oriented compression of FASTQ files. Bioinformatics 30:2213–2215
24. Dekker J (2007) GC- and AT-rich chromatin domains differ in conformation and histone modification status and are differentially modulated by Rpd3p. Genome Biol 8:R116
25. Derrien T, Estelle J, Marco Sola S, Knowles DG, Raineri E, Guigo R, Ribeca P (2012) Fast computation and applications of genome mappability. PLoS One 7:e30377
26. Imakaev M, Fudenberg G, McCord RP, Naumova N, Goloborodko A, Lajoie BR, Dekker J, Mirny LA (2012) Iterative correction of hi-C data reveals hallmarks of chromosome organization. Nat Methods 9:999–1003
27. Yaffe E, Tanay A (2011) Probabilistic modeling of hi-C contact maps eliminates systematic biases to characterize global chromosomal architecture. Nat Genet 43:1059–1065
28. Crane E, Bian Q, McCord RP, Lajoie BR, Wheeler BS, Ralston EJ, Uzawa S, Dekker J, Meyer BJ (2015) Condensin-driven remodelling of X chromosome topology during dosage compensation. Nature 523:240–244
29. Forcato M, Nicoletti C, Pal K, Livi CM, Ferrari F, Bicciato S (2017) Comparison of computational methods for hi-C data analysis. Nat Methods 14(7):679–685
30. Shin H, Shi Y, Dai C, Tjong H, Gong K, Alber F, Zhou XJ (2016) TopDom: an efficient and deterministic method for identifying topological domains in genomes. Nucleic Acids Res 44:e70
31. Mizuguchi T, Fudenberg G, Mehta S, Belton JM, Taneja N, Folco HD, FitzGerald P, Dekker J, Mirny L, Barrowman J, Grewal SI (2014) Cohesin-dependent globules and heterochromatin shape 3D genome architecture in S. pombe. Nature 516:432–435
32. Yang T, Zhang F, Yardimci GG, Song F, Hardison RC, Noble WS, Yue F, Li Q (2017)

HiCRep: assessing the reproducibility of hi-C data using a stratum-adjusted correlation coefficient. Genome Res 27:1939–1949

33. Yan KK, Yardimci GG, Yan C, Noble WS, Gerstein M (2017) HiC-spector: a matrix library for spectral and reproducibility analysis of hi-C contact maps. Bioinformatics 33:2199–2201
34. Baù D, Marti-Renom MA (2012) Genome structure determination via 3C-based data integration by the integrative Modeling platform. Methods 58:300–306
35. Trussart M, Serra F, Bau D, Junier I, Serrano L, Marti-Renom MA (2015) Assessing the limits of restraint-based 3D modeling of genomes and genomic domains. Nucleic Acids Res 43:3465–3477
36. Pettersen EF, Goddard TD, Huang CC, Couch GS, Greenblatt DM, Meng EC, Ferrin TE (2004) UCSF chimera--a visualization system for exploratory research and analysis. J Comput Chem 25:1605–1612
37. Tang B, Li F, Li J, Zhao W, Zhang Z (2018) Delta: a new web-based 3D genome visualization and analysis platform. Bioinformatics 34:1409–1410
38. DeLano WL (2002) The PyMOL molecular graphics system on world wide web. http://www.pymol.org/in
39. Humphrey W, Dalke A, Schulten K (1996) VMD: visual molecular dynamics. J Mol Graph 14:33–38

Chapter 5

Targeted DNase Hi-C

Zhijun Duan

Abstract

Technology advance during the past decade has greatly expanded our understanding of the higher-order structure of the genome. The various chromosome conformation capture (3C)-based techniques such as Hi-C have provided the most widely used tools for interrogating three-dimensional (3D) genome organization. We recently developed a Hi-C variant, DNase Hi-C, for characterizing 3D genome organization. DNase Hi-C employs DNase I for chromatin fragmentation, aiming to overcome restriction enzyme digestion-related limitations associated with traditional Hi-C methods. By combining DNase Hi-C with DNA capture technology, we further implemented a high-throughput approach, called targeted DNase Hi-C, which enables to map fine-scale chromatin architecture at exceptionally high resolution and thereby is an ideal tool for mapping the physical landscapes of cis-regulatory networks and for characterizing phenotype-associated chromatin 3D signatures. Here, I describe a detailed protocol of targeted DNase Hi-C library preparation, which covers experimental steps starting from cell cross-linking to library amplification.

Key words 3C, Hi-C, DNase Hi-C, Chromatin, Chromosome, 3D genome

1 Introduction

The development and widespread application of 3C [1]-based high-throughput techniques [2, 3] such as Hi-C [4] has demonstrated that eukaryotic genomes are hierarchically organized in the nucleus [4–8]. This hierarchy consists of at least four distinct levels: whole chromosome territories (CTs) [9]; large-scale active or repressed compartments (A/B compartments) [4]; chromatin domains, e.g., topologically associated (TAD) [10, 11], lamina associated (LAD) [12, 13], or nucleolus associated (NAD) [14, 15]; and chromatin loops [16], which provide multiple regulatory layers for gene regulation and other genome functions and thereby play pivotal roles in development and disease [6–8]. For example, distal *cis*-elements (e.g., enhancers and super-enhancers (SE)) regulate gene expression through long range physical interactions via chromatin looping, which occurs much more frequently

Beatrice Bodega and Chiara Lanzuolo (eds.), *Capturing Chromosome Conformation: Methods and Protocols*,
Methods in Molecular Biology, vol. 2157, https://doi.org/10.1007/978-1-0716-0664-3_5,

within a TAD than between TADs. TADs are considered to be both building blocks and functional units that guide, constrain, and facilitate promoter-enhancer interactions [6–8]. Each TAD often contains multiple genes and enhancers, allowing for coordinated regulation. Hence, dissecting the 3D genome and its dynamics will provide information essential for us to completely understand gene regulation and the regulation of other nuclear processes.

Mapping the spatial organization of *cis*-regulatory elements in the nucleus will lead to new insights into the mechanisms of how distal enhancers regulate their target genes and how regulatory elements coordinate to achieve temporal and cell type-specific transcriptional regulation during development. Many 3C derivatives, such as ChIA-PET [17], 4C [18, 19], 5C [20], Hi-C [4], Capture-C [21], Capture Hi-C [22], and HiChIP/PLAC-seq [23, 24], can be used to map chromatin interaction landscapes of *cis*-regulatory elements. However, the ultimate resolution of the chromatin interaction maps constructed by these methods is limited by the genomic distribution of restriction enzyme (RE) sites, as these methods all employ RE digestion to fragment chromatins. To overcome this limitation, we recently developed DNase Hi-C [25] and its in situ version in situ DNase Hi-C [26], in which chromatin fragmentation is achieved by sequence-independent digestion of DNase I. By combining DNase Hi-C (or in situ DNase Hi-C) with DNA capture technologies, we subsequently developed targeted DNase Hi-C, a method for mapping fine-scale chromatin architectures of specific genomic regions of interest at high resolution in a massively parallel fashion [25].

In brief, as outlined in Fig. 1, the targeted DNase Hi-C protocol involves cross-linking cells with formaldehyde; chromatin is then randomly fragmented by DNase I. The resulting chromatin fragments are end-repaired and dA-tailed and then marked with a biotinylated internal adaptor, and proximity ligation is carried out in the intact nucleus to favor ligation events between the cross-linked DNA fragments. The resulting DNA sample contains ligation products consisting of chimeric DNA fragments that were originally in close spatial proximity in the nucleus, marked with biotin at the junction. A whole-genome chromatin interaction library is constructed by shearing the DNA, selecting the biotin-containing fragments with streptavidin magnetic beads, and PCR amplification. The chromatin contacts associated with a specific group of loci of interest are then enriched by a bait library targeting the loci via in-solution hybridization-based DNA capture. Finally, the enriched chromatin interactions are quantified by massively parallel deep sequencing. Data processing and further computational analysis can be carried out as previously described [25–27].

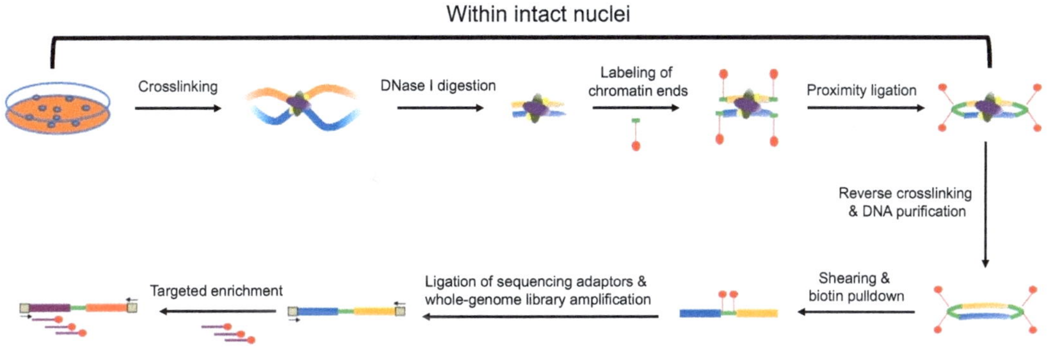

Fig. 1 Targeted DNase Hi-C workflow

2 Materials

2.1 Reagents

1. 37% (vol/vol) formaldehyde.
2. 10× NEBuffer 2 (500 mM NaCl, 100 mM Tris–HCl, 100 mM $MgCl_2$, 10 mM DTT, pH 7.9 at 25 °C).
3. 10% (wt/vol) UltraPure SDS.
4. DNase I, RNase-free (supplied with $MnCl_2$ and reaction buffer) (1 U/μl).
5. RNase A, DNase, and protease-free (10 mg/ml).
6. Klenow fragment (10 U/μl).
7. Klenow fragment (exo–) (5 U/μl).
8. T4 DNA polymerase (5 U/μl)).
9. T4 DNA ligase (5 U/μl) provided with 50% PEG-4000.
10. T4 polynucleotide kinase (10 U/μl).
11. 10× T4 DNA ligase buffer w/ ATP.
12. Agencourt AMPure XP.
13. 2× KAPA HiFi HotStart PCR ReadyMix.
14. Fast DNA End Repair Kit.
15. Proteinase K.
16. FastDigest® BamHI.
17. Dynabeads® MyOne™ Streptavidin C1.
18. dNTP Set (100 mM 4× 0.25 ml).
19. 10 mM dNTP mix.
20. 100 bp DNA ladder.
21. GlycoBlue (15 mg/ml).
22. 3 M sodium acetate (pH 5.2).
23. 100% ethanol.

24. 100% isopropanol.
25. 1 M UltraPure Tris–HCl pH 8.0.
26. 1 M UltraPure Tris–HCl pH 7.5.
27. Buffer EB (10 mM Tris–HCl, pH 8.5).
28. 0.5 M EDTA.
29. Qubit dsDNA HS kit.
30. 100% IGEPAL CA-630.
31. 100% Triton X-100.
32. 1× DPBS.
33. Protease inhibitor tablets.
34. QIAquick PCR Purification Kit.
35. Rapid DNA Ligation Kit.
36. 5 M sodium chloride (NaCl).
37. SeqCap EZ Reagent Kit v2.
38. Zymolyase 20T.

2.2 Buffers

1. Spheroplast buffer: 1 M sorbitol, 100 mM potassium phosphate (PH 7.5).
2. 2.5 M glycine: Bring 9.35 g glycine to 50 ml in ddH_2O and filter sterilize using Steriflip column. Store at RT for up to 6 months.
3. 80% (vol/vol) ethanol: Mix 8 ml 100% ethanol with 2 ml ddH_2O. Make fresh for every day of experiments.
4. 10% (vol/vol) Triton X-100: Mix 1 ml 100% Triton X-100 with 9 ml ddH_2O. Store at RT for up to 6 months.
5. 2% (vol/vol) Tween-20: Mix 1 ml 100% Tween-20 with 49 ml ddH_2O. Store at RT for up to 6 months.
6. 10% IGEPAL CA-630: Mix 5 ml 100% IGEPAL CA-630 with 45 ml ddH_2O.
7. 1× cell lysis buffer: Mix 500 μl 1 M Tris–HCl pH 8.0, 100 μl 5 M NaCl, 150 μl 1 M $MgCl_2$, and 1 ml 10% IGEPAL CA-630, and bring up to 50 ml in ddH_2O. Store at 4 °C for up to 6 months.
8. 1× TE lysis buffer: 50 mM Tris–HCl (pH 7.0), 1 mM EDTA, 1% SDS. Store at RT.
9. 2× B&W buffer: Mix 500 μl 1 M Tris–HCl pH 8.0, 100 μl 0.5 M EDTA, and 20 ml 5 M NaCl, and bring up to 50 ml in ddH_2O. Store at RT for up to 6 months.
10. 1× B&W buffer: Mix 25 ml 2× B&W buffer with 25 ml ddH_2O.

11. 1× B&W buffer with 0.1% Tween-20: Mix 25 ml 2× B&W buffer with 2.5 ml 2% Tween-20 and 22.5 ml ddH_2O.
12. 0.5× DNase SDS buffer: Mix 25 μl 10× DNase digestion buffer with 25 μl of 10 mM $MnCl_2$, 10–20 μl of 10% SDS, and 450 μl of ddH_2O. Use immediately.
13. 0.5× DNase digestion buffer: Mix 25 μl 10× DNase digestion buffer with 25 μl 10 mM $MnCl_2$, 100 μl 10% Triton X-100, and 350 μl of ddH_2O. Use immediately.
14. 20% PEG buffer: Mix 10 g PEG-8000 in 25 ml 5 M NaCl and bring to 50 ml with ddH_2O. Shake vigorously to mix until PEG-8000 has completely gone into solution. Store at 4 °C for up to 6 months.

2.3 Equipment

1. Water bath.
2. Thermocycler.
3. DynaMag Magnetic Rack (e.g., Life Technologies 12321D).
4. NanoDrop 1000.
5. Qubit fluorometer (e.g., Life Technologies Q33216).
6. 0.2 ml PCR tubes (e.g., Fisher 14-230-212).
7. 1.5 ml microcentrifuge tubes (e.g., Fisher 05408129).
8. 6% TBE-PAGE gels (e.g., Life Technologies EC6265BOX).
9. Cell scraper (e.g., Fisher 08-100-241).
10. 50 ml tube (e.g., Fisher 14-432-22).
11. Cell culture plates (e.g., Sigma CLS430167-100EA).
12. Microcentrifuge.
13. Vacuum concentrator.
14. Sonicator (e.g., Covaris S220 focused ultrasonicator).

2.4 Oligos and Primers

1. T-tailed biotinylated bridge adaptor 5′: /5Phos/GCTGAGGGA/iBiodT/C (IDT).
2. Bridge adaptor 3′T: CCTCAGCT (IDT).
3. Blunt bridge adaptor 5′: GCTGAGGGAC (IDT).
4. Blunt bridge adaptor 3′: CCTCAGC (IDT).
5. Illumina_PE_Adapt_F: ACACTCTTTCCCTACACGACGCTCTTCCGATC*T (IDT).
6. Illumina_PE_Adapt_R: PO4-GATCGGAAGAGCGGTTCAGCAGGAATGCCGAG (IDT).
7. Pre-capture PCR primer_F: ACACTCTTTCCCTACACGACG (IDT).

8. Pre-capture PCR primer_R: CGGTCTCGGCATTCCTGCT-GAACC (IDT).
9. Illumina_PE-PCR_primer_F: AATGATACGGCGACCACC-GAGATCTACACTCTTTCCCTACACGAC (IDT).
10. Illumina_PE-PCR_primer_R: CAAGCAGAAGACGGCA-TACGAGAT [8 bp barcode] CGGTCTCGGCATTCCTGCT-GAACCG (IDT).
11. Bait library (e.g., SeqCap® EZ Prime Choice Probes, Roche NimbleDesign).
12. Adaptor-Hi-block: ACACTCTTTCCCTACAC-GACGCTCTTCCGATCT (IDT).
13. Internal adaptor-block: AGCTGAGGGATCCCTCAGCT (IDT).
14. Nbgn-8bp-ID-BL-B: CGGTCTCGGCATTCCTGCT-GAACCGCTCTTCCG AT/3ddC/(IDT).

3 Methods

3.1 Cross-Linking of Cells (See Note 1)

3.1.1 Yeast Cells

1. Yeast cells such as the *Saccharomyces cerevisiae* strain BY4741 (genotype: Mata his3Δ1 leu2Δ0 met15Δ0 ura3Δ0 bar1:: KanMX) are cultured at 30 °C by shaking overnight in 50 ml of YEP media plus 2% glucose.
2. Cultured cells are diluted the next morning to an OD600 = 0.2 in 1 l of YEP plus 2% glucose. Cells are incubated with shaking at 30 °C until reaching an OD600 = 1.0 (about 3–4 h).
3. Cells are treated with 27.7 ml of 37% formaldehyde (final concentration = 1%) for 10 min at room temperature with constant stirring.
4. Fixation is quenched with 52.6 ml of 2.5 M glycine (final concentration = 0.125 M) for 15 min at room temperature with constant stirring.
5. Fixed cells are collected via centrifugation (1500 × *g*—5 min) and resuspended in 50 ml of spheroplast buffer plus 30 mM dithiothreitol (DTT).
6. Fixed cells are recollected via centrifugation (1500 × *g*—5 min) and resuspended in 50 ml of spheroplast buffer plus 1 mM DTT.
7. Fixed cells are converted to spheroplasts with Zymolyase 20T (final concentration = 0.66 g/l) treatment at 30 °C with gentle rotation. Conversion to spheroplasts is confirmed by microscopy.

8. Fixed spheroplasts are collected via centrifugation at 4 °C (1500 × *g* for 5 min).
9. Fixed spheroplasts are washed twice in 50 ml spheroplast buffer and collected via centrifugation at 4 °C (1500 × *g* for 5 min).
10. Fixed spheroplasts are resuspended in 50 ml of 1× NEBuffer 2 and aliquot into 50 1.7 ml microcentrifuge tubes (1 ml per tube, containing about 1–2 × 109 spheroplasts). Collect spheroplasts via centrifugation (2000 × *g* for 5 min) in a refrigerated desktop centrifuge. Spheroplasts can be rapidly frozen in liquid nitrogen and stored at −80 °C until use (up to 1.5 years).

3.1.2 Suspension Cell Cultures

1. Pellet the cells (about 2–3 × 10^7) at 800 × *g* for 5 min at RT.
2. Discard the supernatant and resuspend the pellet in 45 ml of fresh culture medium without serum. Break cell clumps by pipetting up and down.
3. Cross-link the cells by adding 1.25 ml of 37% formaldehyde (1% final concentration). Mix quickly by inverting the tube several times.
4. Incubate at RT for 10 min. Gently invert the tube every 1–2 min.
5. Add 2.5 ml of 2.5 M glycine (0.125 M final concentration) to quench the cross-linking reaction, and mix well.
6. Incubate for 10 min at RT to stop cross-linking.
7. Spin down the cross-linked cells at 800 × *g* for 3 min at RT.
8. Discard the supernatant by aspiration and resuspend the cells with 1× PBS (1 ml PBS per 10^6 cells).
9. Split the cross-linked cell suspension into aliquots of 1–2 × 10^6 cells (in 1.7 ml microtubes).
10. Centrifuge the cross-linked cells at 800 × *g* for 3 min at RT.
11. Discard the supernatant by aspiration.
12. Cells can be snap-frozen in liquid nitrogen and stored at −80 °C until use (up to 1.5 years).

3.1.3 Adherent Monolayer Cell Cultures

1. Aspirate the medium and add 10 ml of fresh medium without serum per 10 cm plate.
2. Cross-link the cells by adding 280 μl of 37% formaldehyde (1% final concentration). Mix gently, immediately after addition of formaldehyde.
3. Incubate at room temperature (RT) for 10 min.
4. Add 560 μl of 2.5 M glycine (0.125 M final concentration) to quench the cross-linking reaction, and mix well.
5. Incubate for 10 min at RT to stop cross-linking completely.

6. Discard the supernatant by aspiration and wash the cross-linked cells with 1× PBS once.
7. Scrape the cells from the plates with a cell scraper and transfer to a 1.7 ml microtube.
8. Split the cross-linked cell suspension into aliquots of 1–2 × 10^6 cells (in 1.7 ml microtubes).
9. Centrifuge the cross-linked cells at 800 × *g* for 3 min at RT.
10. Cells can be snap-frozen in liquid nitrogen and stored at −80 °C until use (up to 1.5 years).

3.1.4 Solid Primary Tissue Cells

1. Place frozen tissue sample (~0.5 g) in 7 ml pre-chilled 7 ml glass homogenizer with 6 ml of cell lysis buffer.
2. Incubate on ice for 5 min.
3. Homogenize on ice until no clumps of cells persist, e.g., dounce 5–10 times with loose pestle followed by 10–15 strokes with tight pestle.
4. Transfer to a 15 ml tube and spin at 800 × *g* for 5 min at 4 °C.
5. Resuspend the cell pellet in 10 ml PBS.
6. Add 280 μl of 37% formaldehyde (1% final concentration), RT for 10 min, with occasional inverting.
7. Quench formaldehyde with 125 mM glycine by adding 500 μl 2.5 M glycine, RT for 5 min.
8. Pellet the cells at 800 × *g* for 5 min.
9. Resuspend cells with 10 ml PBS and aliquot into ten 1.7 ml microtubes, 1 ml per tube.
10. Centrifuge the cross-linked cells at 800 × *g* for 3 min at RT.
11. Cells can be snap-frozen in liquid nitrogen and stored at −80 °C until use (up to 1.5 years).

3.2 Cell Permeabilization and Chromatin Digestion with DNase I

1. Resuspend one cross-linked cell aliquot (1–2 × 10^6 cells) in 1 ml of ice-cold cell lysis buffer containing protease inhibitor cocktail (*see* **Note 2**).
2. Incubate on ice for 30 min.
3. Centrifuge for 1 min at 800 × *g* at RT.
4. Discard the supernatant and resuspend the pellet in 200 μl 0.5× DNase I SDS buffer (*see* **Note 3**).
5. Incubate at 37 °C for 60 min.
6. Add 200 μl of 0.5× DNase I digestion buffer, and mix well.
7. Incubate at 37 °C for 10 min.
8. Add appropriate amount (e.g., 3–6 units) of DNase I, and mix well.

9. Incubate at RT for 5 min.
10. Add 20 μl 0.5 M ethylenediaminetetraacetic acid (EDTA), and mix well. Save 20 μl of lysate for assessing the DNase I digestion efficiency if desired (*see* **Note 4**).
11. Add 220 μl 20% PEG-8000 buffer and 200 μl AMPure XP beads, and mix well (*see* **Note 5**).
12. Incubate at RT for 5 min, and place the tube in a DynaMag-Spin magnet for 2 min.
13. Discard the supernatant and wash the beads twice with 1 ml of 80% ethanol. Briefly spin down the beads and remove the residual ethanol as completely as possible, and then air-dry the beads for no more than 2 min.
14. Resuspend immediately the beads with 100 μl of water, ready for the next end-repair step.

3.3 End Repair, dA-Tailing, and Labelling of Chromatin Ends in Intact Nuclei (See Note 6)

1. Prepare 100 μl end-repair master mix: 70 μl ddH_2O, 20 μl 10× T4 ligase buffer, 5 μl 10 mM dNTPs, 2 μl T4 DNA polymerase (5 U/μl), and 3 μl Klenow (5 U/μl).
2. Add the100 μl end-repair mastermix to the above DNase I digested sample. Mix well by pipetting up and down.
3. Incubate at RT for 1 h.
4. Place the tube in a DynaMag-Spin magnet for 2 min.
5. Discard the supernatant and wash the beads once with 200 μl ddH_2O.
6. Resuspend immediately the beads in 100 μl of ddH_2O, ready for the next dA-tailing step.
7. Prepare 50 μl dA-tailing master: 20 μl ddH_2O, 15 μl 10× NEBuffer 2, 5 μl 20 mM dATP, and 10 μl Klenow (EXO-) (5 U/μl).
8. Add the 50 μl dA-tailing master mix to the end-repaired sample from **step 6**. Mix well by pipetting up and down.
9. Incubate at 37 °C for 1 h.
10. Place the tube in a DynaMag-Spin magnet for 2 min.
11. Discard the supernatant and wash the beads once with 200 μl ddH_2O.
12. Resuspend immediately the beads in 30 μl of ddH_2O, ready for the next step.
13. Prepare 70 μl ligation master mix: 25 μl bridge adaptor w/biotin (40 μM), 20 μl blunt adaptor w/o biotin (50 μM), 10 μl 10× T4 ligase buffer (with 10 mM ATP), 10 μl PEG-4000 (50%), 5 μl T4, and DNA ligase (5 U/μl).
14. Add the 70 μl ligation master mix to the dA-tailed sample from **step 13**. Mix well by pipetting up and down.

15. Incubate at 16 °C overnight.
16. Add 5 μl of 10% SDS to the reaction.
17. Place the tube in a DynaMag-Spin magnet for 2 min.
18. Discard the supernatant.
19. Resuspend the pellet in 200 μl of ddH_2O.
20. Add 165 μl of 20% PEG buffer to the tube, and mix thoroughly by pipetting up and down.
21. Incubate at RT for 5 min, and place the tube in a DynaMag-Spin magnet for 2 min.
22. Discard the supernatant and wash the beads once with 1 ml of 80% ethanol.
23. Briefly spin down the beads and remove the residual ethanol as completely as possible.
24. Resuspend the beads in 200 μl of ddH_2O, and mix thoroughly by pipetting up and down.
25. Add 165 μl of 20% PEG buffer to the tube, and mix thoroughly by pipetting up and down.
26. Incubate at RT for 5 min, and place the three tubes in a DynaMag-Spin magnet for 2 min.
27. Discard the supernatant and wash the beads twice with 1 ml of 80% ethanol. Briefly spin down the beads and remove the residual ethanol as completely as possible, and then air-dry the beads for no more than 2 min.
28. Resuspend immediately the beads in 165 μl of ddH_2O, ready for the next phosphorylation step.

3.4 In Situ Phosphorylation and Proximity Ligation

1. Add 20 μl 10× T4 ligase buffer and 15 μl PNK (10 U/μl) to the 165 μl above adaptor-ligated sample. Mix well.
2. Incubate at 37 °C for 1 h.
3. Add the following reaction to the above tube: 265 μl ddH_2O, 30 μl 10× T4 ligase buffer, and 5 μl T4 DNA ligase (5 U/μl).
4. Incubate at RT for 4 h.

3.5 Reverse Cross-Linking and DNA Purification

1. Centrifuge for 3 min at 800 × g at RT.
2. Resuspend the pellet in 150 μl ddH_2O.
3. Add 20 μl 10× NEBuffer 2, 10 μl 10% SDS, and 20 μl of 20 mg/ml proteinase K.
4. Incubate overnight at 62 °C.
5. Add 150 μl AMPure XP beads, and mix well.
6. Incubate mixture at RT for 5 min, and place the tube in a DynaMag magnet for 2 min.

7. Discard the supernatant and wash the beads twice with 1 ml of 80% ethanol. Briefly spin down the beads and remove residual ethanol as completely as possible, and then air-dry the beads for no more than 2 min.
8. Resuspend the beads in 100 μl nuclease-free water.
9. Incubate beads at RT for 1 min. Collect beads via DynaMag magnet and transfer eluent to fresh 1.5 ml tube.
10. Determine the concentration of the recovered DNA with a NanoDrop spectrophotometer. A typical yield is 3–5 μg if starting with 1–2 × 10^6 cells.

3.6 DNA Sonication, End Repair, dA-Tailing, and Ligation of Sequencing Adaptors (See Note 7)

1. To shear the DNA with a Covaris S2 instrument, and transfer the DNA to Covaris microtube (3–5 μg DNA in 100 μl 1× TE lysis buffer).
2. Shear the DNA to a size of 100–300 bp using the following parameters: duty cycle, 2%; intensity, 5; cycles per burst, 200; set mode, frequency sweeping; continuous degassing; process time, 20 s; and number of cycles, 5.
3. Transfer the 100 μl sonicated DNA solution to a 1.7 ml microtube.
4. Bring the volume to 200 μl by adding ddH_2O.
5. Add 200 μl of AMPure XP beads to the tube, and mix well.
6. Incubate at RT for 5 min.
7. Place the tube in a DynaMag-Spin magnet for 2 min.
8. Discard the supernatant and wash the beads twice with 1 ml of 80% ethanol.
9. Briefly spin down the beads and remove the residual ethanol as completely as possible, and then air-dry the beads for 2 min.
10. Set up the end-repair reaction with the Fast DNA End Repair Kit, 86 μl ddH_2O, 10 μl 10× reaction buffer, and 4 μl enzyme mix, and mix well.
11. Incubate at 16 °C for 10 min.
12. Add 5 μl of 10% SDS to the tube to stop the reaction. Mix well.
13. Add 150 μl of 20% PEG buffer to the tube, and mix well.
14. Incubate at RT for 5 min.
15. Place the tube in a DynaMag-Spin magnet for 2 min.
16. Discard the supernatant and wash the beads twice with 1 ml of 80% ethanol.
17. Briefly spin down the beads and remove the residual ethanol as completely as possible, and then air-dry the beads for 2 min.

18. Set up the dA-tailing reaction by adding 81 μl ddH_2O, 10 μl 10× NEBuffer 2, 4 μl 20 mM dATP, and 5 μl Klenow (exo-) (5 U/μl).
19. Incubate at 37 °C for 30 min.
20. Add 5 μl of 10% SDS to each tube to stop the reaction. Mix well.
21. Add 150 μl of 20% PEG buffer to each tube, and mix thoroughly by pipetting up and down.
22. Incubate at RT for 5 min and place the tubes in a DynaMag-Spin magnet.
23. Discard the supernatant and wash the beads twice with 1 ml of 80% ethanol.
24. Briefly spin down the beads and remove the residual ethanol as completely as possible, and then air-dry the beads for 2 min.
25. Set up the ligation reaction by adding 34 μl ddH_2O, 5 μl 10× T4 ligase buffer (with 10 mM ATP), 5 μl PEG-4000 (50%), 3 μl Illumina PE adaptor (25 μM), and 3 μl T4 DNA ligase (5 U/μl).
26. Incubate at RT for 30 min or 16 °C overnight.
27. Add 5 μl of 10% SDS to each tube to stop the reaction.
28. Add 145 μl of ddH_2O to bring up the volume to 200 μl. Mix well.
29. Add 200 μl of 20% PEG buffer, and mix well.
30. Incubate at RT for 5 min, and place the tube in a DynaMag-Spin magnet for 2 min.
31. Discard the supernatant and wash the beads once with 1.5 ml of 80% ethanol.
32. Briefly spin down the beads and remove the residual ethanol as completely as possible, and then air-dry the beads for 2 min.
33. Elute DNA from the beads with 100 μl of ddH_2O, ready for the next biotin pull-down step.

3.7 Biotin Pull-Down and Amplification of Whole-Genome DNase Hi-C Library

1. Wash 30 μl of MyOne C1 Dynabeads with 100 μl of 1× B&W buffer.
2. Resuspend the beads in 100 μl of 2× B&W buffer.
3. Transfer the 100 μl eluted DNA from above to the tube containing the MyOne C1 beads, and mix well.
4. Incubate the sample for 15 min at RT with rotation.
5. Reclaim beads against the DynaMag-Spin magnet for 1 min, and discard the supernatant.
6. Wash beads four times with 200 μl of 1× B&W buffer containing 0.1% Tween-20.

7. Wash beads twice with 200 μl of EB buffer.
8. Resuspend the beads with 40 μl of EB buffer.
9. To generate DNA templates for targeted DNase Hi-C assays, prepare PCR master mix: 200 μl of 2× KAPA HiFi HotStart DNA Polymerase ReadyMix, 20 μl of 10 μM pre-capture PCR primer pairs, and 140 μl of ddH_2O.
10. Mix PCR master mix with the beads (400 μl in total), divide into ten aliquots of 40 μl, and amplify by PCR using the following conditions: 3 min at 98 °C; 8–9 cycles of 20 s at 98 °C, 30 s at 65 °C, and 30 s at 72 °C; 7 min at 72 °C (*see* **Note 8**).
11. Collect PCR reactions into a new 1.5 ml tube.
12. Purify library by adding 0.8× volumes of AMPure XP beads.
13. Incubate mixture at RT for 5 min and place tube in a DynaMag magnet for 2 min.
14. Discard the supernatant and wash the beads twice with 1 ml of 80% ethanol. Briefly spin down the beads and remove residual ethanol as completely as possible, and then air-dry the beads for 5 min.
15. Elute DNA in 30 μl EB buffer.
16. Determine the DNA concentration using Qubit dsDNA HS kit according to manufacturer's instructions. The yield usually is 500–1500 ng.

3.8 Enrichment of a Subset of Chromatin Interactions of Interest (See Note 9)

1. Design of a bait library to enrich the chromatin interactions associated with specific genomic regions of interest using the design tool of a selected targeted DNA sequencing platform (e.g., the NimbleDesign tool of the SeqCap EZ system). Below is the protocol for the SeqCap EZ system (*see* **Note 10**).
2. Transfer 500 ng to 1 μg of DNase Hi-C library into a new tube, add 1 μl of 1 mM adaptor-Hi-block, 1 μl of 1 mM internal adaptor-block, 1 μl of 1 mM of Nbgn-8bp-ID-BL-B, and 5 μg of human Cot-1 DNA (for human samples) or murine Cot-1 DNA (for mouse samples).
3. Evaporate sample on a vacuum concentrator at 60 °C until dry.
4. Add 7.5 μl of 2× hybridization buffer and 3 μl of hybridization component A to the dried-down sample.
5. Vortex for 10 s and centrifuge at maximum speed for 10 s.
6. Denature DNA at 95 °C for 10 min in a thermocycler, and then centrifuge at maximum speed for 10 s.
7. Add 4.5 μl aliquot of SeqCap EZ bait library to the tube, vortex for 3 s, and centrifuge at maximum speed for 10 s.

8. Incubate in a thermocycler at 47 °C for 48–72 h. The thermocycler's heated lid should be turned on and set to maintain at 57 °C.
9. Set the temperature of a water bath at 47 °C.
10. Dilute 10× wash buffers (I, II, III, and stringent) and 2.5× bead wash buffer from the SeqCap EZ Reagent Kit v2 to create 1× working solutions according to the manufacturer's instruction.
11. Pre-warm 400 μl of 1× stringent wash buffer and 400 μl of 1× wash buffer I at 47 °C in advance (at least 2 h before washing the captured DNA).
12. Add 100 μl of streptavidin Dynabeads M-270 into a new 1.7 ml microtube, place the tube in a DynaMag magnet for 2 min, and remove supernatant.
13. Wash beads with 200 μl of 1× bead wash buffer. Vortex for 10 s. Repeat two more times and discard the supernatant.
14. Transfer the hybridization samples into the tube with the magnetic beads. Mix well.
15. Incubate at 47 °C for 45 min. Mix the samples by vortex for 3 s at 15 min intervals to ensure the beads remain in suspension.
16. Add 200 μl of pre-warmed 1× wash buffer I to the tube. Vortex for 10 s.
17. Reclaim beads on the magnetic separation stand and remove the clear supernatant.
18. Resuspend beads into 200 μl of the pre-warmed 1× stringent wash buffer, and mix well.
19. Incubate for 5 min at 47 °C.
20. Reclaim beads on the magnetic separation stand and remove the clear supernatant. Repeat one more time of washing the beads with 1× stringent wash buffer.
21. Resuspend beads into 200 μl of 1× wash buffer I, and mix by vortexing for 2 min.
22. Place the tubes in the magnetic separation stand and remove the liquid.
23. Resuspend beads into 200 μl of 1× wash buffer II, and mix by vortexing for 1 min.
24. Place the tubes in the magnetic separation stand and remove the liquid.
25. Resuspend beads into 200 μl of 1× wash buffer III, and mix by vortexing for 30 s.
26. Place the tubes in the magnetic separation stand and remove the liquid.
27. Resuspend the beads in 50 μl of EB.

3.9 Library Amplification and Quality Control

1. Prepare PCR master mix: 200 μl of 2× KAPA HiFi HotStart DNA Polymerase ReadyMix, 20 μl of 10 μM Illumina PE forward primer, 20 μl of 10 μM indexed Illumina PE reverse primer, and 120 μl of ddH_2O.
2. Mix PCR master mix with the beads (400 μl in total), divide into ten aliquots of 40 μl, and amplify by PCR using the following conditions: 3 min at 98 °C; 12–18 cycles of 20 s at 98 °C, 30 s at 65 °C, and 30 s at 72 °C; and 7 min at 72 °C (*see* **Note 11**).
3. Collect PCR reactions into a new 1.5 ml tube.
4. Purify library by adding 0.8× volumes of AMPure XP beads.
5. Incubate mixture at RT for 5 min and place tube in a DynaMag magnet for 2 min.
6. Discard the supernatant and wash the beads twice with 1 ml of 80% ethanol. Briefly spin down the beads and remove residual ethanol as completely as possible, and then air-dry the beads for 5 min.
7. Elute DNA in 30 μl EB buffer.
8. Determine the DNA concentration using Qubit dsDNA HS kit according to manufacturer's instructions.
9. Quantitate amount of dsDNA in library using Qubit dsDNA HS kit as per manufacturer's protocols.
10. Digest a small aliquot of the final DNase Hi-C library (50–100 ng) with BamHI to estimate the portion of molecules with valid biotinylated junctions as follows: 1 μl of 10× Fast digestion buffer, 1 μl of targeted DNase Hi-C product, 1 μl of Fast digestion BamHI, and 7 μl ddH_2O. Also set a control reaction without adding the BamHI enzyme.
11. Incubate at 37 °C for 30 min.
12. Run both the BamHI digestion and control reactions side by side on a 6% TBE-PAGE gel. Digested libraries should demonstrate a marked shift in library size distribution, as shown in [26, 27]. If libraries pass this QC metric, proceed to Illumina sequencing (*see* **Note 12**).

4 Notes

1. Different types of cells may require different cross-linking conditions with formaldehyde. Here, the experimental protocols for preparing cross-linked yeast cells, mammalian monolayer and suspension cultures, and solid tissue samples are listed. However, for other cell types, cross-linking conditions might need to be optimized. For example, efficient cross-link of

mouse and human embryonic stem cells, which often aggregate to form large clumps during growth, usually requires one of two alternative approaches: preparing single-cell suspension before carrying out cross-linking under the same condition as for suspension cells or cross-linking the cells using the same protocol as for adherent monolayer cultures but with increased formaldehyde concentration or longer fixation time [27].

2. To achieve optimal complexity and resolution of targeted DNase Hi-C libraries, we recommend starting with 3–5 million cells for a targeted DNase Hi-C assay (or even more cells, depending on the size of the targets and the desired resolution). When starting with more cells, usually three vials of 1–2 million cells can be proceeded in parallel.
3. The concentration of SDS in the 0.5× DNase I SDS buffer can be cell type-specific. When carrying out targeted DNase Hi-C assays on a cell type for the first time, we recommend to titrate the optimal SDS concentration. In general, the SDS concentration should be within 0.2–0.5%.
4. To assess DNase I digestion efficiency, add 70 μl of 1× TE lysis buffer and 10 μl of proteinase K (20 mg/ml) to 20 μl of lysed cells from this step. Incubate for 30 min at 65 °C. Purify DNA by the QIAquick PCR Purification Kit. Check size distribution of the DNA fragments by running the sample on 1% agarose gel. In general, efficient DNase I digestion will result in the majority of DNA fragments with a size less than 1 kb [26, 27]. This QC step is recommended when applying DNase Hi-C protocol to a new cell type.
5. In the protocols of DNase Hi-C, in situ DNase Hi-C, and targeted DNase Hi-C, SPRI (solid-phase reversible immobilization) magnetic beads (e.g., AMPure XP) are employed to immobilize nuclei/chromatin complexes, which enables efficient removal of DNase I and the DNA modifying enzymes used in one of the protocols, including the Klenow enzyme, T4 DNA polymerase, Klenow fragment ($3' \rightarrow 5'$ exo-), and T4 DNA ligase [27]. These beads are also essential for the efficient removal of the low-molecular-weight DNA fragments that escape the nucleus following cell lysis and chromatin digestion free, and more importantly, for the efficient removal of unligated internal bridge adaptors following bridge-adaptor ligation. If not removed, the unligated internal bridge adaptors will severely interfere the downstream in situ proximity ligation step and lead to high percentage of dangling ends in the final DNase Hi-C libraries. Plus, when starting with a low number of cells (e.g., less than 1 million cells), the SPRI beads (usually with a brown color) can help visualize the cell/nuclei pellets throughout the protocol.

6. The efficiency of these three enzymatic reactions in intact nuclei, Klenow and T4 DNA polymerase enzyme-catalyzed DNA end-repair, Klenow (Exo-)-mediated dA-tailing, and ligation of the biotinylated bridge adaptors to the dA-tailed chromatin ends by T4 DNA ligase, largely determines the success of an in situ DNase Hi-C or targeted DNase Hi-C assay. Hence, a control ligation reaction in parallel to the experimental ligation of the biotinylated bridge adaptors can be performed to examine the efficiency of these reactions. Briefly, a small portion (~5–10%) of dA-tailed sample can be taken and ligated to the dT-tailed Illumina sequencing Y-adaptors. After the ligation, genomic DNA is then purified and the ligation efficiency examined by performing quantitative PCR (qPCR) with the appropriate Illumina PCR primers [27]. If amplification signal appears before ten PCR cycles using 10 ng of genomic DNA as a template, it suggests that the efficiency of the upstream end-repair and dA-tailing steps is acceptable. This QC step is recommended when applying the DNase Hi-C protocol to a new cell type.
7. When the amount of the starting DNase Hi-C genomic DNA is low (e.g., less than 1 μg), these four steps can be replaced by using the tagmentation-based library construction protocol with the Illumina Nextera DNA Library Preparation Kit.
8. The number of PCR cycles for amplifying the whole-genome library to generate sufficient templates for the downstream hybridization-mediated capture assay is dependent on both the amount of starting materials and the efficiency of the upstream enzymatic reactions. To determine the appropriate number of PCR cycles, trial PCR reactions with 6, 8, 10, or 12 cycles can be carried out as follows: 3 μl ddH_2O, 5 μl 2× HotStart ReadyMix, 1 μl 10 μM pre-capture PCR primer pairs, and 1 μl of DNA-bound streptavidin beads. Use the following PCR program: 3 min at 98 °C; indicated numbers of cycles of: 20 s at 98 °C, 30 s at 65 °C, and 30 s at 72 °C; and 7 min at 72 °C. Then run the 10 μl of each PCR reaction product on a 2% agarose gel to determine the appropriate number of cycles and amount of input beads for the large-scale library amplification.
9. In principle, the enrichment of a subset chromatin interactions of interest can be achieved by the various array-based or in-solution-based DNA capture techniques, such as the Agilent SureSelect and the Roche NimbleGen SeqCap EZ technologies [27]. The protocol described here is for using the SeqCap EZ technology.
10. Targeted DNase Hi-C can be used to enrich chromatin interactions associated with any genomic region or regions of

interest. The target regions can be continuous genomic segments (even an entire chromosome) or discrete cis-regulatory elements (e.g., promoters, enhancers, transcription factors' binding sites and DNase I hypersensitive sites, etc.). In principle, the size of the targeted regions can also vary, ranging from a few kilobases to hundreds of megabases.

11. The number of PCR cycles for amplifying the targeted DNase Hi-C library to generate sufficient materials for Illumina sequencing is largely dependent on the overall size of the target genomic regions. The bigger the overall size of the targeted regions, the more chromatin interactions they associate and the fewer PCR cycles required. It is recommended to carry out trial PCR reactions to determine the appropriate number of PCR cycles.
12. Before the purified targeted DNase Hi-C libraries being subjected to high-throughput sequencing, the capture efficiency of the libraries can be estimated using quantitative PCR (qPCR) assays. PCR primers to specific target regions can be designed and qPCR carried out by using the targeted DNase Hi-C library or the parent whole-genome DNase Hi-C library as templates. The enrichment of a specific target can be assessed by comparing its amplification rate from the target DNase Hi-C library with that from the parent whole-genome DNase Hi-C library [27]. Since the enrichment efficiency of a specific target in a targeted DNase Hi-C assay is determined by multiple factors and is likely to be uneven between individual targets, it is recommended to examine enrichment at multiple individual targets simultaneously using this qPCR assay to ensure the accuracy of the assessment.

Acknowledgments

This work was supported by the UW Bridge Fund (ZD), ASH Bridge Grant (ZD), and the NIH Common Fund U54DK107979.

References

1. Dekker J, Rippe K, Dekker M et al (2002) Capturing chromosome conformation. Science 295:1306–1311
2. De Wit E, De Laat W (2012) A decade of 3C technologies: insights into nuclear organization. Genes Dev 26:11–24
3. Denker A, De Laat W (2016) The second decade of 3C technologies: detailed insights into nuclear organization. Genes Dev 30:1357–1382
4. Lieberman-Aiden E, Van Berkum NL, Williams L et al (2009) Comprehensive mapping of long-range interactions reveals folding principles of the human genome. Science 326:289–293
5. Duan Z, Andronescu M, Schutz K et al (2010) A three-dimensional model of the yeast genome. Nature 465:363–367
6. Duan Z, Blau CA (2012) The genome in space and time: does form always follow function? How does the spatial and temporal

organization of a eukaryotic genome reflect and influence its functions? BioEssays 34:800–810

7. Krumm A, Duan Z (2019) Understanding the 3D genome: emerging impacts on human disease. Semin Cell Dev Biol 90:62–77
8. Ramani V, Shendure J, Duan Z (2016) Understanding spatial genome organization: methods and insights. Genomics Proteomics Bioinformatics 14:7–20
9. Cremer T, Cremer M (2010) Chromosome territories. Cold Spring Harb Perspect Biol 2: a003889
10. Dixon JR, Selvaraj S, Yue F et al (2012) Topological domains in mammalian genomes identified by analysis of chromatin interactions. Nature 485(7398):376–380
11. Nora EP, Lajoie BR, Schulz EG et al (2012) Spatial partitioning of the regulatory landscape of the X-inactivation centre. Nature 485 (7398):381–385
12. Guelen L, Pagie L, Brasset E et al (2008) Domain organization of human chromosomes revealed by mapping of nuclear lamina interactions. Nature 453:948–951
13. Peric-Hupkes D, Meuleman W, Pagie L et al (2010) Molecular maps of the reorganization of genome-nuclear lamina interactions during differentiation. Mol Cell 38:603–613
14. Nemeth A, Conesa A, Santoyo-Lopez J et al (2010) Initial genomics of the human nucleolus. PLoS Genet 6:e1000889
15. Van Koningsbruggen S, Gierlinski M, Schofield P et al (2010) High-resolution whole-genome sequencing reveals that specific chromatin domains from most human chromosomes associate with nucleoli. Mol Biol Cell 21:3735–3748
16. Rao SS, Huntley MH, Durand NC et al (2014) A 3D map of the human genome at kilobase resolution reveals principles of chromatin looping. Cell 159:1665–1680
17. Fullwood MJ, Liu MH, Pan YF et al (2009) An oestrogen-receptor-alpha-bound human chromatin interactome. Nature 462:58–64
18. Simonis M, Klous P, Splinter E et al (2006) Nuclear organization of active and inactive chromatin domains uncovered by chromosome conformation capture-on-chip (4C). Nat Genet 38:1348–1354
19. Zhao Z, Tavoosidana G, Sjolinder M et al (2006) Circular chromosome conformation capture (4C) uncovers extensive networks of epigenetically regulated intra- and interchromosomal interactions. Nat Genet 38:1341–1347
20. Dostie J, Richmond TA, Arnaout RA et al (2006) Chromosome conformation capture carbon copy (5C): a massively parallel solution for mapping interactions between genomic elements. Genome Res 16:1299–1309
21. Hughes JR, Roberts N, Mcgowan S et al (2014) Analysis of hundreds of cis-regulatory landscapes at high resolution in a single, high-throughput experiment. Nat Genet 46:205–212
22. Mifsud B, Tavares-Cadete F, Young AN et al (2015) Mapping long-range promoter contacts in human cells with high-resolution capture hi-C. Nat Genet 47:598–606
23. Fang R, Yu M, Li G et al (2016) Mapping of long-range chromatin interactions by proximity ligation-assisted ChIP-seq. Cell Res 26:1345–1348
24. Mumbach MR, Rubin AJ, Flynn RA et al (2016) HiChIP: efficient and sensitive analysis of protein-directed genome architecture. Nat Methods 13:919–922
25. Ma W, Ay F, Lee C et al (2015) Fine-scale chromatin interaction maps reveal the cis-regulatory landscape of human lincRNA genes. Nat Methods 12:71–78
26. Ramani V, Cusanovich DA, Hause RJ et al (2016) Mapping 3D genome architecture through in situ DNase hi-C. Nat Protoc 11:2104–2121
27. Ma W, Ay F, Lee C et al (2018) Using DNase hi-C techniques to map global and local three-dimensional genome architecture at high resolution. Methods 142:59–73

Chapter 6

C-HiC: A High-Resolution Method for Unbiased Chromatin Conformation Capture Targeting Small Locus

Jérôme D. Robin

Abstract

Within the nucleus, precise DNA folding and organization is mandatory for a tight control of gene expression. In the past 20 years, a wealth of molecular approaches has unraveled the existence of DNA territories. With the emergence of affordable deep-sequencing approaches, "Cs" techniques such as 4C, 5C, and HiC, to name a few, are now routinely performed by the scientific community in a large number of model systems. We have modified the HiC approach to a capture probe-based version named C-HiC. This updated assay has resulted in an improved throughput analysis, reduced input material, and good repeatability. The protocol described below details our procedure and notes for a C-HiC approach, designed to target only specific portion of a given genome.

Key words Higher-order organization, HiC, Chromatin conformation capture, Chromatin looping, Chromatin domain

1 Introduction

Higher order organization of chromatin allows DNA to be compacted into separated compartments (e.g., A and B) and territories. This organization sets the chromatin into defined structured loops and discrete domains allowing precise gene expression patterns that can be modulated by protein complexes controlling long-distance interactions between promoters and *cis*-regulatory regions [1–3].

First investigated by contrast microscopy [4] then, more recently by chromatin conformation capture (i.e., 3C) [5], chromosomal territories and corresponding 3D organization constitute a wide field of investigation that aims at understanding the relation between chromatin states, dynamics, and influence on gene expression in various contexts (i.e., pathologies, cancers, differentiation). The rise of techniques used to monitor chromatin landscapes has established the existence of major chromosomal domains of different scales. Topologically associated domains (TADs) [6] are the

Beatrice Bodega and Chiara Lanzuolo (eds.), *Capturing Chromosome Conformation: Methods and Protocols*, Methods in Molecular Biology, vol. 2157, https://doi.org/10.1007/978-1-0716-0664-3_6,

most studied and represent domains of 500Kb to few Mbs. Associations between similar TADs (based on dynamic gene expression) form the higher scale of organization, dividing the DNA into chromosomal compartments (A and B). A being the open transcriptionally active chromatin (e.g., euchromatin), found mostly internally, whereas B compartments are associated with silenced chromatin (e.g., heterochromatin) and localized at the nuclear periphery [7, 8]. At smaller scales, other domains have been detailed such as those associated with nuclear organelles known as lamina-associated domains (LADs) [9] and nucleolus-associated domains (NADs) [1, 10]. The formation of chromosomal domains is thought to be driven by a combination of sequence specificity, genomic localization, and epigenetic states. Our understanding of the 3D structure of DNA arises from a continuum of complementary techniques, performed over the past 30 years, that have set our views on DNA folding from yeast to human nuclei. Indeed, the most recent studies in yeast revealed that the sequences per se seem to hold the code of this organization rather than its genomic localization [11, 12].

The complete folding of the human genome was released a decade ago, thanks to a "C-like" technique named HiC (for high-resolution capture) that was adapted to a deep sequencing design [7]. Since then, the description of the higher-order organization of the human genome has been explored in different normal and pathological contexts [13–15]. Importantly, the core of C-based techniques relies on the determination of frequency of interactions detected by PCR or sequencing. If the average of 3D structures can be drawn from these data, it is not clear yet how one can detect discrete *per cell* information, underlying the need of controls (*see* Subheading 3.12). Additionally, the lack of resolution (e.g., data are limited to 500 kb to 1 Mb scale) and the requirement of large amounts of materials to achieve appropriate depth of sequencing have led to innovative approaches combining already existing techniques.

Among many, tethered HiC (T-HiC) [16, 17] and captured-based HiC (C-HiC) [18–22] seem to be the easiest on-hand assays. The first one allows for a protein-centered vision of the 3D chromatin, using an immunoprecipitation step (i.e., emphasis on *trans*-regulatory interactions). The second uses a probe capture step, hence focusing on a specific region of the genome (5–10 Mb), leading to a higher resolution otherwise not solved by classical HiC [19–21]. Importantly, C-HiC techniques with additional steps not only filter out regions that are not of interest but also increase the depth of sequencing of the targeted region, decreasing at the same time the complexity of the data to be handled. Last, those approaches require less material, an important factor when dealing with rare samples (e.g., biopsies or pathological specimen) [21, 22]. The C-HiC aims to provide regulatory information (in *cis*) of targeted loci relevant to specific clinical diseases [19, 22], ultimately reducing cost of sequencing and increasing the potential treatment of side-by-side samples (patients analyzed per flow cell).

The protocol provided below describes all materials, reagents, and steps needed to perform a complete C-HiC assay along with important controls and notes that one should consider when designing such experiments (Fig. 1).

2 Materials

Prepare all solutions following appropriate safety rules (i.e., fume hoods, gloves). Set up all amplification and PCR steps using typical PCR precautions.

2.1 Equipment

Rocking shaker.

Dounce homogenizer.

Thermoshaker.

Sonicator/Covaris/Bioruptor.

DNA LoBind tubes.

Magnet holder.

Rotation wheel.

Thermocycler.

2.2 Chemicals

16% formaldehyde (w/v), methanol-free.

2.5 M glycine.

PBS 1×.

Protease inhibitors.

1% SDS.

10% Triton X-100.

Phenol pH 8.0.

Chloroform.

3 M Na-acetate.

0.5 M EDTA pH 8.0.

100% ethanol.

UltraPure BSA.

0.1 M NaOH.

1 M Tris–HCl pH 7.5.

1 M Tris–HCl pH 8.0.

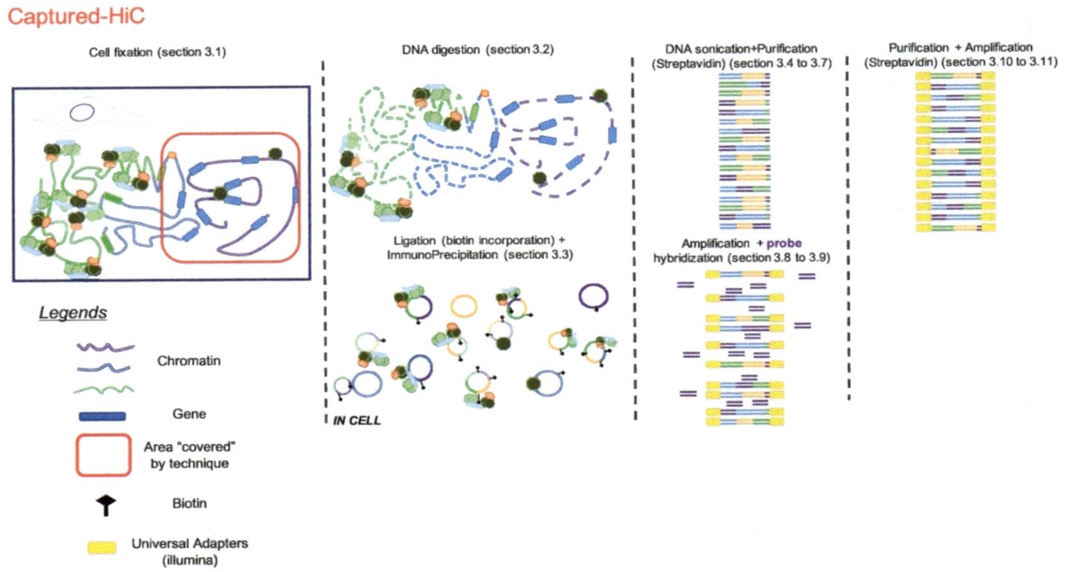

Fig. 1 Captured HiC (C-HiC) overview. Schematic representation of the main steps of a typical C-HiC procedure, where the target locus is represented in purple (red square region). Importantly, all steps up to ligation (Subheading 3.1, **steps 1–9**) are performed in cellulo. Matching steps' numbers are reported accordingly to the protocol provided

2.3 Enzymes

*Hind*III, *Nhe*I restriction enzyme.

T4 DNA ligase.

T4 DNA polymerase.

T4 polynucleotide kinase.

DNA polymerase I large fragment (Klenow).

Klenow (exo-).

Proteinase K.

RNase A.

NEBNext® Ultra™ II Q5® Master Mix (NEB).

2.4 Buffers

Fixation buffer: Cell culture media (depending on cell culture) complemented with formaldehyde (i.e., 10 ml media +0.625 μl formaldehyde; 1% final).

Lysis buffer: 10 mM Tris–HCl pH 8.0, 10 mM NaCl, 0.2% NP40.

Lysis buffer$^+$: 10 mM Tris–HCl pH 8.0, 10 mM NaCl, 0.2% NP40, 1× protease inhibitors.

CutSmart® Buffer 10×: 500 mM potassium acetate, 200 mM Tris-acetate, 100 mM magnesium acetate, 100 μg/ml BSA.

NEBuffer 2 10×: 500 mM NaCl,100 mM Tris–HCl pH 7.9, 100 mM $MgCl_2$, 10 mM DTT.

Ligation buffer 10×: 500 mM Tris–HCl pH 7.5, 100 mM $MgCl_2$, 50 mM DTT.

TE: 10 mM Tris–HCl pH 8.0, 1 mM EDTA.

Tween wash buffer 1× (TWB): 5 mM Tris–HCl pH 8.0, 0.5 mM EDTA, 1 M NaCl, 0.05% Tween.

Binding buffer 2× (BB): 10 mM Tris–HCl pH 8.0, 1 mM EDTA, 2 M NaCl.

RNABait wash buffer 1 (RWB1): 1× SSC, 0.1% SDS.

RNABait wash buffer 2 (RWB2): 0.1× SSC, 0.1% SDS.

MySelectHYB#1: 20× SSPE.

MySelectHYB#3: 50× Denhardt's.

2.5 Nucleic Acids, Nucleotides, and Oligos

dNTPs (separated mix; 100 mM each).

0.4 mM biotin-14-dCTP (Life Technologies).

100 mM ATP.

1 μg/ml human *Cot1* DNA (Life Technologies).

1 μg/ml Salmon sperm DNA (Life Technologies).

Dynabeads MyOne Streptavidin C1 beads (Life Technologies).

Biotinylated-RNA capture probe library (MYcroarray-MYselect, MI) or SureSelectXT RNA library (Agilent).

Control-*Hind*III *-1* GTTCATCTTGCTGCCAGAAATGCCGAG CCTG.

Control-*Hind*III*-2* ATCCCAGCTGTCTGTAGCTTTAGAAAG TGGG.

Illumina Paired End adapters.

Illumina PE PCR Primer 1.0.

Illumina PE PCR Primer 2.0.

MYSelect purification Kit (MYcroarray).

Gel extraction kit (Qiagen).

MinElute purification Kit (Qiagen).

dsDNA QuBit assay (Life Technologies).

3 Methods

3.1 Cross-Linking

1. Grow 5 dishes of five million cells in recommended media to 70% of confluence, and wash cells once with fresh media at 37 °C.
2. Fix cells using a fresh fixation buffer (*see* **Note 1**).
3. Incubate at room temperature for 10 min on a rocking platform.
4. Add 580 μl of 2.5 M glycine solution (0.1 M final), and incubate at room temperature for 5 min on a rocker.
5. Incubate dishes on ice for 5 min.
6. Wash cells with ice-cold PBS1×, twice.
7. Using 200 μl of ice-cold PBS1×, scrap cell from dish and transfer them to a 1.5 ml epi tube.
8. Centrifuge at 250 × *g* for 10 min at 4 °C.
9. Discard supernatant.
10. Proceed with next steps or flash freeze cells and store at −80 °C.

3.2 Lysis-Chromatin Digestion

1. Resuspend the pellet by adding 250 μl lysis buffer complemented by 50 μl of protease inhibitors per cross-linked cell pellets.
2. Incubate on ice for at least 15 min.
3. Centrifuge at 2500 × *g* for 10 min at 4 °C, and discard the supernatant.
4. Wash pellets once with ice-cold lysis buffer^{+}.

5. To remove uncross-linked proteins, resuspend pellet in 50 µl lysis buffer^{+} complemented with 0.5% SDS (add 2.5 µl of a 10% SDS solution).
6. Avoid foam formation and incubate 8 min at 62 °C in a thermoshaker.
7. To quench the reaction, add 150 µl of Triton X-100 mixture (20 µl of Triton X-100 in 130 µl water). Mix gently to avoid foam formation.
8. Incubate 15 min at 37 °C—Take at this step an aliquot of 10 µl of each sample. Store and label tubes as undigested control.
9. Add 22.5 µl of CutSmart Buffer 10× and 400 units of *Hind*III.
10. Incubate overnight at 37 °C in a thermoshaker to digest the chromatin—Take an aliquot of 10 µl of each sample. Store and label tubes as digested control.

3.3 DNA Ends Labeling and In-Cell Ligation

1. Inactivate the restriction enzyme at 62 °C for 20 min in a thermoshaker.
2. Centrifuge at 2500 × *g* for 10 min at 4 °C, and discard the supernatant.
3. Resuspend the cells in 300 µl NEBuffer2 1.1×, and split samples in 3 epi tubes.
4. Keep 1 tube on ice as a 3C control.
5. Add 50 µl of fill in mix to each of the two remaining tubes as prepared below:

   1.5 µl of 10 mM dATP.

   1.5 µl of 10 mM dGTP.

   1.5 µl of 10 mM dTTP.

   37.5 µl of 0.4 mM biotin-14-dCTP (Invitrogen).

   8 µl of 5 U/µl Klenow (NEB).
6. Incubate at 37 °C for 45 min in a thermoshaker.
7. Put all tubes on ice, and retrieve 3C control tube.
8. Add 1 ml of complete ligation mix 1× prepared as below:

   130 µl of ligation buffer 10×.

   130 µl of 10% Triton X-100.

   13 µl of 10 mg/ml BSA.

   1.3 µl of 100 mM ATP.

   1 µl of T4 DNA ligase—use 0.1 µl for the 3C control.

   qsp 1 ml water.
9. Incubate at 16 °C for 8 h on a rotating wheel. Add 1.3 µl of 100 mM ATP every 2 h.

3.4 Reverse Cross-link-DNA Purification

1. Centrifuge at 2500 × *g* for 10 min at RT, and discard the supernatant (*see* **Note 2**).
2. Resuspend the cells in 200 μl TE.
3. Add 20 μl of proteinase K, and incubate at 60 °C overnight in a thermoshaker.
4. Add an extra 5 μl of proteinase K, and incubate at 60 °C for 1 h in a thermoshaker.
5. Cool the reaction mixture to RT.
6. Extract DNA by standard phenol/chloroform procedure:
   Add 300 μl of phenol (pH 8.0) to each tube.

   Vortex.

   Centrifuge at 20,000 × *g* for 10 min at RT.

   Transfer supernatants to new epi tubes.

   Add 300 μl of phenol/chloroform (1:1 mix) to each tube.

   Vortex.

   Centrifuge at 20,000 × *g* for 10 min at RT.

   Transfer supernatants to new epi tubes.

   Add 300 μl of chloroform to each tube.

   Vortex.

   Centrifuge at 20,000 × *g* for 10 min at RT.

   Transfer sups to new epi tubes.

   Add 300 μl of chloroform to each tube.

   Vortex.

   Centrifuge at 20,000 × *g* for 10 min at RT.

   Transfer supernatants to new epi tubes.
7. Bring the volume of each tubes to 200 μl with TE.
8. Precipitate DNA by adding 20 μl of 3 M Na-acetate to each sample.
9. Mix well and add 500 μl of ice-cold 100% ethanol.
10. Invert tubes several times, and incubate at −80 °C for 1 h.
11. Centrifuge at 20,000 × *g* for 20 min at 4 °C and carefully remove the supernatant.
12. Wash pellets five times with 500 μl 70% ethanol to remove salts.
13. Resuspend the pellets in 50 μl TE.
14. Degrade RNA by adding 1 μl of RNAse A (1 mg/ml) to each tube.
15. Incubate at 37 °C for 25 min.
16. Add 150 μl of TE and extract DNA by standard phenol/chloroform procedure (re-do Subheading 3.4, **steps 6–13**, without the phenol-only step).

17. Resuspend the pellets in 50 μl TE, and quantify DNA using a dsDNA QuBit assay. For now on, only use DNA LoBind tubes (*see* **Note 2**).

3.5 Quality Controls (Fig. 2)

1. Digestion-ligation control: run a 0.8% agarose gel, and load 1 μl of each sample aside with the undigested and digested controls (from Subheading 3.2, **steps 8** and **10**).
2. HiC ligation control (*see* **Note 3**):
   Perform a PCR reaction using 1 μl of each sample (HiC and 3C) as template and Control-*Hind*III primers.

   Run PCR using the following program:

95 °C for 3 min	
40 cycles at:	95 °C for 45 s 60 °C for 45 s 72 °C for 90 s
7 min final extension at 72 °C	

   Digest PCR products by either *Hind*III or *Nhe*I and run products on a 1% agarose gel.

   Efficient procedure is verified by a yield of 70% of digestion by *Nhe*I.

3. Pursue steps only for successful samples; store the 3C control sample at −80 °C.

3.6 Biotin-End Removal, DNA Shearing, and Repair

1. Remove biotin from unligated ends by preparing 100 μl reactions as follows:
   5 μg of HiC sample.

   1 μl of 10 mg/ml BSA.

   10 μl of 10× T4 DNA polymerase buffer.

   1 μl 10 mM dATP.

   2 U of T4 DNA polymerase.
2. Incubate at 20 °C for 1 h in a thermoshaker.
3. Stop reaction by adding 2 μl of 0.5 M EDTA pH 8.0.
4. Pool tubes and sonicate DNA to a 300–500 bp size range.
5. Using a Bioruptor: 30/90 s cycles on/off for five cycles for a 250 μl sample (*see* **Note 4**).
6. Control sonication by running 1 μl aliquots on a 1% agarose gel. Adjust conditions accordingly.
7. Purify DNA using a MinElute kit. Elute DNA in 2× 50 μl TE and quantify DNA by dsDNA QuBit assay.
8. Repair DNA ends by adding to a 100 μl sample:

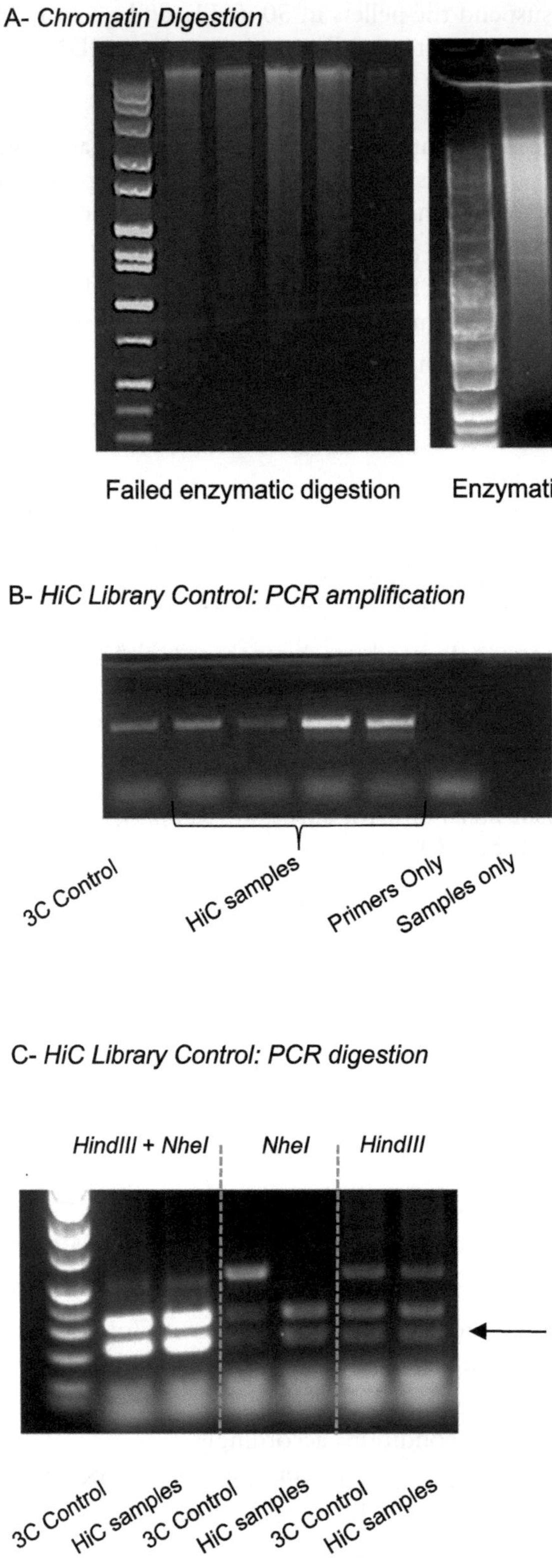

Fig. 2 C-HiC controls. Representative agarose gel (2%) of a typical HiC procedure. (**a**) Enzymatic restriction of chromatin. Failed digestion (left) results in a non-migrating DNA stuck at the top of the gel along with a very

14 μl of 10× ligation buffer.

2 μl of 10 mM dNTPs.

2 U T4 DNA polymerase.

2 U T4 polynucleotide kinase.

1 μl Klenow DNA polymerase.

9. Incubate for 30 min at 20 °C in a thermoshaker.
10. Purify DNA using a MinElute kit. Elute DNA in 2× 15 μl TE.
11. Add an "A" to the 3′ end using Klenow (exo-) mix in a 50 μl final solution:
 5 μl of 10x Klenow buffer.
 1 μl of 10 mM dATP.
 2U Klenow (exo-).
12. Incubate at 37 °C for 30 min in a thermoshaker, and inactivate at 65 °C for 20 min.

3.7 First Biotin Pull-Down

1. Prepare the Dynabeads MyOne Streptavin C1 beads, for each sample:
 Add 400 μl TWB to 6 μl of beads.
 Retain beads by holding tubes against a magnet for 1 min, and remove supernatant.
 Repeat the wash for a total of two TWB washes.
 Resuspend beads in 200 μl BB 2×, and use fresh tubes.
2. Combine the beads in BB 2× mixture to HiC samples mix in a 400 μl final volume (*see* **Note 5**).
3. Incubate at RT for 15 min on a rotation wheel.
4. Hold sample against a magnet and resuspend in 400 μl 1× BB.
5. Transfer to new tubes, hold against magnet, and resuspend in 100 μl ligation buffer.
6. Transfer to new tubes, hold against magnet, and resuspend in 50 μl ligation buffer.

3.8 PE Adapter Ligation: HiC Library

1. Complement each tube with 6 pmol of PE adapters (can be indexed) per μg of DNA (measured in Subheading 3.6, **step** 7).
2. To each tube, add 1 μl of 10 mM ATP and 1U of T4 DNA ligase.

Fig. 2 (continued) faint smear. Successful digestion (right) results in a bright smear, with an ideal average size at 4 kb. (**b**) PCR amplification of a known loop, used to control for biotin's integration (Subheading 3.5, **step 2**). (**c**) Digestion of the PCR product. Successful HiC preparation (biotin integration) will translate into a PCR product that is digested by *Nhe*I (Subheading 3.5, **step 3**). A 3C control is shown as control of digestion. Digested products are indicated by the arrow

3. Incubate at 16 °C overnight on a rotation wheel.
4. Add 400 μl of TWB, hold sample against a magnet for 3 min, remove supernatant, and resuspend beads in 400 μl of TWB.
5. Transfer samples to new tubes, hold against magnet, remove supernatant, and resuspend in 200 μl TWB.
6. Transfer samples to new tubes, hold against magnet, remove supernatant, and resuspend in 200 μl BB 1×.
7. Transfer samples to new tubes, hold against magnet, remove supernatant, and resuspend in 200 μl NEB 2.
8. Transfer samples to new tubes, hold against magnet, remove supernatant, and resuspend in 50 μl NEB 2.
9. Repeat Subheading 3.8, **step 8**.
10. Amplify the library by PCRs, and assess the appropriate number of amplification cycles:
    Set a series of PCR library mixture (5 points):
    1.5 pm of PE PCR primers 1.0.
    1.5 pm of PE PCR primers 2.0.
    10 μl NEBNext® Ultra™ II Q5® Master Mix.
    0.6 μl HiC DNA-bound beads.
11. Run the PCR using the following program:

98 °C for 1 min	
Do *X* cycles of:	98 °C for 10 s 65 °C for 30 s 72 °C for 30 s
Final extension at 72 °C for 7 min	

where $X = 5, 8, 10, 12$, and 18 cycles.

12. Run PCR products on a 2% agarose gel (Fig. 3).
13. Amplify the library accordingly by large-scale PCRs, in at least five separate tubes to maximize amplification heterogeneity.
14. Clear the DNA from the beads, heat mixture to 98 °C, hold against magnet, and transfer supernatant to a fresh tube. Rinse the beads once with 50 μl TE, hold against a magnet, and retrieve the supernatant. Discard beads.
15. Bring volume to 250 μl with TE.
16. Purify library with AMPure XP beads following Illumina's recommendations.
17. Add 175 μl of RT AMPure beads to the 250 μl library, mix well, and incubate at RT for 15 min.

Library Amplification

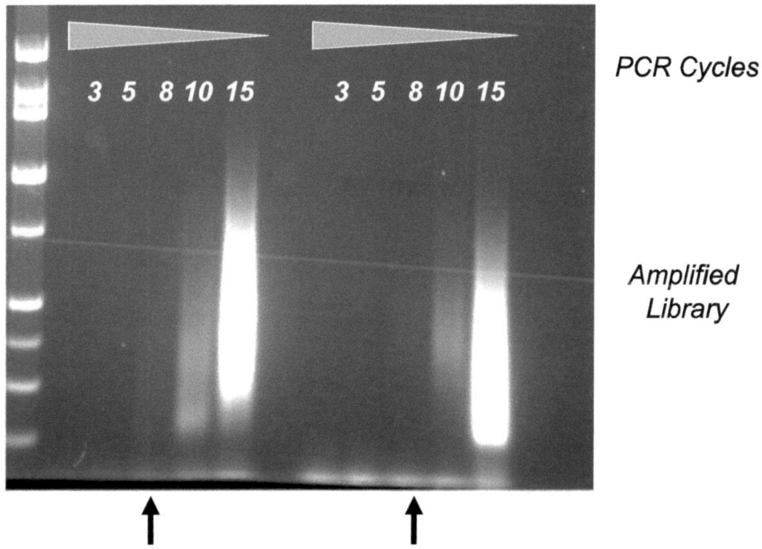

Fig. 3 Library amplification. PCR amplification of the HiC and C-HiC libraries (Subheadings 3.8, **step 12** and 3.11, **step 2**, respectively) using the PE primers and PCR amplification profile according to manufacturers'/sequencing core instructions (Illumina). From 0.6 to 1 μl of library are used for the assessment of the minimal cycles' number needed for amplification. Hence, appropriate number is determined by a visible smear on a 2% agarose gel (black arrow)

18. Hold sample against a magnet, and remove the supernatant.
19. Keep the sample against the magnet and wash twice with 700 μl of freshly made 70% ethanol. Do not mix.
20. Dry the beads by leaving the tube open at RT for 5 min.
21. Add 25 μl of TE to elute DNA. Mix by pipetting.
22. Incubate at RT for 5 min, hold sample against a magnet, and transfer supernatant to a fresh tube. Quantify DNA by dsDNA QuBit assay.

3.9 RNA Probe Capture (Fig. 4)

Use between 50 and 250 ng of HiC library for each capture reaction. Do all capture in triplicate to ensure good reproducibility.

1. Set the following profile on a thermocycler that can hold 0.5 ml epi tubes.
   95 °C for 5 min.
   65 °C for 3 min.
   65 °C for 2 min.
   65 °C for 48 h.

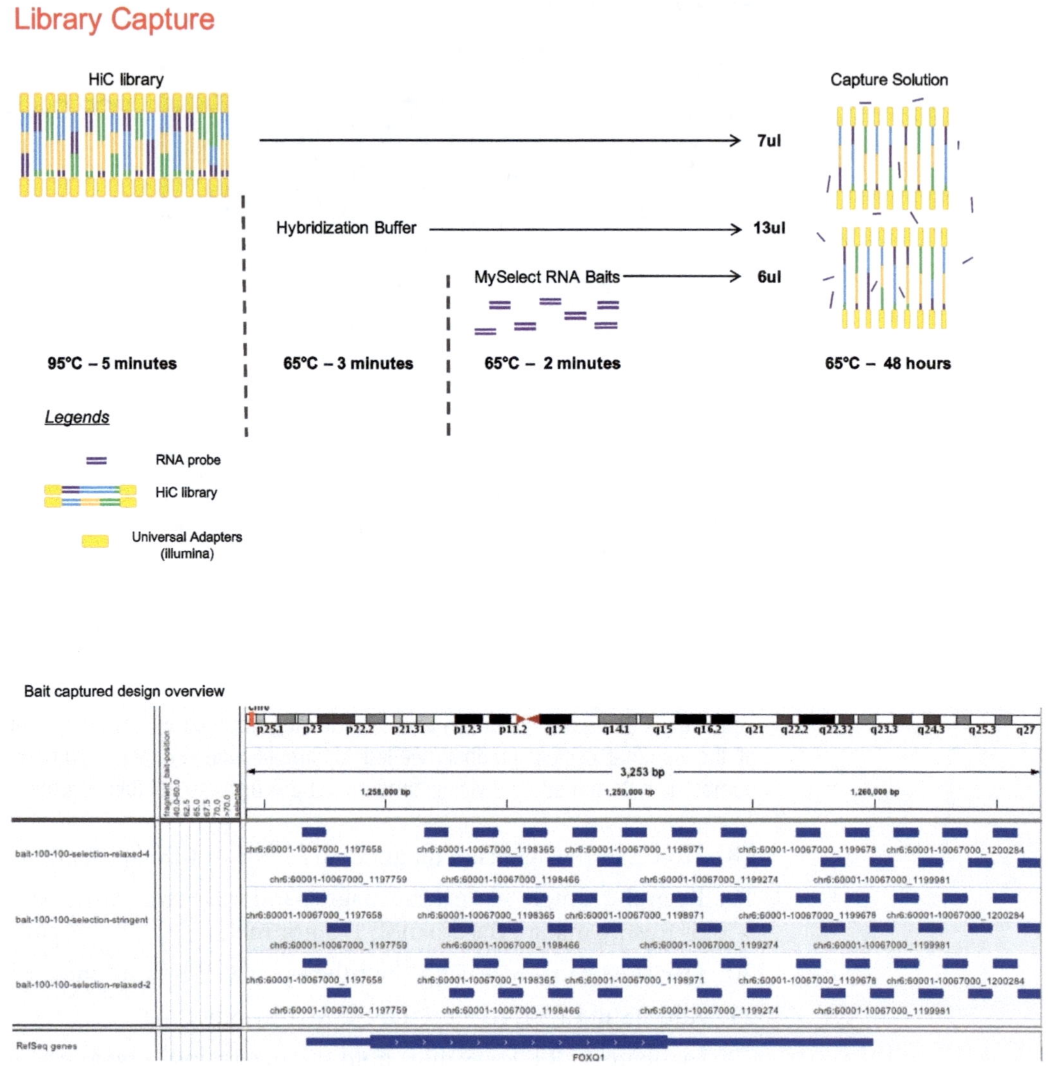

Fig. 4 Locus capture. Schematic representation of the library capture steps, preparation of mixes (top), and localization of baits (bottom). From an amplified Hi-C library, DNA is denatured, mixed with a warmed solution and finally hybridized with biotinylated RNA for 48 h at 65 °C (Subheading 3.9, **steps 1–8**). Baits are designed to overlap the full loci to capture, each bait covering a unique specific sequence. Example is shown for the *FOXQ1* gene, located at the 6p25 locus

2. Prepare the following mix:

Library mix
2.5 μl of 1 μg/μl Cot1 DNA
2.5 μl of 1 μg/μl Salmon sperm DNA
0.6 μl of MySelect blocking buffer
3.4 μl of HiC library

(continued)

Hybridization mix	
	20 μl of 20× SSPE
	0.8 μl 0.5 M EDTA
	8 μl of 50× Denhardt's
	8 μl of 1% SDS
Capture mix	
	5 μl of RNA baits (*see* **Note 6**)
	1 μl of RNA block

3. Transfer the library mix to the thermocycler, and start program.
4. When reaching **step 2**, briefly put the library mix on ice.
5. Transfer the library mix and hybridization mix to the thermocycler.
6. Add the capture mix to the thermocycler for the last 2 min at 65 °C.
7. Open the thermocycler, leaving all tubes at 65 °C, and add 7 μl of library mix and 13 μl of hybridization mix to the capture mix tube. Briefly mix by pipetting.
8. Hybridize solution for 48 h.

3.10 Second Biotin Pull-Down: C-HiC Sample

1. Prepare Dynabeads MyOne Streptavin C1 beads as described in Subheading 3.7, **step 1**.
2. Hold tubes against a magnet and resuspend in 200 μl BB 1×.
3. Transfer the hybridized solution to the beads.
4. Incubate at RT for 30 min on a rotation wheel.
5. Hold tubes against a magnet, remove supernatant, and resuspend in 400 μl RNABait wash buffer 1 (RWB1).
6. Incubate at RT for 15 min on a rotation wheel.
7. Hold tubes against a magnet, remove supernatant, and resuspend in 400 μl RWB2 (pre-warmed at 65 °C).
8. Incubate at 65 °C for 10 min in a thermoshaker.
9. Repeat Subheading 3.10, **steps** 7 and **8**.
10. Hold tubes against a magnet, remove supernatant, and resuspend in 100 μl RWB2 (pre-warmed at 65 °C).
11. Incubate at 65 °C for 10 min in a thermoshaker.
12. Hold tubes against a magnet, remove supernatant, and resuspend in 50 μl of 0.1 M NaOH to degrade the RNA.

13. Vortex and incubate at RT for 10 min.
14. Pellet the beads, and transfer the supernatant to a fresh epi tube labelled C-HiC.
15. Add 70 μl of 1 M Tris–HCl pH 7.5 to neutralize the NaOH.
16. Purify DNA using a MinElute kit. Elute DNA in 2× 15 μl TE.

3.11 C-HiC Libraries

1. Amplify the library by PCR. Assess the number of amplification cycles in order to generate enough material as required by the sequencing core—usually between 8 and 12 pM per lane is required [23]:
   Set a series of PCR library mixture (5 points).
   1.5 pm of PE PCR primers 1.0.
   1.5 pm of PE PCR primers 2.0.
   10 μl NEBNext® Ultra™ II Q5® Master Mix.
   0.6 μl C-HiC.
2. Run the PCR using the following program:

98 °C for 1 min	
Do *X* cycles of:	98 °C for 10 s 65 °C for 30 s 72 °C for 30 s
72 °C for 7 min	

   where $X = 5, 8, 10, 12$, and 18 cycles.
3. Run PCR products on a 2% agarose gel.
4. Amplify the library accordingly by large-scale PCRs, in at least five separate tubes to maximize amplification heterogeneity. This will decrease possible PCR bias and allow the preservation of discrete events.
5. Quantify library, and process with sequencing (*see* **Note 7**).

3.12 Controls

1. Once treated, take advantage of the generated 3C control sample and assess one or two newly found interactions thanks to C-HiC by 3C. As mentioned previously, C-techniques are based on the detection of interactions by de novo ligation after digestion. As frequency interaction/population-based data can mask discrete events, it is important to test the most important interaction by additional approaches such as in situ hybridization. 3D FISH assays should be performed for the same interactions (as 3C) to confirm and consolidate the data.

4 Notes

1. Use fume hood and appropriate handling of formaldehyde. We found more consistent results when cross-linking was done in dish rather than after treatment with trypsin.
2. In case salts were not removed, an agarose dialysis step can be added to the procedure.
   (a) Make an agarose gel in the desired exchange buffer (TE or H_2O). Concentration doesn't really matter. At this point, samples are digested and ligated genomic DNA, hence big structures that will not easily diffuse in the gel.
   (b) Pour agarose in a 24-well plate, 2 ml/well.
   (c) Overlay for >20 min with exchange buffer and let the agarose be saturated with the buffer (this is to prevent the transfer of some of your samples into the gel).
   (d) Discard the buffer from the agarose (shake your wells).
   (e) Overlay the wells with your samples (200 μl), and let stand for 30 min at 4 °C with a smooth shake.
   (f) Retake your DNA samples from the wells; you may lose some of your volumes but *not* your DNA.
3. Confirm that the ligation process worked by taking advantage of the fact that successful ligation creates a *Nhe*I restriction site (AAGCTT) in lieu of *Hind*III (GCTAGC). Use PCR to amplify a ligation product formed by two nearby fragments (use the proximity effect of chromatin conformation capture). Determine that 70% of amplicons can be cut only by *Nhe*I (Fig. 3).
4. We routinely use a Bioruptor (Diagenode). However, any sonication instrument can be used. To note, profile for using a Misonix Sonicator 3000 is two pulses of 30 W for 90s separate by a 90s rest on ice, using a 500 μl sample in TE. For a Covaris setup, settings are as follows: fill level 10, duty cycle 15, PIP 500, and cycles/burst 200 for 60 s.
5. The binding buffer mix can be complemented with 0.1% Triton X-100 + 1% serum to block beads and ensure better stringency.
6. The capture probe library is a set of 200,000 digitally synthesized 120 nt biotinylated RNA probes covering a 10 Mb area.
7. Quality of raw reads is checked using FastQC. Reads are aligned individually to the most recent UCSC hg19 assembly of the human genome reference using Bowtie2, and HOMER (Hypergeometric Optimization of Motif EnRichment) was used for all remaining analysis. Importantly, parameters are set to filter out self-ligation; include only reads within 1000 bp of a *Hind*III restriction site, and ensure a minimum

distance of 20 kb between read pairs. A 100 kb resolution background model was generated using all paired reads, and false discovery rate was controlled using a Benjamini correction.

References

1. Robin JD, Magdinier F (2016) Physiological and pathological aging affects chromatin dynamics, structure and function at the nuclear edge. Front Genet 7:153
2. Ferreira J, Paolella G, Ramos C, Lamond AI (1997) Spatial organization of large-scale chromatin domains in the nucleus: a magnified view of single chromosome territories. J Cell Biol 139:1597–1610
3. Dekker J, Mirny L (2016) The 3D genome as moderator of chromosomal communication. Cell 164:1110–1121
4. Belmont AS, Braunfeld MB, Sedat JW, Agard DA (1989) Large-scale chromatin structural domains within mitotic and interphase chromosomes in vivo and in vitro. Chromosoma 98:129–143
5. Dekker J, Rippe K, Dekker M, Kleckner N (2002) Capturing chromosome conformation. Science 295:1306–1311
6. Dixon JR et al (2012) Topological domains in mammalian genomes identified by analysis of chromatin interactions. Nature 485:376–380
7. Lieberman-Aiden E et al (2009) Comprehensive mapping of long-range interactions reveals folding principles of the human genome. Science 326:289–293
8. Fortin J-P, Hansen KD (2015) Reconstructing A/B compartments as revealed by Hi-C using long-range correlations in epigenetic data. Genome Biol 16:289
9. Guelen L et al (2008) Domain organization of human chromosomes revealed by mapping of nuclear lamina interactions. Nature 453:948–951
10. Robin JD, Magdinier F (2017) Handbook of Epigenetics. Elsevier, Amsterdam, pp 73–92. https://doi.org/10.1016/B978-0-12-805388-1.00006-7
11. Pretorius IS, Boeke JD (2018) Yeast 2.0-connecting the dots in the construction of the world's first functional synthetic eukaryotic genome. FEMS Yeast Res 18:55
12. Mercy G et al (2017) 3D organization of synthetic and scrambled chromosomes. Science 355:eaaf4597
13. Rao SSP et al (2015) A 3D map of the human genome at kilobase resolution reveals principles of chromatin looping. Cell 162:687–688
14. Rao SSP et al (2014) A 3D map of the human genome at kilobase resolution reveals principles of chromatin looping. Cell 159:1665–1680
15. Tang Z et al (2015) CTCF-mediated human 3D genome architecture reveals chromatin topology for transcription. Cell 163:1611–1627
16. Kalhor R, Tjong H, Jayathilaka N, Alber F, Chen L (2011) Genome architectures revealed by tethered chromosome conformation capture and population-based modeling. Nat Biotechnol 30:90–98
17. Fullwood MJ et al (2009) An oestrogen-receptor-alpha-bound human chromatin interactome. Nature 462:58–64
18. Robin JD et al (2014) Telomere position effect: regulation of gene expression with progressive telomere shortening over long distances. Genes Dev 28:2464–2476
19. Robin JD et al (2015) SORBS2 transcription is activated by telomere position effect-over long distance upon telomere shortening in muscle cells from patients with facioscapulohumeral dystrophy. Genome Res 25:1781–1790
20. Dryden NH et al (2014) Unbiased analysis of potential targets of breast cancer susceptibility loci by Capture Hi-C. Genome Res 24:1854–1868
21. Jäger R et al (2015) Capture Hi-C identifies the chromatin interactome of colorectal cancer risk loci. Nat Commun 6:6178
22. Baxter JS et al (2018) Capture Hi-C identifies putative target genes at 33 breast cancer risk loci. Nat Commun 9:1028
23. Robin JD, Ludlow AT, LaRanger R, Wright WE, Shay JW (2016) Comparison of DNA quantification methods for next generation sequencing. Sci Rep 6:24067

Chapter 7

Computational Analysis of Hi-C Data

Mattia Forcato and Silvio Bicciato

Abstract

The chromatin organization in the 3D nuclear space is essential for genome functionality. This spatial organization encompasses different topologies at diverse scale lengths with chromosomes occupying distinct volumes and individual chromosomes folding into compartments, inside which the chromatin fiber is packed in large domains (as the topologically associating domains, TADs) and forms short-range interactions (as enhancer-promoter loops). The widespread adoption of high-throughput techniques derived from chromosome conformation capture (3C) has been instrumental in investigating the nuclear organization of chromatin. In particular, Hi-C has the potential to achieve the most comprehensive characterization of chromatin 3D structures, as in principle it can detect any pair of restriction fragments connected as a result of ligation by proximity. However, the analysis of the enormous amount of genomic data produced by Hi-C techniques requires the application of complex, multistep computational procedures that may constitute a difficult task also for expert computational biologists. In this chapter, we describe the computational analysis of Hi-C data obtained from the lymphoblastoid cell line GM12878, detailing the processing of raw data, the generation and normalization of the Hi-C contact map, the detection of TADs and chromatin interactions, and their visualization and annotation.

Key words Hi-C data, 3D chromatin structure, Topologically associating domains, Chromatin interactions, Bioinformatics

1 Introduction

1.1 Generalities

In the last decade, methods based on chromosome conformation capture (3C [1]), such as 4C [2], 5C [3], and Hi-C [4], complemented fluorescence-based microscopy in the investigation of genome architecture and chromatin structures. These methods share the capacity to translate the three-dimensional information on spatial proximity of DNA into biochemical events, quantifiable by either PCR or next-generation sequencing. The primary steps, common to all 3C-based techniques, are cross-linking with formaldehyde and digestion with a restriction enzyme, with subsequent

Mattia Forcato and Silvio Bicciato contributed equally to this work.

Beatrice Bodega and Chiara Lanzuolo (eds.), *Capturing Chromosome Conformation: Methods and Protocols*, Methods in Molecular Biology, vol. 2157, https://doi.org/10.1007/978-1-0716-0664-3_7,

filling and ligation of the resulting ends to obtain a chimeric molecule comprising two spatially proximal loci. In Hi-C, fragment ends are filled with biotinylated nucleotides and ligated; the resulting molecules are sheared, precipitated with streptavidin, and sequenced in paired-end mode [5]. Typically, Hi-C data consist of hundreds of millions, or even billions, of read pairs generated by the ligation of pairs of DNA fragments that were proximal in the three-dimensional genome space. Read pairs are processed to generate a genome-wide contact map in which each entry accounts for the number of ligation products observed between two genomic regions. The width of such regions, which determines the resolution of the data, is selected based on the sequencing depth and on the restriction enzyme used to fragment the genome. Hi-C contact maps are analyzed to extract patterns and associate signal intensities to different chromatin conformations, as active and inactive compartments [4], regions of high self-interaction named topologically associating domains (TADs [6, 7]), and local contact enrichments associated with chromatin interactions. TADs appear as highly interacting regions that tend to segregate with respect to the neighboring chromatin, and their boundaries are particularly enriched in binding of architectural and insulator proteins as CTCF and cohesin [7, 8]. Chromatin interactions are evidenced by signal spikes in the contact map and may indicate the interaction between regulatory regions, as in the case of promoters and enhancers [9].

Hi-C data are commonly analyzed using multistep computational pipelines that, exploiting different algorithms and tools, allow the alignment and filtering of the sequence reads, the generation and normalization of the contact maps, and the identification and visualization of chromatin structures (as compartments, TADs, and interactions). Most of these methods have been reviewed in [10–12], and their performances have been compared in [13, 14].

1.2 Calling TADs and Interactions

In this chapter, we describe a pipeline for the analysis of raw Hi-C data and the identification, visualization, and annotation of TADs and chromatin interactions (Fig. 1). A detailed description on how to calculate active and inactive compartments from processed Hi-C data is reported in [15]. Here, the processing of raw Hi-C data is performed using HiC-Pro, a pipeline that, starting from sequencing reads (FASTQ files), executes the steps of sequence alignment and filtering and of binning and normalization to generate an Hi-C contact map [16]. For the identification of TADs we use Insulation Score [17] and Arrowhead [18, 19], while interactions are identified using Fit-Hi-C [20] and HiCCUPS [18, 19]. Although a variety of tools is available for the identification of TADs and chromatin interactions [10–12], no algorithm can be considered the gold standard to identify chromatin structures [13, 14]. Most approaches differ not only in terms of the implemented algorithm

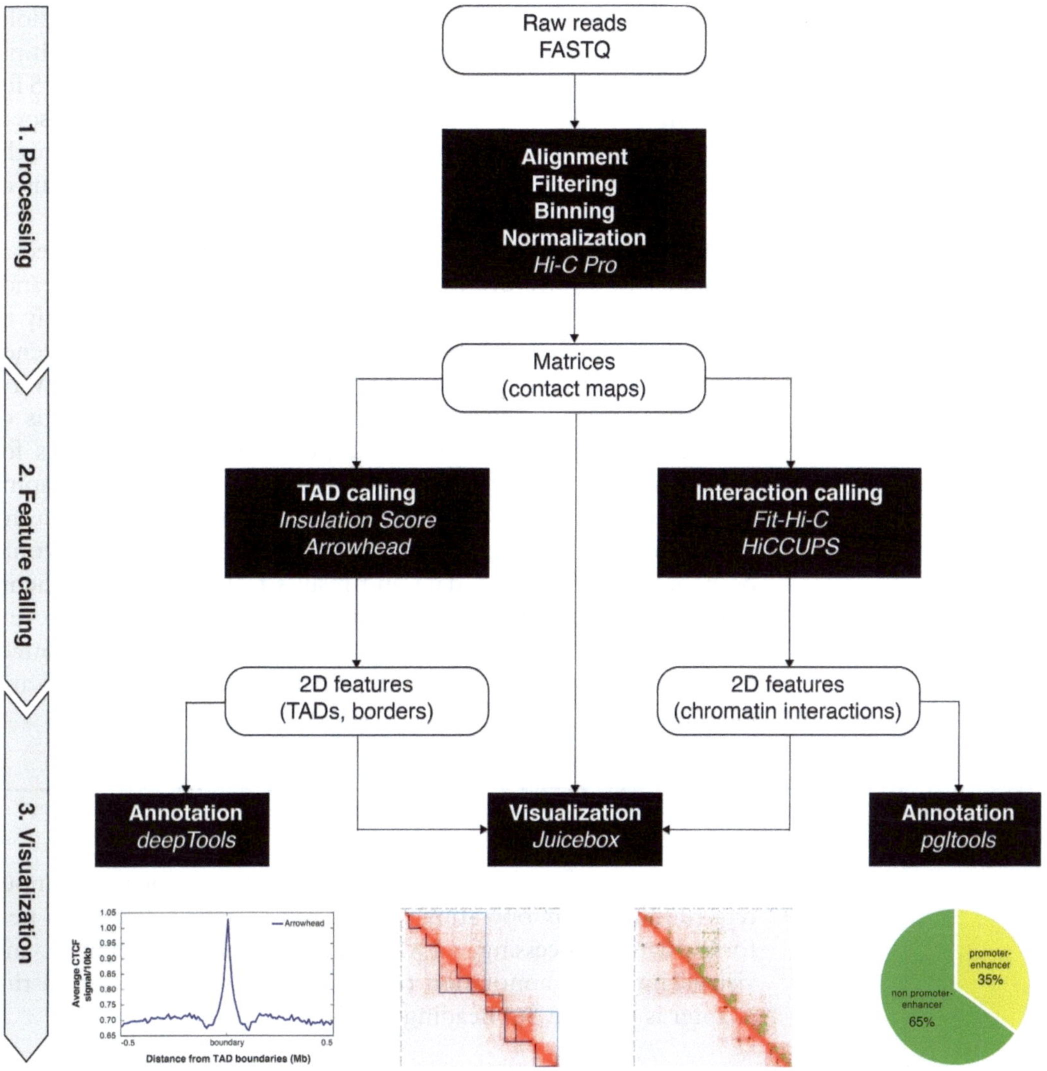

Fig. 1 Workflow of the Hi-C data analysis presented in this chapter. The white shapes contain the raw reads, matrices, and features, while the black boxes highlight the analysis steps. The tools used to perform each step are shown in italic. Examples of annotation results and visualization of TADs and interactions are shown in the lower part of the workflow

but also in the definition of the structures that are investigated and in usability. Therefore, on one hand, the choice of the method may impact the quantity and characteristic attributes of the identified TADs and interactions, and, on the other hand, the lack of large sets of true domains and interactions hinders the rational selection of the most robust method [13]. For these reasons, in this chapter, the identification of TADs and chromatin interactions is conducted using two different approaches for each type of chromatin structure. Specifically, for TADs we use Insulation Score, a method based on a one-dimensional representation of the contact matrix and that returns a single level of TADs, and Arrowhead, a tool that exploits the 2D structure of the Hi-C contact map and identifies multiple levels of TADs. For interactions we use Fit-Hi-C and HiCCUPS for the identification of globally or locally enriched contacts, respectively. Arrowhead and HiCCUPS are only applicable to high-resolution Hi-C data sets accounting for hundreds of millions of contacts.

Contact matrices are commonly represented and visualized as triangular or square 2D heatmaps using web interfaces or standalone software that can project TADs and interactions directly to the heatmap and complement the analysis results with other genomic data, as ChIP-seq signals. Most of the visualization software, while providing elegant interfaces and efficient combinations of diverse types of genomic data, adopt different data formats for loading contact matrices that complicate the exchange of inputs and outputs among the various tools. Considering the increasing production of data, the adoption of common data formats would be beneficial for sharing data and results from Hi-C experiments [21, 22]. Here, visualization is performed using the desktop version of Juicebox [23], a tool that displays the entire 2D contact matrix and provides browsing capabilities, as zooming and panning.

2 Materials

Materials used for analysis of Hi-C data primarily include computers running Unix operating system and specific software to perform data processing, TAD and interaction calling, and visualization and annotation of results. The software used in this chapter is listed in Subheading 2.1.2.

2.1 Necessary Resources

2.1.1 Hardware

UNIX operating systems with AWK package installed, standard CPUs or GPUs (*see* **Note 1**).

2.1.2 Software

The following software and tools are required to analyze the example data:

SRA Toolkit 2.9.0: https://trace.ncbi.nlm.nih.gov/Traces/sra/sra.cgi?view=toolkit_doc.

HiC-Pro 2.10.0: http://nservant.github.io/HiC-Pro/.

Juicer tools 1.8.9: https://github.com/aidenlab/juicer/wiki/Feature-Annotation.

Juicebox 1.8.8: https://github.com/aidenlab/Juicebox/wiki.

Cworld 0.01: https://github.com/dekkerlab/cworld-dekker.

Fit-Hi-C 2.0.5: https://github.com/ay-lab/fithic.

deepTools 2.5.3: https://deeptools.readthedocs.io/en/develop/index.html.

pgltools 2.1.4: https://github.com/billgreenwald/pgltools.

All listed tools run on Unix via the command line. For Juicebox, we use the desktop version for Windows. We assume that tools are installed in the /Software folder (either as binary or from source code), and programs are added to the user's path. For installation refer to the help page of each tool.

2.2 Data Sets

2.2.1 Description of the Example Data

In this chapter, the Hi-C example data are from the lymphoblastoid cell line GM12878 and correspond to sample HIC003 of [18]. Data have been obtained using the in situ Hi-C protocol and the restriction enzyme MboI (a four-cutter that recognizes the GATC sequence) and comprise 486 million paired-end reads with a length of 101 bps.

2.2.2 Download of the Data

The example data is available at the Sequence Read Archive (SRA) under the accession number SRR1658572 (https://www.ncbi.nlm.nih.gov/sra/?term=SRR1658572). Raw reads can be downloaded and directly converted into FASTQ format using SRA Toolkit:

```
fastq-dump --split-files SRR1658572 -O Rao_GM12878_HIC003
--gzip
```

where `--split-files` is used to generate two separate FASTQ files (having the data been obtained using paired-end sequencing), `SRR1658572` is the accession number of the example data, and `-O` specifies the folder where data will be saved, i.e., `Rao_GM12878_HIC003`.

2.2.3 Download of Support Data

A set of genomic annotation files are necessary to perform data processing and result annotation. The processing of the raw reads requires the sequence in FASTA format and the Bowtie2 indexes of the reference genome. The reference genome sequence, i.e., the assembly of the human genome UCSC hg19, and the Bowtie2 indexes of the same genome version can be both retrieved from the Illumina's iGenomes collection at the Bowtie 2 website (http://bowtie-bio.sourceforge.net/bowtie2/index.shtml). In this chapter, TADs and interactions will be annotated using the ENCODE project CTCF ChIP-seq and the chromatin states of the GM12878 cell line. Both data sets can be downloaded from ENCODE and UCSC using the following commands:

```
wget https://www.encodeproject.org/files/ENCFF364OXN/@@download/ENCFF364OXN.bigWig
wget http://hgdownload.soe.ucsc.edu/goldenPath/hg19/encodeDCC/wgEncodeBroadHmm/wgEncodeBroadHmmGm12878HMM.bed.gz
```

and stored in an annotation folder, as, for instance, `/home/Annotation`.

3 Methods

3.1 Processing of Raw Data Using HiC-Pro

3.1.1 HiC-Pro Input and Configuration Files

HiC-Pro is a pipeline for the processing of Hi-C data that takes in input sequencing reads (FASTQ files); performs alignment, filtering, binning, and normalization; and returns in output raw and normalized contact matrices along with diagnostic and quality control information [16]. Raw reads of all samples to be processed must be stored in a directory containing one folder for each sample. Thus, we first move the `Rao_GM12878_HIC003` folder with the FASTQ data to the `/home/data` folder:

```
mv Rao_GM12878_HIC003 /home/data
```

Before running HiC-Pro, it is advisable to perform quality controls on raw reads as commonly done for any other experiment involving sequencing data (*see* **Note 2**).

HiC-Pro requires three annotation files, i.e., the Bowtie 2 indexes, a table with the chromosome sizes and a file with the coordinates of the restriction fragments after digestion, and a configuration file. All annotation files must refer to the same version of the reference genome, e.g., hg19. The installation folder of HiC-Pro contains the table with the chromosome sizes for hg19 and the restriction fragment files for some enzymes. In the case the restriction fragment file of a given enzyme is not available, as is for MboI, it is possible to create it using the *digest_genome.py* script located in the `HiC-Pro-2.10.0/bin/utils` folder. *digest_genome.py* needs in inputs the sequence cut by the restriction enzyme (`GATC` for MboI) and the sequence of the reference genome in FASTA format (here `genome.fa`) and returns the restriction fragment file (here, `MboI_resfrag_hg19.bed`):

```
/Software/HiC-Pro-2.10.0/bin/utils/digest_genome.py -r ^GATC
-o MboI_resfrag_hg19.bed /home/Annotation/Homo_sapiens/UCSC/
hg19/Sequence/WholeGenomeFasta/genome.fa
```

Once created, the restriction fragment file has to be moved to the HiC-Pro `/annotation` folder:

```
mv MboI_resfrag_hg19.bed /Programs/HiC-Pro_2.10.0/annotation/
```

HiC-Pro configuration file contains all settings and paths for the analysis. A pre-compiled configuration file, named `config-hicpro.txt`, is available in the HiC-Pro folder. Although most of the parameters can be left as default, some others need to be modified to correctly run the analysis. Specifically, (a) in the data section, we must specify the suffixes that distinguish the FASTQ file names of the two mates, here identified as `_1` and `_2`; (b) in the alignment options section, we must provide the path to the Bowtie 2 indexes; (c) in the annotation files section, we indicate the name of the reference genome, i.e., the same that appears as the prefix of the Bowtie 2 index files; (d) in the digestion Hi-C processing section, we have to specify the name of the restriction fragments file and the ligation site. Note that the ligation site is the sequence resulting after ligation of two digested fragments and depends on the fill-in strategy adopted in the experiment. For the data used in this chapter, the ligation site is `GATCGATC`. The updated fields are as follows:

```
PAIR1_EXT = _1
PAIR2_EXT = _2
BOWTIE2_IDX_PATH = /home/Annotation/Homo_sapiens/UCSC/hg19/
Sequence/Bowtie2Index
REFERENCE_GENOME = genome
GENOME_FRAGMENT = MboI_resfrag_hg19.bed
 LIGATION SITE = GATCGATC
```

Finally, the edited file can be saved as `config-Rao.txt` in the `/home` folder (*see* **Note 3**).

3.1.2 Alignment

Starting from the `/home` folder, HiC-Pro can perform the entire processing workflow (i.e., alignment, filtering, binning, and normalization) in a single run or can be operated in sequential mode, i.e., performing only a subset of steps at the time. Here to check the quality of the results and eventually re-run some steps after modifying the configuration file, we will run HiC-Pro in sequential mode specifying the `-s` option. The alignment of all samples contained in the data folder, using the settings specified in config-Rao.txt and controlling the alignment quality (`-s quality_checks`), is performed through the following command:

```
HiC-Pro -c config-Rao.txt -i data -o hicpro_results_-
Rao_GM12878_HIC003 -s mapping -s quality_checks
```

which will create a folder named `hicpro_results_-Rao_GM12878_HIC003` containing the results of the alignments in the `bowtie2_results` subfolder and the alignment statistics in the `hic_results/pic` subfolder.

HiC-Pro adopts a two-step alignment strategy using Bowtie 2. First, reads for the two mates that align on the genome are saved in BAM format in the `bwt2_global` folder. Then, unmapped

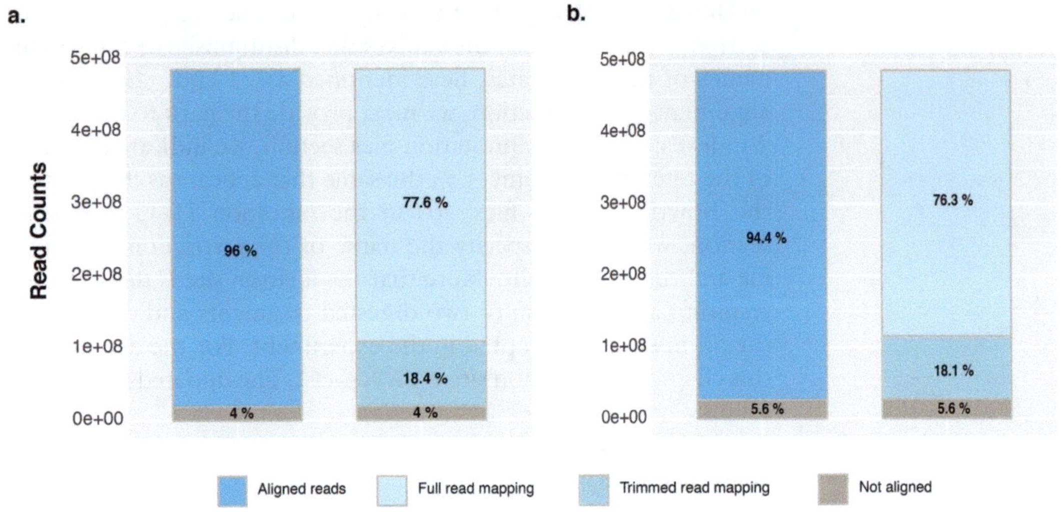

Fig. 2 Results of HiC-Pro two-step alignment for each of the two mates. (**a**) In the left bar is reported the total percentage of aligned (96%) and not aligned (4%) reads after the first (full read mapping) and the second (trimmed read mapping) alignment step for mate 1. The right bar reports the percentage of full read mapping (77.6%), trimmed read mapping (18.4%), and not aligned reads (4%) for mate 1. (**b**) Same as in (**a**) for mate 2

reads are trimmed according to the ligation site indicated in the configuration file, Bowtie 2 is run again on the trimmed reads, and alignments are saved in the `bwt2_local` folder. Finally, both alignments are combined in the `bwt2` folder and the statistics of the alignment plotted and saved as `plotMapping_Rao_GM12878_HIC003.pdf` in the `hic_results/pic` folder (Fig. 2). Since the two mates are aligned independently, statistics are reported separately for each mate. The plot of Fig. 2 indicates that a large percentage of the reads maps on the genome (more than 95% on average) and that about 18% of reads is mapped after trimming. This means that 18% of the reads were chimeric, i.e., spanned the ligation junction. The proportion of chimeric reads depends on the size of the fragments selected during sample preparation and on the read length. We expect the number of unmapped reads to be low (less than 5% on average), while a high number of unmapped reads could indicate problems with the library (e.g., read-through adapters or contamination) or that the wrong ligation site was given as input to HiC-Pro. Considering the alignment performance summarized in Fig. 2, we can proceed with filtering.

3.1.3 Filtering

The filtering step takes as input the configuration file and the alignment results contained in `hicpro_results_Rao_GM12878_HIC003/bowtie_results/bwt2` and performs the pairing of the reads and their mapping to restriction fragments (`-s proc_hic`) through the following command:

```
HiC-Pro -c config-Rao.txt -i hicpro_results_-
Rao_GM12878_HIC003/bowtie_results/bwt2 -o hicpro_results_-
Rao_GM12878_HIC003 -s proc_hic -s merge_persample -s
quality_checks
```

where `-s merge_persample` indicates to merge BAM files if multiple sequencing runs are present for the same sample and `-s quality_checks` allows saving several diagnostic plots in the `hic_results/pic` folder. This command generates a `data/` subfolder inside the `hic_results/` folder with the coordinates of all valid interaction products (`Rao_GM12878_HIC003_allValidPairs` file). Since HiC-Pro performs all the filters defined in the configuration file, we can check the results and, in case, re-run the filtering step with a set of modified filters. In the first predefined filter, read pairs are filtered based on their alignment and pairing. Results are saved in the `plotMappingPairing_Rao_GM12878_HIC003.pdf` file and displayed as in Fig. 3a. The plot summarizes the percentage of all reads retained after pairing and the percentage of not reported pairs, i.e., pairs with multiple alignments, pairs with singleton (only one mate is mapped), and pairs with low-quality alignments. These filters based on the alignment quality can be controlled editing the `RM_MULTI`, `RM_SINGLETON`, and `MIN_MAPQ` fields of the configuration file. Then, pairs are mapped on the restriction fragments and are classified as valid and invalid pairs based on their position and orientation on the restriction fragments. Results from this filter are saved in the `plotHiCFragment_Rao_GM12878_HIC003.pdf` and displayed as in Fig. 3b. Valid 3C interaction pairs are classified according to the orientation of the reads on the restriction fragments (ligation classes; FF and RR, same strand; FR, inward; RF, outward), and, since digested fragments are ligated randomly, the expected result is an equal proportion of the four categories. Invalid pairs are classified into different categories (as dangling end, self-circle, dumped pair, or re-ligation pairs) and are all discarded. In this experiment, almost 94% of all pairs are deemed as valid pairs. It is expected that a successful Hi-C experiment, and especially one obtained using the in situ protocol which reduces the frequency of spurious contacts, results in a low number of invalid pairs. A higher proportion of invalid pairs or an imbalance in the ligation classes of valid pairs indicate issues in the digestion or ligation step during the sample preparation. We can also apply a distance filter to remove pairs whose reads have a distance from the restriction sites outside the expected range (determined by the experimental size selection step). This is can be implemented setting the options `MIN_INSERT_SIZE` and `MAX_INSERT_SIZE` in the configuration file. Reads failing this filter may represent nonspecific ligation events due to random breakage of the chromatin or unspecific enzyme activity and will be labeled as dumped pairs. The absolute numbers

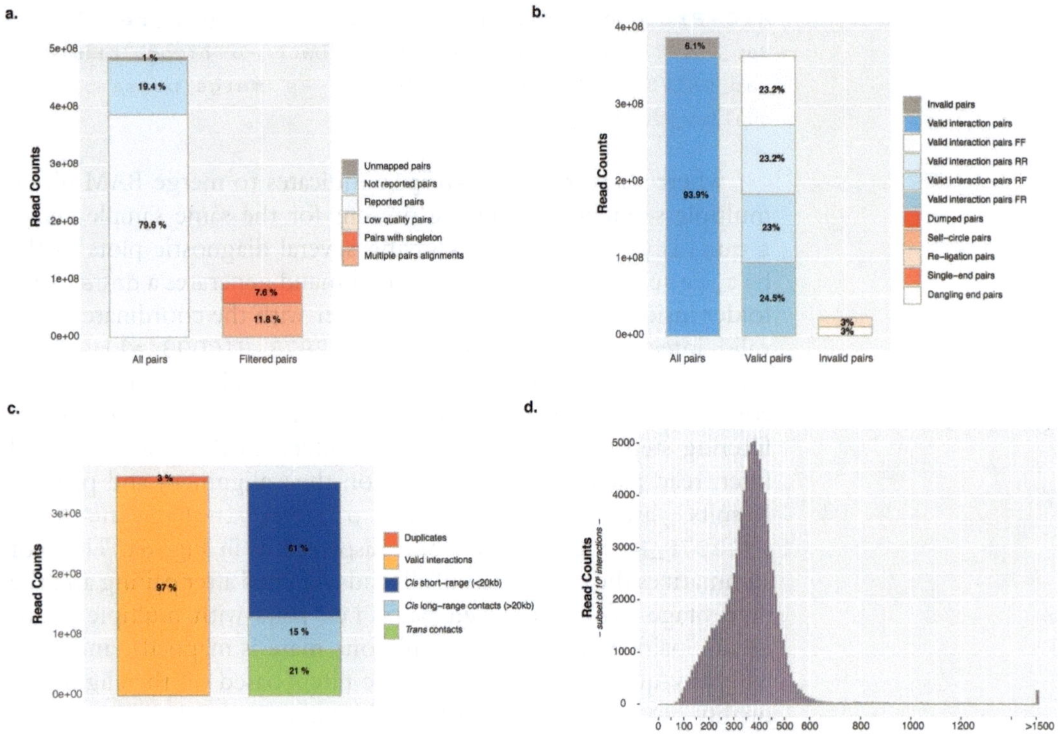

Fig. 3 Results of HiC-pro pairing and filtering steps. (**a**) After being independently mapped, reads from the two BAM files are paired. The *not reported pairs* in the left bar (19.4% of mapped reads) comprise the categories shown in the right bar (i.e., *low-quality pairs*, *pairs with singleton*, and *multiple pairs alignments*). In this analysis, the percentage of *low-quality pairs* is zero because reads were not filtered on mapping quality. (**b**) *Reported pairs* from panel (**a**) are mapped to the restriction fragments and divided into *valid* and *invalid interactions pairs*. *Valid pairs* are classified based on the orientation of the reads (F and R stand for forward and reverse, respectively). *Invalid pairs* include read pairs falling on (i) same fragment with the same strand orientation (*dumped pairs*; 0%), (ii) same fragment with inward orientation (*dangling end pairs*; 3%), (iii) same fragment with outward orientation (*self-circle pairs*; 0.1%), (iv) adjacent fragments (*re-ligation pairs*; 3%), and, if not filtered in the previous step, *single-end pairs* (0%). (**c**) *Valid pairs* from panel (**b**) are deduplicated and split according to their distance and interaction type. (**d**) A subset of *cis* valid interactions is used to estimate the fragment size (in bp)

of pairs falling in each category are reported in the `Rstat` files contained in the `hic_results/data/Rao_GM12878_HIC003` folder. The last filter removes PCR duplicates. Results of this filtering step are reported in the `plotHiCContactRanges_Rao_GM12878_HIC003.pdf` file and plotted as in Fig. 3c. This graph shows the number of duplicates and the proportion of deduplicated valid pairs classified into short (<20 kb) and long (>20 kb) range intra-chromosomal (cis) and inter-chromosomal (trans) contacts. According to [18], many trans contacts and a small proportion of long-range cis contacts (<15%) are indicators of an unsuccessful experiment with mostly random ligation products. Finally, the `plotHiCFragmentSize_Rao_GM12878_HIC003.`

pdf file contains the fragment size distribution derived from a subset of valid deduplicated cis contacts (Fig. 3d). The plot indicates that the distribution is centered around 380 bp which is compatible with the fragment size selected to generate the sample.

3.1.4 Binning and Normalization

To generate the contact matrices, HiC-Pro default settings comprise five bin sizes, i.e., 20 kb, 40 kb, 150 kb, 500 kb, and 1 Mb, as specified in the configuration file. Since the filtering steps allowed retaining a consistent number of valid pairs (>350 million), we can also produce the contact matrices at 5 and 10 kb resolution. To specify these two additional bin sizes, we add 5000 and 10000 to the BIN_SIZE option in the configuration file. HiC-Pro performs normalization with a fast implementation of the iterative correction method (ICE [24]). Binning (-s build_contact_maps) and normalization (-s ice_norm) of the matrices are performed using the following command:

```
HiC-Pro -c config-Rao.txt -i hicpro_results_-
Rao_GM12878_HIC003/hic_results/data -o hicpro_results_-
Rao_GM12878_HIC003 -s build_contact_maps -s ice_norm
```

HiC-Pro bins the valid interaction products stored in the hic_results/data folder and stores the matrices in the hic_results/matrix folder. The matrix folder is divided into two subfolders (raw and iced) that contain subfolders corresponding to the various bin sizes (i.e., the raw matrix binned at 10 kb is saved in the Rao_GM12878_HIC003_10000.matrix file contained in the hicpro_results_Rao_GM12878_HIC003/hic_results/matrix/Rao_GM12878_HIC003/raw/10000/ subfolder). Matrices are saved in the triple sparse or COO format, where only nonzero entries of the upper triangular matrix are saved. To save space, bins are indicated with a bin identifier, and the correspondence between bin identifiers and genomic coordinates is reported in a BED file. The iced subfolders also contain the ICE calculated biases.

3.1.5 Visualization of the Contact Maps

Once generated the matrices, it is possible to start exploring the results. To make this step faster, rename the hicpro_results_-Rao_GM12878_HIC003 result folder as WORKDIR and enter it:

```
WORKDIR=/home/hicpro_results_Rao_GM12878_HIC003
 cd $WORKDIR
```

The easiest way to explore the contact matrices is through visualization. Hi-C matrices are normally visualized as square symmetric heatmaps (contact maps), with a color scale corresponding to raw or normalized counts. Heatmaps can be generated using tools such as HiTC, an R package able to produce a static image of a chromosome or of selected regions [25]. However, it is more attractive the use of an interactive approach where the matrix can be browsed in real time, zoomed in and out, and where the annotations such as TADs and interactions (see next sections) can be easily overlaid on the heatmap. There are several tools that can be used to perform these tasks, such as Juicebox [23] and HiGlass [26]. HiC-Pro conveniently provides utilities to convert results into formats compatible with these tools. Indeed, text formats are not efficient to smoothly load and browse big matrices and are commonly replaced by binary compressed formats. Here, we will use the `hicpro2juicebox.sh` script, a wrapper for Juicer tools, to load valid interaction pairs, bin matrices at several resolutions, and perform matrix normalization. In `hicpro2juicebox.sh`, the ICE normalization is replaced by Knight and Ruiz (KR) normalization [27], a different implementation of matrix balancing with faster convergence, and the output is a file in the HIC format (*see* **Note 4**). This format, supported by Juicebox for visualization, will also be used as input to Arrowhead and HiC-CUPS for features calling. The HIC file, here named `Rao_GM12878_HIC003_allValidPairs.hic`, is generated as follows:

```
/Software/HiC-Pro-2.10.0/bin/utils/hicpro2juicebox.sh -i hi-
c_results/data/Rao_GM12878_HIC003/Rao_GM12878_HIC003_allVa-
lidPairs -g hg19 -j /Software/juicer_tools.1.8.9_jcuda.0.8.jar
```

Here, the Hi-C matrix will be visualized using Juicebox desktop client. In Juicebox, the map is loaded selecting "File/open," "Local," and the HIC file from the pop-up window. At first, cis and trans maps for all chromosomes are displayed. The chromosomes menu or double-clicking on one of the chromosomes can be used to display the chromosome cis Hi-C heatmap. The map can be inspected using mouse to zoom in (or zoom out), thus highlighting the presence of structures such as compartments. By zooming over the diagonal, it is possible to visualize potential TADs and focal points of interactions (Fig. 4a, b).

3.2 Identification of TADs

3.2.1 Insulation Score

Insulation Score slides a square (insulation square) along the diagonal and aggregates the signal inside the square to calculate an Insulation Score for each bin [17]. The slope of the Insulation Score (delta vector) is approximated calculating the difference between insulation in a window (delta window) to the left and to the right of the central bin. Boundaries are called at bins where the delta vector crosses zero at Insulation Score minima. The coordinates of TADs are then retrieved as regions between boundaries. An implementation of this method is available in cworld, a collection of scripts developed at Dekker Lab (https://github.com/dekkerlab/cworld-dekker). The input to cworld scripts is a normalized matrix

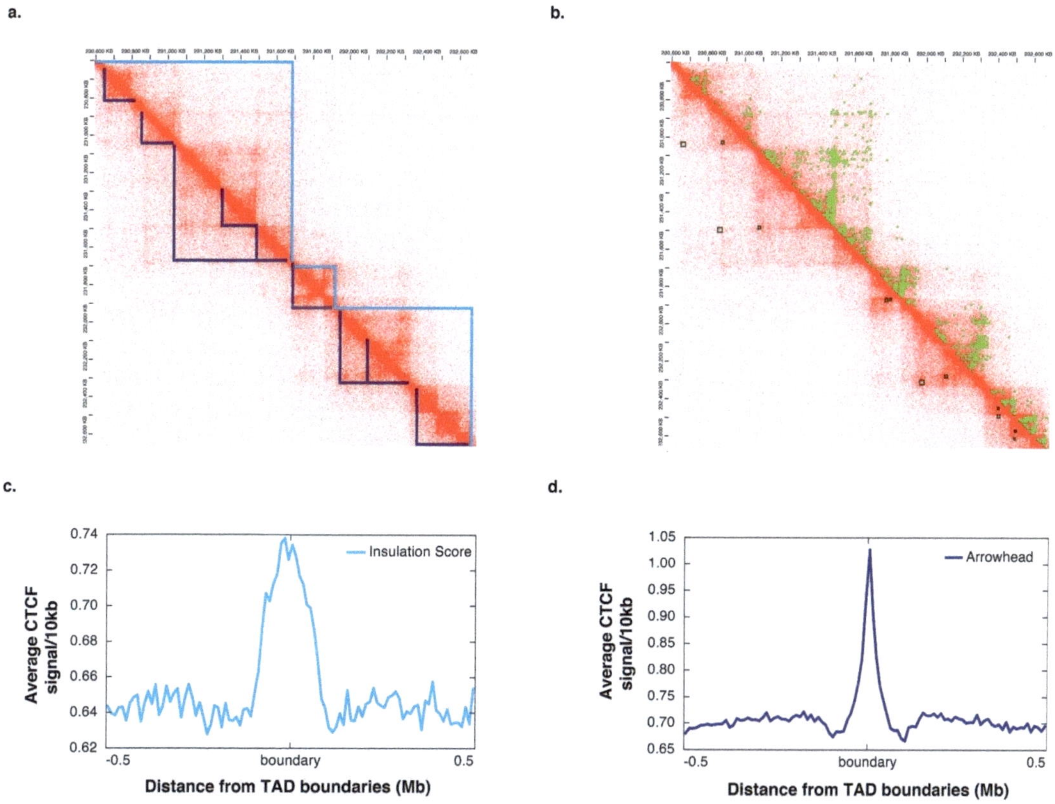

Fig. 4 Visualization and annotation of the results. (**a**) Heatmap produced with Juicebox showing the KR normalized counts for region chr2:230,600,000-232,700,000. In the upper part of the map are reported the TADs identified by Insulation Score (light blue lines); in the lower part the TADs called by Arrowhead (dark blue lines). (**b**) The same heatmap as in (**a**) displaying in the upper part the interactions found by Fit-Hi-C (light green squares) and in the lower part those identified by HiCCUPS (dark green squares). (**c**) and (**d**) Enrichment of CTCF binding (signal from CTCF ChIP-seq in GM12878 cell line) in a window of 1 Mb around the TAD boundaries found by Insulation Score (left panel) and Arrowhead (right panel)

in dense format. To convert the HiC-Pro ICE normalized matrix, we use `sparseToDense.py`, a script contained in the HiC-Pro / `utils` folder, that converts the HiC-Pro ICE normalized matrix from sparse to dense. The following command generates a dense matrix at 10 kb resolution for each chromosome:

```
/Software/HiC-Pro-2.10.0/bin/utils/sparseToDense.py -b hi-
c_results/matrix/Rao_GM12878_HIC003/raw/10000/Rao_GM12878_HI-
C003_10000_abs.bed hic_results/matrix/Rao_GM12878_HIC003/
iced/10000/Rao_GM12878_HIC003_10000_iced.matrix --perchr --di
```

For the file to be fully compliant with the cworld format, the IDs in first column and first row have to be formatted as `name|assembly|chr:start-end`. In doing this, we also create a subfolder named `insulationScore_results/` inside the hicpro_results_Rao_GM12878_HIC003/ folder to store intermediate files and results:

```
mkdir insulationScore_results
cd insulationScore_results/
chr_list=$(egrep -v '(_|Y|M)' /Software/HiC-Pro-2.10.0/anno-
tation/chrom_hg19.sizes | cut -f1)
for chr in $chr_list; do
awk '{print "bin"NR"|hg19|"$1":"$2"-"$3,$0}' OFS="\t" $WORK-
DIR/hic_results/matrix/Rao_GM12878_HIC003/iced/10000/
Rao_GM12878_HIC003_10000_iced_${chr}_dense.matrix > tmp.txt
cut -f 1 tmp.txt | awk 'BEGIN{ORS="\t"}{print}' | awk '{print
"",$0}' OFS="\t" > Rao_GM12878_HIC003_10000_iced_${chr}_den-
se_Cworld.matrix
NF=$(awk 'NR==1{print NF}' tmp.txt)
cut -f 1,5-$NF tmp.txt >> Rao_GM12878_HIC003_10000_iced_
${chr}_dense_Cworld.matrix
rm tmp.txt
 done
```

To run Insulation Score, we will use two scripts, i.e., matrix2insulation.pl which calculates boundaries and insulation2tads.pl that derives the TADs. This is done through a cycle that runs the scripts on all chromosomes and then concatenates results in a single file named Rao_GM12878_HIC003_10000_IS.bed (*see* **Note 5**):

```
for chr in $chr_list; do
perl /Software/cworld-dekker/scripts/perl/matrix2insulation.
pl -i Rao_GM12878_HIC003_10000_iced_${chr}_dense_Cworld.ma-
trix --is 500000 --ids 200000 --bmoe 0
boundaries=$(ls *.boundaries | grep "$chr"_)
insulation=$(ls *.insulation | grep "$chr"_)
perl /Software/cworld-dekker/scripts/perl/insulation2tads.pl
-i $insulation -b $boundaries --mbs 0.1
done
for file in *tads.bed; do
awk 'NR>1{print}' OFS="\t" $file >> Rao_GM12878_HI-
C003_10000_IS.bed
 done
```

Here, the size of the Insulation Square (--is) and the insulation delta window (--ids) are set equal to 500 Mb and 200 Mb, respectively, as done in the original paper when using 10 kb resolution matrices [17]. Moreover, the boundary margin of error (--bmoe) is set equal to zero and the minimum boundary strength (--mbs) to 0.1. Among various output files, the Insulation Score scripts save, for each chromosome, a plot with the profiles of the

Insulation Score and the delta vector along the chromosome (a PNG file with `insulation.debug` as name suffix). Using the data of this example, Insulation Score identifies 4460 domains with an average size of 611 kb.

3.2.2 Arrowhead

Differently from Insulation Score, Arrowhead algorithm adopts image processing techniques to directly analyze the 2D matrix, i.e., without reducing the data matrix to a one-dimensional vector [18, 19]. Arrowhead is based on a matrix transformation that converts the square pattern of TADs into an easy-to-detect arrowhead-shaped feature, with the arrow pointing the upper-left corner of the domain. Subsequently, a corner score matrix is obtained from the arrowhead matrix using a heuristic scoring algorithm; high scoring pixels corresponding to pixel with high likelihood of being domain corners are identified using dynamic programming. While Insulation Score almost continuously partitions the genome into a single layer of TADs, Arrowhead can instead identify overlapping and nested non-adjacent TADs. This is an important feature since the visual inspection of high-resolution Hi-C maps suggests the presence of TADs hierarchies. Arrowhead is embedded in Juicer tools [19] and can be run on the same HIC file used for the matrix visualization in Juicebox:

```
cd $WORKDIR
 java -Xmx20g -jar /Software/juicer_tools.1.8.9_jcuda.0.8.jar
arrowhead Rao_GM12878_HIC003_allValidPairs.hic arrowhead_re-
sults
```

Results are saved in a folder named `arrowhead_results`.

Since the HIC file contains KR normalized matrices binned with various bin sizes, Arrowhead first evaluates if the matrix contains enough contacts (sparsity check; *see* **Note 6**) and then, depending on the number of contacts, calls TADs on the matrix binned at 5 or 10 kb resolution. Since the sample used in this example successfully passes the sparsity check as a medium resolution map, Arrowhead is run on the matrix binned at 10 kb. The result file `10000_blocks.bedpe` contains 4006 domains with a median size of 325 kb.

3.3 Identification of Interactions

3.3.1 Fit-Hi-C

Fit-Hi-C identifies significant chromatin interactions comparing the observed contact count of a given pair to a global background [20]. Briefly, Fit-Hi-C calculates the contact probability of a given pair using a binomial distribution and estimates the parameters of the distribution with a monotonic spline fit of the contact counts versus the genomic distance. The method can perform an additional refinement by removing outliers and calculating a second spline fit. Fit-Hi-C can incorporate biases learned by ICE in the confidence estimation and therefore can be applied on the raw matrices. Here, Fit-Hi-C is applied to the matrix binned at 5 kb resolution. First input files for Fit-Hi-C are generated using a HiC-Pro script:

```
cd $WORKDIR
 /Software/HiC-Pro-2.10.0/bin/utils/hicpro2fithic.py -i hi-c_results/matrix/Rao_GM12878_HIC003/raw/5000/Rao_GM12878_HIC003_5000.matrix -b hic_results/matrix/Rao_GM12878_HIC003/raw/5000/Rao_GM12878_HIC003_5000_abs.bed -s hic_results/matrix/Rao_GM12878_HIC003/iced/5000/Rao_GM12878_HIC003_5000_iced.matrix.biases -r 5000 -o fithi-c_results
```

where `-i`, `-b`, and `-s` represent the raw interaction counts, the bin coordinates, and the ICE biases, respectively. This script generates three Fit-Hi-C input files in the `fithic_results` folder. Fit-Hi-C is then run using the following command (see Note 7):

```
cd fithic_results/
 fithic -f fithic.fragmentMappability.gz -i fithic.interactionCounts.gz -r 5000 -o ./ -t fithic.biases.gz -l fithic_5kb -U 5000000 -L 10000 -v
```

It is recommended to specify an upper and lower bound on the range of the intra-chromosomal interactions by setting the `-U` and `-L` parameters. By default, this version of Fit-Hi-C reports only cis interactions, but both cis and trans interactions can be retrieved setting the `-x` parameter to `All`. In case a small number of interactions are returned, the null model can be refined by adding a second spline fit setting the `-p` option to 2. FDR values for all detected interactions are returned in the `fithic_5kb.spline_pass1.res5000.significances.txt` file. The FDR threshold can be set inspecting the `fithic_5kb.spline_pass1.qplot.png` file which displays the number of called interactions as a function of the FDR threshold. This plot can be generated specifying the `-v` option. Interactions with an FDR < 0.01 can be retrieved with the command:

```
awk '$7<0.01 || $7=="q-value" {print}' fithic_5kb.spline_pass1.res5000.significances.txt > fithic_5kb.q0.01.txt
```

Setting this FDR threshold, we obtain 337,644 interactions at a median distance of 115 kb.

3.3.2 HiCCUPS

HiCCUPS identifies regions that are enriched with respect to a local background [18, 19]. In particular, HiCCUPS searches for pixels with a signal (number of contacts) that is significantly higher with respect to four local neighborhoods, represented by different portions of a donut-shaped area surrounding the pixels under investigation. The results are focal peaks, clearly visible on the contact map, which are also called loops. As Arrowhead, HiCCUPS is encoded in Juicer tools and can be applied only on matrices with a high number of reads (*see* **Note 6**). HiCCUPS estimates the resolution of the data set (medium or high) in terms of number of contacts and then calls significant interactions at multiple resolutions. Nearby enriched pixels are clustered, and then results are merged across resolutions in such a way that if the same loop is identified at multiple bin sizes, the highest resolution version of the peak is reported. In this case (medium resolution map), loops are called on KR normalized matrices binned at 5 kb, 10 kb, and 25 kb and then merged. Since the current version of HiCCUPS run on GPUs using CUDA, the path to JCUDA has to be specified in launch command of HiCCUPS (*see* **Note 8**):

```
cd $WORKDIR
  java -Xmx20g -Djava.library.path=jcuda -jar /Software/juicer_tools.1.8.9_jcuda.0.8.jar hiccups Rao_GM12878_HIC003_allValidPairs.hic hiccups_results
```

The list of significant interactions is saved in the `merged_loops.bedpe` file inside the `hiccups_results` folder. In the example data, HiCCUPS identifies 6226 loops with a median distance of 250 kb.

3.4 Visualization and Annotation of Analysis Results

3.4.1 Data Format Conversion for Visualization

Juicebox visualizes 2D annotations (squares) saved in BEDPE format, i.e., a tab delimited file where the first six columns are the *x* (chr, start, end) and *y* (chr, start, end) coordinates of each annotation. In the case of TADs, *x* will be equal to *y*, whereas for interactions they will be different. TADs and interactions called by Arrowhead and HiCCUPS are already saved in this format and can be directly loaded in Juicebox (*see* **Note 9**). Instead, TADs found by Insulation Score and interactions called by Fit-Hi-C need to be converted in a BEDPE file using the following commands:

```
cd $WORKDIR
awk '{print $1,$2,$3,$1,$2,$3}' OFS="\t" insulationScore_results/Rao_GM12878_HIC003_10000_IS.bed > insulationScore_results/Rao_GM12878_HIC003_10000_IS.bedpe
  awk 'NR>1{print $1,$2-2500,$2+2500,$3,$4-2500,$4+2500}' OFS="\t" fithic_results/fithic_5kb.q0.01.txt > fithic_results/fithic_5kb.q0.01.bedpe
```

3.4.2 Visualization of TAD and Interactions

To load annotations in Juicebox, open the annotation panel and select the 2D annotations tab. Then, choose if we want to display pre-loaded list of loops and domains or to load a local file. By selecting "Add Local" we can load the domains obtained with Insulation Score and Arrowhead. In the Annotations Layer Panel, we can change the colors and visibility of the tracks. Figure 4a displays the TADs identified by Insulation Score (light blue lines in the upper part of the matrix) and by Arrowhead (dark blue lines in the lower part of the matrix). We can see how Insulation Score captures one level of TADs organization, often the one corresponding to wider TADs, whereas Arrowhead returns nested structures and some of the visible small domains. However, some TADs seem to be missed by both approaches. Regardless of TADs, the identified boundaries are quite conserved among the two tools. TADs can be temporarily hidden by clicking on the eye symbol in the annotation panel and interactions loaded following the same steps used for loading TADs. Figure 4b shows the interactions found by Fit-Hi-C (light green squares in the upper part of the matrix) and those called by HiCCUPS (dark green squares in the lower part of the matrix). From the figure, it can be noted how the approaches implemented by the two methods return completely different results: i.e., HiCCUPS finds a limited number of focal peaks, often involving one or two TAD boundaries, whereas Fit-Hi-C identifies many more interactions also occurring inside the domains.

3.4.3 Result Integration for TADs

The identified TADs can be further characterized investigating the enrichment of genomic signals around TADs and TAD boundaries. For instance, through the annotation panel of Juicebox clicking on 1D annotations and "Add Local," we can load the BIGWIG of the CTCF signal in GM12878 cell line downloaded from ENCODE and evaluate the presence of CTCF binding sites at TAD boundaries. The CTCF track will appear on either side of the heatmap revealing that many TAD borders correspond to high CTCF signal. Moreover, we can evaluate this enrichment at a genome-wide level by plotting the average CTCF signal around all boundaries using deepTools. To use deepTools, we need a BED file with the boundary coordinates and a BIGWIG file with the genomic signal we want to plot (i.e., here CTCF but, in general, any transcription factor or histone modification). First, concatenate the boundary files generated by Insulation Score (for a total of 4507), and extract the list of 7857 boundaries found by Arrowhead with the following commands:

```
cd insulationScore_results/
for file in *boundaries.bed; do
awk 'NR>1 {if($5>0.1)print}' OFS="\t" $file >> Rao_GM12878_-
HIC003_10000_IS_boundaries.bed
done
```

```
cd $WORKDIR/arrowhead_results/
 awk 'NR>1{print "chr"$1,$2-5000,$2+5000"\nchr"$1,$3-5000,$3
+5000}' OFS="\t" 10000_blocks.bedpe | awk '!_[$0]++' >
10000_blocks_boundaries.bed
```

Then, use deepTools as follows:

```
cd $WORKDIR
computeMatrix reference-point -R insulationScore_results/
Rao_GM12878_HIC003_10000_IS_boundaries.bed -S /home/Annota-
tion/ENCFF364OXN.bigWig --referencePoint center --binSize
10000 --upstream 500000 --downstream 500000 -o CTCF_avera-
ge_IS.matrix.gz
plotProfile --matrixFile CTCF_average_IS.matrix.gz --outFile-
Name CTCF_average_IS.profile.pdf --refPointLabel boundary --
samplesLabel CTCF --regionsLabel InsulationScore
computeMatrix reference-point -R arrowhead_results/
10000_blocks_boundaries.bed -S /home/Annotation/ENCFF364OXN.
bigWig --referencePoint center --binSize 10000 --upstream
500000 --downstream 500000 -o CTCF_average_arrowhead.matrix.gz
 plotProfile --matrixFile CTCF_average_arrowhead.matrix.gz --
outFileName CTCF_average_arrowhead.profile.pdf --refPointLa-
bel boundary --samplesLabel CTCF --regionsLabel Arrowhead
```

to define a 1 Mb window centered around each boundary and a bin size (length of bin for averaging the score) of 10 kb to calculate the score and then plot the profile.

In Fig. 4c, d are shown the enrichments in CTCF signal around the lists of boundaries identified by Insulation Score (Fig. 4c) and Arrowhead (Fig. 4d). The peak around the boundaries identified by Insulation Score is less sharp, and this is in line with the fact that, in the original publication, the final boundary zones were defined with a boundary margin of error of ±30 kb.

3.4.4 Result Integration for Interactions

Chromatin interactions can be used to assign regulatory regions or SNPs to their target genes, in a 3D aware annotation process. An interesting class of chromatin interactions is the one involving promoters and enhancers. Such interactions can be used to annotate, e.g., ChIP-seq peaks falling in distal enhancer regions. In Juicebox, we can load several 1D annotation such as Refseq genes, p300 peaks, and files containing peaks from ChIP-seq experiments and investigate the interactome of genes or regions of interest. At a genome-wide level, we can count the number and proportion of interactions (identified by either Fit-Hi-C or HiC-CUPS) which involve promoters and enhancers. To do this, we use the list of GM12878-specific promoters and enhancers (extracted by the chromatin states defined by ChromHMM software [28, 29] and downloaded from ENCODE) and generate two BED files containing the coordinates of promoters and enhancers as follows:

```
awk '{print $1,$2,$3,$4}' OFS="\t" /home/Annotation/wgEncodeB-
roadHmmGm12878HMM.bed | grep Promoter > /home/Annotation/
GM12878_Promoters.bed
 awk '{print $1,$2,$3,$4}' OFS="\t" /home/Annotation/wgEnco-
deBroadHmmGm12878HMM.bed | grep Enhancer > /home/Annotation/
GM12878_Enhancers.bed
```

To classify interactions, we then use some functions of the pgltools package [30] to convert BEDPE into PGL (paired genomic loci) files, calculate the overlap of Fit-Hi-C interactions with promoters and enhancers, and intersect the results:

```
cd fithic_results/
awk '{print $1,$2+1,$3,$4,$5+1,$6}' OFS="\t" fithic_5kb.q0.01.
bedpe > fithic.bedpe
pgltools formatbedpe fithic.bedpe > fithic.pgl
pgltools intersect1D -a fithic.pgl -b /home/Annotation/
GM12878_Promoters.bed -wa | awk '!_[$0]++' > promoters_-
fithic.pgl
pgltools intersect1D -a fithic.pgl -b /home/Annotation/
GM12878_Enhancers.bed -wa | awk '!_[$0]++' > enhancers_-
fithic.pgl
pgltools intersect -a promoters_fithic.pgl -b enhancers_-
fithic.pgl -allA | awk '!_[$0]++'| pgltools merge -stdInA -o
distinct,distinct -c 7,8 > annotated_PE_fithic.pgl
 awk 'NR>1{if(!($7=="A" && $8=="A")&&!($7=="B" && $8=="B"))
{print}}' annotated_PE_fithic.pgl | wc -l
```

Columns 7 and 8 in the final `annotated_PE_fithic.pgl` table report the interacting bins containing at least one promoter and at least one enhancer, respectively. We can finally count the number of promoter-enhancer interactions excluding those where both bins overlap the same category.

The same code can be run on interactions identified by HiCCUPS with the addition of removing the BEDPE file header as follows:

```
cd $WORKDIR/hiccups_results/
 awk 'NR>1{print "chr"$1,$2+1,$3,"chr"$4,$5+1,$6}' OFS="\t"
merged_loops.bedpe > merged_loops_noheader.bedpe
```

We find that 68,678 out of 337,644 interactions (20%) identified by Fit-Hi-C and 2169 out of 6226 interactions (35%) called by HiCCUPS are classified as promoter-enhancer interactions.

4 Notes

1. While most tools run on CPUs, the stable version of HiCCUPS requires the use of GPUs. However, the developmental version of HiCCUPS runs also on CPUs (*see* also **Note 8**).
2. Before running the alignment step, it is a good practice to control the quality of FASTQ files using, for instance, FastQC (https://www.bioinformatics.babraham.ac.uk/projects/fastqc/). FastQC performs various quality checks and can mark the presence of contamination, adapters, quality loss, and high duplication. For sake of space, we do not include this step in our workflow, but we encourage users to do it.
3. A section of the configuration file regards the system used for the analysis, where the user can specify the number of CPUs and the parameters for running HiC-Pro on a computer cluster.
4. Since `hicpro2juicebox.sh` takes in input valid interaction pairs, the script performs binning and normalization. However, using juicer tools `pre`, it is possible to create the HIC file starting directly from the matrix binned and normalized with HiC-pro. This latter approach requires the conversion of the matrix from the sparse to the *Short with score* format and results a HIC file at a single resolution.
5. Insulation Score was reported to run very fast on 40 kb binned samples [13]. However, using the example data binned at 10 kb, Insulation Score takes almost 28 h to call TADs on all chromosomes.
6. The authors recommend running Arrowhead (and HiCCUPS) only if the matrix contains a large number of contacts (e.g., more than 300 million contacts; sparsity check). If the data do not pass the sparsity check, the algorithm exits with a warning. This control can be overridden adding the `--ignore_sparsity` flag to the command line.
7. Fit-Hi-C is also available in Bioconductor as a R package (https://bioconductor.org/packages/release/bioc/html/FitHiC.html). However, using the R package, memory issues can be encountered when analyzing samples at 5 kb resolution.
8. The development version of Juicer tools (1.9.8) comprises an implementation of HiCCUPS that can run on CPUs but restricts the search of loops within 8 Mb of the diagonal. Since this impacts the calculation of FDR, results obtained using the GPU- or the CPU-based versions are not identical. With the example data used here, the GPU and the CPU versions identified 6226 and 5939 loops, respectively. The

95.7% of the interactions identified with the CPU version is included in the interactions called by the GPU version.

9. Although the extension of these files is BEDPE, the BEDPE format, as defined at https://bedtools.readthedocs.io/en/latest/content/general-usage.html, does not include the header.

Acknowledgments

This work was supported by Bando Ricerca Finalizzata 2016 grant GR-2016-02362451 (to M.F.) and by AIRC Special Program Molecular Clinical Oncology "5 per mille" grant 10016 and CNR-MIUR Epigenetics Flagship project (to S.B.). We thank Martina Dori for collaboration on the analysis of example data and critical feedback on the manuscript.

References

1. Dekker J, Rippe K, Dekker M, Kleckner N (2002) Capturing chromosome conformation. Science 295:1306–1311. https://doi.org/10.1126/science.1067799
2. Simonis M, Klous P, Splinter E et al (2006) Nuclear organization of active and inactive chromatin domains uncovered by chromosome conformation capture-on-chip (4C). Nat Genet 38:1348–1354. https://doi.org/10.1038/ng1896
3. Dostie J, Richmond TA, Arnaout RA et al (2006) Chromosome conformation capture carbon copy (5C): a massively parallel solution for mapping interactions between genomic elements. Genome Res 16:1299–1309. https://doi.org/10.1101/gr.5571506
4. Lieberman-Aiden E, van Berkum NL, Williams L et al (2009) Comprehensive mapping of long-range interactions reveals folding principles of the human genome. Science 326:289–293. https://doi.org/10.1126/science.1181369
5. Denker A, de Laat W (2016) The second decade of 3C technologies: detailed insights into nuclear organization. Genes Dev 30:1357–1382. https://doi.org/10.1101/gad.281964.116
6. Nora EP, Lajoie BR, Schulz EG et al (2012) Spatial partitioning of the regulatory landscape of the X-inactivation centre. Nature 485:381–385. https://doi.org/10.1038/nature11049
7. Dixon JR, Selvaraj S, Yue F et al (2012) Topological domains in mammalian genomes identified by analysis of chromatin interactions. Nature 485:376–380. https://doi.org/10.1038/nature11082
8. Rao SSP, Huang S-C, Glenn St Hilaire B et al (2017) Cohesin loss eliminates all loop domains. Cell 171:305–320.e24. https://doi.org/10.1016/j.cell.2017.09.026
9. Jin F, Li Y, Dixon JR et al (2013) A high-resolution map of the three-dimensional chromatin interactome in human cells. Nature 503:290–294. https://doi.org/10.1038/nature12644
10. Ay F, Noble WS (2015) Analysis methods for studying the 3D architecture of the genome. Genome Biol 16:183. https://doi.org/10.1186/s13059-015-0745-7
11. Schmitt AD, Hu M, Ren B (2016) Genome-wide mapping and analysis of chromosome architecture. Nat Rev Mol Cell Biol 17:743–755. https://doi.org/10.1038/nrm.2016.104
12. Nicoletti C, Forcato M, Bicciato S (2018) Computational methods for analyzing genome-wide chromosome conformation capture data. Curr Opin Biotechnol 54:98–105. https://doi.org/10.1016/j.copbio.2018.01.023
13. Forcato M, Nicoletti C, Pal K et al (2017) Comparison of computational methods for Hi-C data analysis. Nat Methods 14:679–685. https://doi.org/10.1038/nmeth.4325
14. Dali R, Blanchette M (2017) A critical assessment of topologically associating domain prediction tools. Nucleic Acids Res 45:2994–3005. https://doi.org/10.1093/nar/gkx145

15. Miura H, Poonperm R, Takahashi S, Hiratani I (2018) Practical analysis of Hi-C data: generating A/B compartment profiles. Methods Mol Biol 1861:221–245. https://doi.org/10.1007/978-1-4939-8766-5_16
16. Servant N, Varoquaux N, Lajoie BR et al (2015) HiC-Pro: an optimized and flexible pipeline for Hi-C data processing. Genome Biol 16:259. https://doi.org/10.1186/s13059-015-0831-x
17. Crane E, Bian Q, McCord RP et al (2015) Condensin-driven remodelling of X chromosome topology during dosage compensation. Nature 523:240–244. https://doi.org/10.1038/nature14450
18. Rao SSP, Huntley MH, Durand NC et al (2014) A 3D map of the human genome at kilobase resolution reveals principles of chromatin looping. Cell 159:1665–1680. https://doi.org/10.1016/j.cell.2014.11.021
19. Durand NC, Shamim MS, Machol I et al (2016) Juicer provides a one-click system for analyzing loop-resolution Hi-C experiments. Cell Syst 3:95–98. https://doi.org/10.1016/j.cels.2016.07.002
20. Ay F, Bailey TL, Noble WS (2014) Statistical confidence estimation for Hi-C data reveals regulatory chromatin contacts. Genome Res 24:999–1011. https://doi.org/10.1101/gr.160374.113
21. Dekker J, Belmont AS, Guttman M et al (2017) The 4D nucleome project. Nature 549:219–226. https://doi.org/10.1038/nature23884
22. Marti-Renom MA, Almouzni G, Bickmore WA et al (2018) Challenges and guidelines toward 4D nucleome data and model standards. Nat Genet 50:1352–1358. https://doi.org/10.1038/s41588-018-0236-3
23. Durand NC, Robinson JT, Shamim MS et al (2016) Juicebox provides a visualization system for Hi-C contact maps with unlimited zoom. Cell Syst 3:99–101. https://doi.org/10.1016/J.CELS.2015.07.012
24. Imakaev M, Fudenberg G, McCord RP et al (2012) Iterative correction of Hi-C data reveals hallmarks of chromosome organization. Nat Methods 9:999–1003. https://doi.org/10.1038/nmeth.2148
25. Servant N, Lajoie BR, Nora EP et al (2012) HiTC: exploration of high-throughput "C" experiments. Bioinformatics 28:2843–2844. https://doi.org/10.1093/bioinformatics/bts521
26. Kerpedjiev P, Abdennur N, Lekschas F, et al (2018) HiGlass: web-based visual exploration and analysis of genome interaction maps. Genome Biol 19:125. https://doi.org/10.1186/s13059-018-1486-1
27. Knight PA, Ruiz D (2013) A fast algorithm for matrix balancing. IMA J Numer Anal 33:1029–1047. https://doi.org/10.1093/imanum/drs019
28. Ernst J, Kellis M (2012) ChromHMM: automating chromatin-state discovery and characterization. Nat Methods 9:215–216. https://doi.org/10.1038/nmeth.1906
29. Ernst J, Kellis M (2017) Chromatin-state discovery and genome annotation with ChromHMM. Nat Protoc 12:2478–2492. https://doi.org/10.1038/nprot.2017.124
30. Greenwald WW, Li H, Smith EN et al (2017) Pgltools: a genomic arithmetic tool suite for manipulation of Hi-C peak and other chromatin interaction data. BMC Bioinformatics 18:207. https://doi.org/10.1186/s12859-017-1621-0

Chapter 8

Profiling Chromatin Landscape at High Resolution and Throughput with 2C-ChIP

Xue Qing David Wang, Christopher J. F. Cameron, Dana Segal, Denis Paquette, Mathieu Blanchette, and Josée Dostie

Abstract

Chromatin immunoprecipitation (ChIP) is used to probe the presence of proteins and/or their posttranslational modifications on genomic DNA. This method is often used alongside chromosome conformation capture approaches to obtain a better-rounded view of the functional relationship between chromatin architecture and its landscape. Since the inception of ChIP, its protocol has been modified to improve speed, sensitivity, and specificity. Combining ChIP with deep sequencing has recently improved its throughput and made genome-wide profiling possible. However, genome-wide analysis is not always the best option, particularly when many samples are required to study a given genomic region or when quantitative data is desired. We recently developed carbon copy-ChIP (2C-ChIP), a new form of the high-throughput ChIP analysis method ideally suited for these types of studies. 2C-ChIP applies ligation-mediated amplification (LMA) followed by deep sequencing to quantitatively detect specified genomic regions in ChIP samples. Here, we describe the generation of 2C-ChIP libraries and computational processing of the resulting sequencing data.

Key words Epigenomics, Chromatin immunoprecipitation, Transcription, Gene expression, Posttranslational modification, Deep sequencing

1 Introduction

Chromatin immunoprecipitation (ChIP) is an important method for profiling the protein landscape of chromatin. Over the years, it has shaped much of what we know about transcription and earned a place squarely at the center of investigations that study nuclear processes. ChIP is often performed together with Hi-C and other related chromosome conformation capture techniques (e.g., 5C) to provide valuable information about chromatin landscape and molecularly describe potential underlying mechanisms responsible for changes in genome organization. Characterizing the occupancy of chromatin architectural proteins (e.g., CTCF, cohesin, BEAF-32) can give insight into location and maintenance mechanisms of

Beatrice Bodega and Chiara Lanzuolo (eds.), *Capturing Chromosome Conformation: Methods and Protocols*, Methods in Molecular Biology, vol. 2157, https://doi.org/10.1007/978-1-0716-0664-3_8,

Topologically Associating Domain (TAD) and subTAD boundaries as well as chromatin loops and enhancer-promoter interactions. Furthermore, ChIP against transcriptional regulatory factors (e.g., Lamins, HP1, H3K9me3) can provide information on heterochromatin and nuclear chromosomal organization. Combining ChIP with high-throughput sequencing (ChIP-seq) [1] radically improved the method to enable genome-wide profiling in single experiments. Although powerful, the standard genome-wide ChIP-seq is not always the appropriate analysis method, particularly for studies seeking information depth at given genomic regions.

The carbon copy-ChIP (2C-ChIP) method was developed to quantitatively profile multiple chromatin marks and/or associated proteins within defined genomic domains in many samples simultaneously [2]. During 2C-ChIP, chemically fixed chromatin is first sheared into short fragments ranging in size between 300 and 500 bp and then immunoprecipitated with an antibody of choice, and the resulting genomic DNA (gDNA) is purified along with a sample of the corresponding input material (Fig. 1). The purified input and ChIP gDNA are next separately annealed to a pool of 2C-ChIP library primers (Forward and Reverse) designed against defined genomic region(s). Forward and Reverse 2C-ChIP primers annealed adjacent to each other on the same DNA strand are then specifically ligated together with Taq DNA ligase. This step

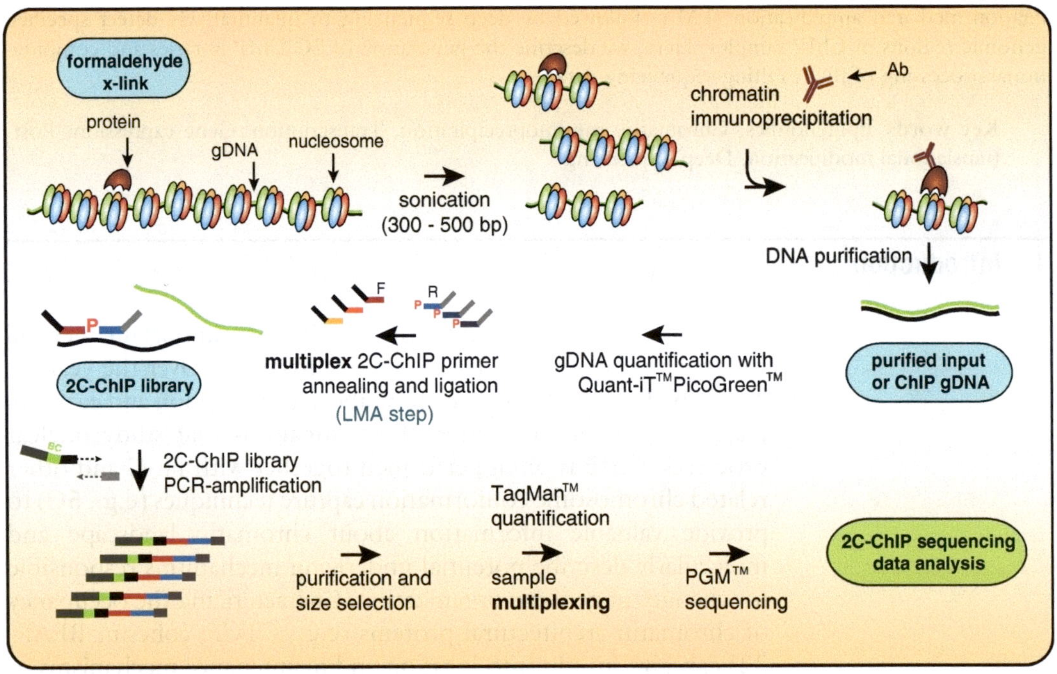

Fig. 1 Overview of the 2C-ChIP protocol. The method proceeds in the order according to the arrows. *gDNA* genomic DNA, *Ab* antibody, *F* Forward 2C-ChIP primer, *R* Reverse 2C-ChIP primer, *LMA* ligation-mediated amplification

produces the 2C-ChIP library, which is then PCR-amplified using 2C-ChIP amplification primers designed against the universal 3′ and 5′ tails of ligation products. When different 2C-ChIP libraries are multiplexed prior to sequencing (e.g., from the input or ChIP), one amplification primer must contain a barcode. Our example shows the Forward 2C-ChIP amplification primer containing the barcode because it uses the PGM™ sequencing system. The position and nature of the barcode should be selected according to the sequencing platform. Amplified 2C-ChIP libraries should also be purified to remove unincorporated primers as background associated with them has been observed. When samples are multiplexed prior sequencing, the 2C-ChIP libraries should be combined at equimolar ratios based on TaqMan™ quantification. The Ion Torrent PGM™ sequencing system outputs a FASTQ file that can be processed with the Ligation-mediated Amplified and Multiplexed Primer-pair Sequence (LAMPS) analysis pipeline [3], which processes sequencing output for each sample into bedGraph format.

The protocol described in this article explains how to design and produce 2C-ChIP libraries from ChIP material and outlines how to process 2C-ChIP sequencing data. Before proceeding with 2C-ChIP, we recommend controlling for the quality of the ChIP material by ChIP quantitative PCR (ChIP-qPCR) [4–6].

2 Materials

2.1 2C-ChIP Library Production

1. Input of ChIP genomic DNA template.
2. 2C-ChIP library primers multiplexed diluted at 200 pM individual concentrations.
3. NEBuffer™ 4 (10×).
4. Salmon Sperm DNA (10 mg/mL).
5. Taq DNA ligase.
6. Taq DNA ligase buffer (10×).
7. Quant-iT™ PicoGreen™ dsDNA Assay Kit (Thermo Fisher Scientific). This kit is optional (*see* Subheading 3.2).

2.2 Quality Control of 2C-ChIP Libraries

2.2.1 2C-ChIP Libraries' PCR Reactions

1. 2C-ChIP library.
2. HIFI PCR buffer (10×).
3. 50 mM Mg_2SO_4.
4. 25 mM dNTP mix.
5. 20 μM Forward and Reverse 2C-ChIP amplification primer stocks.
6. Taq DNA polymerase.

2.2.2 Resolving 2C-ChIP Library PCR Controls

1. Agarose powder (DNAse/RNAse-free molecular biology grade).
2. 10× TBE buffer (*see* **Note 1**).
3. 10 mg/mL ethidium bromide solution (*see* **Note 2**).
4. DNA gel loading buffer (10×).

2.3 2C-ChIP Library Amplification

1. 2C-ChIP library.
2. HIFI PCR buffer (10×).
3. 50 mM Mg_2SO_4.
4. 25 mM dNTP mix.
5. 20 μM barcoded Forward 2C-ChIP amplification primer stocks (A-keys).
6. 20 μM Reverse 2C-ChIP amplification primer (P1-key).
7. Taq DNA polymerase.

2.4 2C-ChIP Library Purification

1. 2C-ChIP library.
2. Sera-Mag SpeedBeads™ (Thermo Fisher Scientific).
3. Ethanol (80%).
4. 1× TE.

2.5 Quantifying 2C-ChIP Libraries Prior to Deep Sequencing

1. Purified 2C-ChIP library.
2. Ion Library TaqMan™ Quantitation Kit (Thermo Fisher Scientific).

2.6 Equipment

1. Thermocycler.
2. Agarose gel electrophoresis apparatus.
3. Gel documentation system.
4. Real-time quantitative PCR machine.
5. Magnetic stir bar.
6. Microcentrifuge.
7. High-throughput DNA sequencing platform (can be outsourced).
8. Computer with Internet access.
9. P10, P20, P200, and P1000 micropipettes.

2.7 Disposables

1. 1.7 mL microtubes.
2. PCR tubes or plates.
3. Filter micropipette tips.

2.8 Solutions

1. 1× Tris-EDTA (1× TE): 10 mM Tris–HCl pH 8.0, 1 mM EDTA.

2. 10× HIFI PCR buffer: 600 mM Tris-SO_4 pH 8.9, 180 mM $(NH_4)_2SO_4$.
3. NEBuffer™ 4 (New England BioLabs).
4. Taq DNA ligase buffer.
5. 50 mM $MgSO_4$.
6. 10× TBE.
7. 10 mg/mL ethidium bromide.

3 Methods

This protocol was optimized to generate sufficient 2C-ChIP products per library for sequencing on a Personal Genome Machine (PGM™; Thermo Fisher Scientific) sequencer at a final DNA concentration of 9 pM (*see* **Note 3**). Adjustments may be required if sequencing is performed on a different platform. Before generating 2C-ChIP libraries, we recommend verifying the quality of ChIP samples by performing ChIP-qPCR tests at control regions.

3.1 Designing 2C-ChIP primers

2C-ChIP quantifies genomic sequences by ligating pairs of 2C-ChIP library primers annealed onto the same DNA strand (Fig. 2a). Each pair consists of one Forward and one Reverse primer designed to anneal contiguously onto the target genomic sequence (no space in between). All Forward primers should feature a 5′ end universal sequence, and all Reverse primers should include a 3′ end universal sequence distinct from the one used in Forward primers

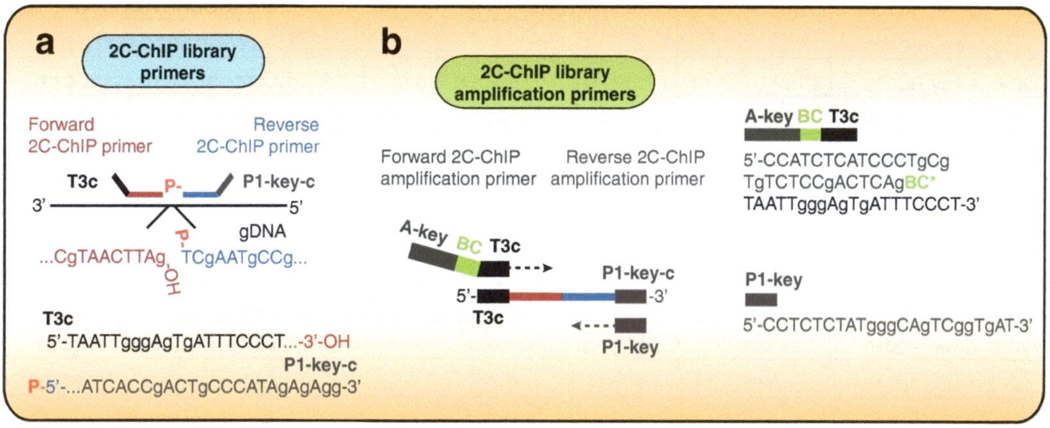

Fig. 2 Primers used in 2C-ChIP experiments. Two types of primers are used in 2C-ChIP experiments. The **"2C-ChIP library primers"** are those containing genomic homology regions that will be multiplexed, annealed, and ligated onto genomic DNA. The **"2C-ChIP library amplification primers"** are those used to PCR-amplify the 2C-ChIP ligation products from the universal 5′ and 3′ tails. The "A-key" and "P1-key" sequences are those used by the PGM™ sequencing system and can be replaced by any other sequence according to the platform of choice

(*see* **Note 4**). In our example, we show Forward primers with a 'T3c' sequence and Reverse primers with the complementary P1-key (P1-key-c) sequence used in sequencing with the PGM™ system (*see* **Note 5**). All 2C-ChIP Reverse primers must also be 5′ end phosphorylated (*not* the Forward oligos) for specific ligation of annealed Forward-Reverse (F-R) primer pairs by Taq DNA ligase (*see* below and **Note 6**). 2C-ChIP primers are designed according to the biological questions addressed (*see* **Note 7**).

"2C-ChIP library amplification primers" are used to generate sufficient 2C-ChIP library amounts for sequencing (Fig. 2b, Table 1). Forward oligos feature the A1-key sequence used in the Ion PGM™ sequencing system. They also contain a barcode and a T3-complementary (T3c) region that will anneal to the T3 sequence in the Forward 2C-ChIP library primers. The Reverse 2C-ChIP amplification oligo consists of the P1-key sequence from the Ion PGM™ system. All of these sequences can be customized according to the sequencing platform selected.

Table 1
List of barcode sequences used for 2C-ChIP sequencing with the PGM™ sequencing system

name	primer sequence (5' - 3')
2C-C_A-key_BC001*	CCATCTCATCCCTGCGTGTCTCCGACTCAGCTAAGGTAAC**TAATTGGGAGTGATTTCCCT**
2C-C_A-key_BC002	CCATCTCATCCCTGCGTGTCTCCGACTCAGTAAGGAGAAC**TAATTGGGAGTGATTTCCCT**
2C-C_A-key_BC003	CCATCTCATCCCTGCGTGTCTCCGACTCAGAAGAGGATTC**TAATTGGGAGTGATTTCCCT**
2C-C_A-key_BC004	CCATCTCATCCCTGCGTGTCTCCGACTCAGTACCAAGATC**TAATTGGGAGTGATTTCCCT**
2C-C_A-key_BC005	CCATCTCATCCCTGCGTGTCTCCGACTCAGCAGAAGGAAC**TAATTGGGAGTGATTTCCCT**
2C-C_A-key_BC006	CCATCTCATCCCTGCGTGTCTCCGACTCAGCTGCAAGTTC**TAATTGGGAGTGATTTCCCT**
2C-C_A-key_BC007	CCATCTCATCCCTGCGTGTCTCCGACTCAGTTCGTGATTC**TAATTGGGAGTGATTTCCCT**
2C-C_A-key_BC008	CCATCTCATCCCTGCGTGTCTCCGACTCAGTTCCGATAAC**TAATTGGGAGTGATTTCCCT**
2C-C_A-key_BC009	CCATCTCATCCCTGCGTGTCTCCGACTCAGTGAGCGGAAC**TAATTGGGAGTGATTTCCCT**
2C-C_A-key_BC010	CCATCTCATCCCTGCGTGTCTCCGACTCAGCTGACCGAAC**TAATTGGGAGTGATTTCCCT**
2C-C_A-key_BC011	CCATCTCATCCCTGCGTGTCTCCGACTCAGTCCTCGAATC**TAATTGGGAGTGATTTCCCT**
2C-C_A-key_BC012	CCATCTCATCCCTGCGTGTCTCCGACTCAGTAGGTGGTTC**TAATTGGGAGTGATTTCCCT**
2C-C_A-key_BC013	CCATCTCATCCCTGCGTGTCTCCGACTCAGTCTAACGGAC**TAATTGGGAGTGATTTCCCT**
2C-C_A-key_BC014	CCATCTCATCCCTGCGTGTCTCCGACTCAGTTGGAGTGTC**TAATTGGGAGTGATTTCCCT**
2C-C_A-key_BC015	CCATCTCATCCCTGCGTGTCTCCGACTCAGTCTAGAGGTC**TAATTGGGAGTGATTTCCCT**
2C-C_A-key_BC016	CCATCTCATCCCTGCGTGTCTCCGACTCAGTCTGGATGAC**TAATTGGGAGTGATTTCCCT**
2C-C_A-key_BC017	CCATCTCATCCCTGCGTGTCTCCGACTCAGTCTATTCGTC**TAATTGGGAGTGATTTCCCT**
2C-C_A-key_BC018	CCATCTCATCCCTGCGTGTCTCCGACTCAGAGGCAATTGC**TAATTGGGAGTGATTTCCCT**
2C-C_A-key_BC019	CCATCTCATCCCTGCGTGTCTCCGACTCAGTTAGTCGGAC**TAATTGGGAGTGATTTCCCT**
2C-C_A-key_BC020	CCATCTCATCCCTGCGTGTCTCCGACTCAGCAGATCCATC**TAATTGGGAGTGATTTCCCT**
2C-C_P1-key	CCACTACGCCTCCGCTTTCCTCTCTATGGGCAGTCGGTGAT

*The Ion PGM(™) "A-key" (5′-CCATCTCATCCCTGCGTGTCTCCGACTCAG-3′) is colored dark green in Forward PCR amplification primers. Barcodes (BC) are in light green. The P1-key sequence is found in the Reverse PCR amplification primer

3.2 Diluting 2C-ChIP Primers

1. All 2C-ChIP primers should be ordered desalted and either lyophilized or already resuspended in 1× TE (*see* **Note 8**). Only Reverse 2C-ChIP primers must also be 5′-phosphorylated. All primers should be resuspended at 80 μM concentration in 1× TE upon receipt and kept at −80 °C for long-term storage (at least 2 years; *see* **Note 9**).
2. To perform 2C-ChIP, library primers must be diluted from the stocks and mixed at a final individual concentration of 200 pM in the "diluted primer mix" (34 pM final concentration in annealing reactions; *see* Subheading 3.2 below and **Note 10**). Primer pools can be prepared on the same day of the experiment or kept at −80 °C for long-term storage. Prepare all pools on ice and quick-freeze on dry ice before transferring to the −80 °C freezer (*see* **Note 11**).
3. On the day of the experiment, dilute primer stocks or primer pools in *water* (the EDTA from 1× TE will inhibit PCR reactions). Keep the Forward and Reverse primer dilutions separate until immediately before use. Do not keep the diluted primers past the day of the experiment since oligos are not stable in water (*see* **Note 12**).

3.3 Generating 2C-ChIP Libraries

2C-ChIP libraries are generated in two steps: the 2C-ChIP primers are first annealed to the gDNA overnight. The annealed 2C-ChIP primers are ligated on the following day with Taq DNA ligase. A general procedure is described below.

1. Anneal 2C-ChIP library primers to purified genomic DNA (from input or ChIP) overnight at 55 °C in a thermocycler. Dilute and mix your primer pools in water at a final concentration of 200 pM each as described in Subheading 3.1, and then prepare the 2C-ChIP "primer master mix" as in Table 2.
2. Set up the 2C-ChIP annealing reactions as in Table 3. We recommend setting up at least two annealing reactions per sample to account for the "no ligase" controls that will be included in the experiment on the next day.
3. Annealed 2C-ChIP primers are ligated on the next day with Taq DNA ligase. Using filtered tips, prepare a (+) and (−) "ligation master mix" as in Table 4.

Table 2
2C-ChIP library "primer master mix"

	Per reaction
Forward + Reverse "diluted primer mix" (200 pM)	1.7 μL
NEBuffer 4™ (10×)	1 μL

Table 3
2C-ChIP annealing reactions

	Per reaction
DNA sample∗	up to 5.8 μL
Salmon sperm DNA (1/10 dilution)∗∗	1.5 μL
2C-ChIP "primer master mix"∗∗∗	2.7 μL
Water (if required)	(complete to 10 μL)
Total reaction volume	10 μL

∗*see* **Note 13**
∗∗∗*see* **Note 14**

Table 4
Ligation master mixes

	Per reaction (+)/(−)
Taq DNA ligase	0.25 μL/0 μL
Taq DNA ligase buffer (10×)	2.5 μL
Water	17.25 μL/17.5 μL
Total master mix volume	20 μL

4. Quickly spin down the reaction tubes or plates to collect any condensation that might have formed overnight (quick spin). Do not let the samples reach room temperature to reduce chances of nonspecific annealing. Immediately add 20 μL of "ligation master mix" to each tube and return the sample to the thermocycler to initiate ligation at 55 °C for 1 h. Always dispense the "no ligase" ((−) ligase) samples first. Do not mix by pipetting to avoid cross-contamination.
5. Place the 2C-ChIP libraries on ice after reactions are completed. 2C-ChIP libraries can be kept frozen at −20 °C for at least 1 month.

3.4 Quality Control of 2C-ChIP Libraries

Before moving ahead with large-scale PCR amplification of 2C-ChIP libraries for sequencing, we suggest verifying the presence of only one band of the correct size in (+) ligase samples and the absence of PCR products in the (−) ligase control.

1. Set up PCR reactions according to Table 5. To minimize background issues, prepare samples on ice by adding first "master mix 1" to the tubes (A), then the 2C-ChIP library

Table 5
2C-ChIP PCR quality control

A. master mix 1	*per* reaction
10X PCR HIFI buffer	2.5 μL
50 mM Mg_2SO_4	2.0 μL
25 mM dNTP mix	0.2 μL
20 μM A-key (Forward primer)*	0.5 μL
20 μM P1-key (Reverse primer)	0.5 μL
volume master mix 1	**5.7** μL
B. 2C-ChIP library	3.0 μL ***added individually***
C. master mix 2	
NEB Taq DNA polymerase	0.2 μL
water	16.1 μL
volume master mix 2	**16.3** μL
total reaction volume	25 μL

**see* Note 15

Table 6
2C-ChIP quality control PCR amplification program

No. of PCR cycles	Denature	Anneal	Extend
1	95 °C, 5 min		
26	95 °C, 30 s	60 °C, 30 s	72 °C, 15 s*
1	95 °C, 30 s	60 °C, 30 s	72 °C, 3 min

**see* Note 16

samples (B), and lastly the "master mix 2" (C). Do not forget a "water control" for the PCR analysis that will not contain any 2C-ChIP library to control for the specificity of the reagents.

2. Amplify the samples using the PCR program in Table 6.
3. Prepare a 2.5% agarose gel in 1× TBE and contain 0.5 μg/mL ethidium bromide as described in [7]. Resolve 10 μL of each PCR reaction on this gel for approximately 45 min to 1 h at 120 V. Image the results using a gel documentation system.
4. Figure 3 shows the kind of results expected from an input and ChIP sample. In this example, input and ChIP genomic DNA was annealed to 320 2C-ChIP library primers designed against the human *HOXA* gene cluster as we have published previously (Table 7, [2]). As shown in this example, a single band migrating at ~160 bp should be present specifically in the (+) ligase

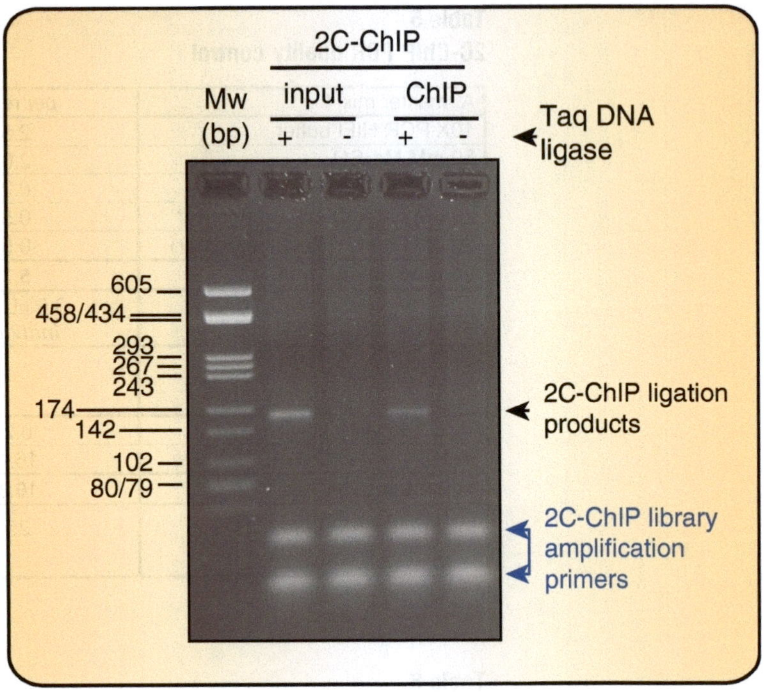

Fig. 3 Quality control of 2C-ChIP libraries by end-point PCR. 2C-ChIP libraries from input and ChIP (anti-H3K4me3, Abcam; ab8580) samples were PCR-amplified and resolved on agarose gel as described in Subheading 3.3. The 2C-ChIP libraries were generated with 320 2C-ChIP library primers designed against the human *HOXA* gene cluster as was previously done (Table 7, [2]). ChIP was from five million NT2-D1 cells and 2C-ChIP was performed on volumes equivalent to approximately 6000 and 30,000 cells for the input and ChIP samples, respectively. The top and bottom bands of the "2C-ChIP amplification primers" correspond to the Forward and Reverse oligos, respectively

samples. If this is observed and if no PCR products appear in the (−) ligase controls, proceed with the large-scale PCR amplification of 2C-ChIP libraries for sequencing. Take note of the library that yields the weakest band on gel. Use this sample for PCR optimization as described in Subheading 3.4.

3.5 Amplifying 2C-ChIP Libraries for Deep Sequencing

Sequencing on the PGM™ system requires that barcoded 2C-ChIP libraries be mixed at final individual concentrations of 9 pM after pooling. This PCR amplification step should therefore yield sufficient sequencing material using the least possible number of cycles to limit amplification biases. The number of cycles will depend on the 2C-ChIP primer design (complexity of the library), the number of cells collected to generate the libraries, and the antibody used for ChIP. We recommend first testing different cycle numbers ranging from 12 to 24 before selecting this experimental condition. We outline below PCR conditions that

Table 7
List of 2C-ChIP library primers used in this protocol

Name	Sequence with adapter sequences (5′–3′)*
2C-C_HoxA_T3_F1	TAATTGGGAGTGATTTCCCTGGAGGCAAGAAAATGGTTTGAATCCCTAAT
2C-C_HoxA_T3_F2	TAATTGGGAGTGATTTCCCTGGAAAGGTAAAATTAAGACAAGAAGCAACATCT
2C-C_HoxA_T3_F3	TAATTGGGAGTGATTTCCCTTCTTCCATATGCCTCCTTACCCTCATCTGC
2C-C_HoxA_T3_F4	TAATTGGGAGTGATTTCCCTGCATGCAAAACATAGGTACAGAGAACCCAGAAG
2C-C_HoxA_T3_F5	TAATTGGGAGTGATTTCCCTGGTAATGGTATCTACCTGGTCAAAGTGCTG
2C-C_HoxA_T3_F6	TAATTGGGAGTGATTTCCCTCATCTAATCAGTATTCTGTGAATCTGTTGAAAG
2C-C_HoxA_T3_F7	TAATTGGGAGTGATTTCCCTCCAGAAGCTATCTTTGGGATCTTCATAGTTTCT
2C-C_HoxA_T3_F8	TAATTGGGAGTGATTTCCCTCATCCCTCCTGTTCATCTGTACATATCTTAAG
2C-C_HoxA_T3_F9	TAATTGGGAGTGATTTCCCTTGGTGTTGAATTTAGGATCACAAAGGTTTG
2C-C_HoxA_T3_F10	TAATTGGGAGTGATTTCCCTTAGATCTGACTGCCACGCTTTTCATTTTTCAAG
2C-C_HoxA_T3_F11	TAATTGGGAGTGATTTCCCTAGGAGGAAACACAAACAGCAGTGACCATTT
2C-C_HoxA_T3_F12	TAATTGGGAGTGATTTCCCTGAGGGATGGAACTATAATTGTTGCCTTTAG
2C-C_HoxA_T3_F13	TAATTGGGAGTGATTTCCCTGGAGTCAATGGCATGCAAATGACACTTTAA
2C-C_HoxA_T3_F14	TAATTGGGAGTGATTTCCCTGTTATGTGGCATGGAATTAGAACTGTGGAT
2C-C_HoxA_T3_F15	TAATTGGGAGTGATTTCCCTCACTGATGTGAGGAAAGACTGACTGAAGAC
2C-C_HoxA_T3_F16	TAATTGGGAGTGATTTCCCTCCGAGAAGAAAGGGGTAGAGATTGGGAAAG
2C-C_HoxA_T3_F17	TAATTGGGAGTGATTTCCCTGACTTCTCTGAGGATTCCTCGGCCTTCTCG
2C-C_HoxA_T3_F18	TAATTGGGAGTGATTTCCCTGGTGATGCTGGACCATGGGAGATGAGAGAT
2C-C_HoxA_T3_F19	TAATTGGGAGTGATTTCCCTGCAGTGATGGATCACCGTTTTAGTGGCATT
2C-C_HoxA_T3_F20	TAATTGGGAGTGATTTCCCTCCTCTCGTCATTAGCATGGCAATGAGGAGT
2C-C_HoxA_T3_F21	TAATTGGGAGTGATTTCCCTGTCCTGTTCCTCATCACCCTACTTCCCTGA
2C-C_HoxA_T3_F22	TAATTGGGAGTGATTTCCCTGCGGCTTCCTTGGGAGCACCAGCTTCCAGC
2C-C_HoxA_T3_F23	TAATTGGGAGTGATTTCCCTATCCTCTTACGTTATTTGCCGGGGA
2C-C_HoxA_T3_F24	TAATTGGGAGTGATTTCCCTCTCAACTTGCTAACCTGACTTCGAAGCATT
2C-C_HoxA_T3_F25	TAATTGGGAGTGATTTCCCTCAGGACTGTCATTGTTTAGGCCAGCTCCAC
2C-C_HoxA_T3_F26	TAATTGGGAGTGATTTCCCTATGAAGAGGTTGGGGGGAGCCACAGGCATGAA
2C-C_HoxA_T3_F28	TAATTGGGAGTGATTTCCCTTGGTGTGGGTGGGGATGTCCCGGAGTACGT
2C-C_HoxA_T3_F29	TAATTGGGAGTGATTTCCCTAGAAGTTGATGGCGCAGGAAGGTCGGGAAG
2C-C_HoxA_T3_F29b	TAATTGGGAGTGATTTCCCTCCAAAGGAAGTCTGGGCGCGATCAATCTTG
2C-C_HoxA_T3_F30	TAATTGGGAGTGATTTCCCTCGTGTTCAAATAGTGATATCCCTGAAGGAT
2C-C_HoxA_T3_F31	TAATTGGGAGTGATTTCCCTCTGTTTCGCTCCGCGTGCAGATTTTTGGAG
2C-C_HoxA_T3_F32	TAATTGGGAGTGATTTCCCTCTGACTGTTCACCAGCATACACACACGGAA
2C-C_HoxA_T3_F33	TAATTGGGAGTGATTTCCCTGCTCTTGTCGCCAGCGCAGCTTTCGCCTGC

(continued)

Table 7
(continued)

Name	Sequence with adapter sequences (5′–3′)*
2C-C_HoxA_T3_F34	TAATTGGGAGTGATTTCCCTCCAGGAGATCTTTGGCTGCTTTCATTGTAC
2C-C_HoxA_T3_F35	TAATTGGGAGTGATTTCCCTCCGAGCTGTCGTAGTAGGTCGCTTTTTGCA
2C-C_HoxA_T3_F36	TAATTGGGAGTGATTTCCCTGGGTCATTTGGCTCCCGACGAGGGATGGAA
2C-C_HoxA_T3_F37	TAATTGGGAGTGATTTCCCTATCCCACCAAAATCCTTCCTATTGTTCGAG
2C-C_HoxA_T3_F38	TAATTGGGAGTGATTTCCCTGGGCAGAGGAATACGAAGTTTCTCAAACGAA
2C-C_HoxA_T3_F39	TAATTGGGAGTGATTTCCCTTCAGCGGGCAGAAGTGGGGCTCCGACCTCA
2C-C_HoxA_T3_F40	TAATTGGGAGTGATTTCCCTGGGTGTTTTAGAGCACTTGAAAGGGCCTCA
2C-C_HoxA_T3_F41	TAATTGGGAGTGATTTCCCTTGGCAGCGCATCTGCACATTTCCCTCATCG
2C-C_HoxA_T3_F42	TAATTGGGAGTGATTTCCCTGGGAGGGTAGGTGGGGAACCTTGACCAGCA
2C-C_HoxA_T3_F43	TAATTGGGAGTGATTTCCCTTGGAGAATGGGTCTTCAGCTGGCTATGCC TGAG
2C-C_HoxA_T3_F44	TAATTGGGAGTGATTTCCCTGCAGGCAACCAAGGGTACACACCGTTAGCC
2C-C_HoxA_T3_F45	TAATTGGGAGTGATTTCCCTAGGAAAAGGAAACGCCAAGACATAGAAAAC
2C-C_HoxA_T3_F46	TAATTGGGAGTGATTTCCCTGGAAGCCTTGTCTGGGAGCATTCAATCACG
2C-C_HoxA_T3_F47	TAATTGGGAGTGATTTCCCTCTCCAAGTCTAATATTCATGGCTGCTGTAC
2C-C_HoxA_T3_F48	TAATTGGGAGTGATTTCCCTCAGGTCAAGTGTAAGCTTAGATGCATACCT
2C-C_HoxA_T3_F49	TAATTGGGAGTGATTTCCCTCCATCTATCTAGGTAATCTTCCCATGGCTC
2C-C_HoxA_T3_F50	TAATTGGGAGTGATTTCCCTAAGATGCTGTGTCCTTGCCTGAGACCACATT
2C-C_HoxA_T3_F51	TAATTGGGAGTGATTTCCCTACCCAGTTTGGCTATTTTCCAGCAGTAAAC
2C-C_HoxA_T3_F52	TAATTGGGAGTGATTTCCCTGGAGTTTATTCTTAGCACATGGCTTTCTAT
2C-C_HoxA_T3_F53	TAATTGGGAGTGATTTCCCTCTGAGTTTGTGCTTTCCCTGGTGGGCCGGC
2C-C_HoxA_T3_F54	TAATTGGGAGTGATTTCCCTCAGAGGCTTCAGGCCATGTGTCTTTGAAG
2C-C_HoxA_T3_F55	TAATTGGGAGTGATTTCCCTTACCGGCGCTGACATGGATCTTCTTCATCC
2C-C_HoxA_T3_F56	TAATTGGGAGTGATTTCCCTGGGCTCGATGTAGTTGGAGTTTATCAAAAA
2C-C_HoxA_T3_F57	TAATTGGGAGTGATTTCCCTGTGTCCTTCAGGGGTTCTAGGCTAACAGGC
2C-C_HoxA_T3_F58	TAATTGGGAGTGATTTCCCTGTGCGATTGTGAGAGAGAAGGTGGCCCAAG
2C-C_HoxA_T3_F59	TAATTGGGAGTGATTTCCCTGAGCCACAGACAGAAATCAACACTGAGTGT
2C-C_HoxA_T3_F60	TAATTGGGAGTGATTTCCCTCTGAAGTTTACCAGCAGCATCTCGCCTGAG
2C-C_HoxA_T3_F61	TAATTGGGAGTGATTTCCCTGCAGAATTTCAAAACGCCTGATGGTGGCAT
2C-C_HoxA_T3_F62	TAATTGGGAGTGATTTCCCTGGATTTCACCTTTCTGCATTCCCTACAGTC
2C-C_HoxA_T3_F63	TAATTGGGAGTGATTTCCCTAGGTCATTCAACCACTAGGGTTCACCTGGA
2C-C_HoxA_T3_F64	TAATTGGGAGTGATTTCCCTGTAAATGCAGAGAAGATAAATCTGCACACCCT
2C-C_HoxA_T3_F65	TAATTGGGAGTGATTTCCCTTAGTTCAGAAATATTGTCTCTTCAGCTTGT
2C-C_HoxA_T3_F66	TAATTGGGAGTGATTTCCCTCTGGGATCAAACAGAAAGAGCAACTAACAA
2C-C_HoxA_T3_F67	TAATTGGGAGTGATTTCCCTAAGTCATTAACATCCGCGGTTGTGCTGCAA

(continued)

Table 7
(continued)

Name	Sequence with adapter sequences (5′–3′)*
2C-C_HoxA_T3_F68	TAATTGGGAGTGATTTCCCTCGTGACGTTTTATAGCCAGTGAGCCGATCT
2C-C_HoxA_T3_F69	TAATTGGGAGTGATTTCCCTTGCAAAGGGTCCTATAAAGGCACGCAGGGA
2C-C_HoxA_T3_F70	TAATTGGGAGTGATTTCCCTGCTTTTCGCCTGGGTTCCCTGCTCA TGACCCAAG
2C-C_HoxA_T3_F71	TAATTGGGAGTGATTTCCCTTTCCAACGTCTAAACTGTCCCAGAGAACGC
2C-C_HoxA_T3_F72	TAATTGGGAGTGATTTCCCTTTATGATGAATTATGGAAATGACTGGGACA
2C-C_HoxA_T3_F73	TAATTGGGAGTGATTTCCCTGTACTGACCGTCCTGCCAGCAGCTCTGAAT
2C-C_HoxA_T3_F74	TAATTGGGAGTGATTTCCCTAGCTCAGAGACACTAGCACAGGAGCCCCAG
2C-C_HoxA_T3_F75	TAATTGGGAGTGATTTCCCTCAGTAGCATCCTCAGAGAGACCACTATGAA
2C-C_HoxA_T3_F76	TAATTGGGAGTGATTTCCCTAGTTCATCCGCTGCATCCAAGGGTAAACCG
2C-C_HoxA_T3_F77	TAATTGGGAGTGATTTCCCTCATCGAACTGGTTTGTTTCTGGATGGCCGAA
2C-C_HoxA_T3_F78	TAATTGGGAGTGATTTCCCTAGAAAGGACCGACAGAGCAGTGGCCGTGAG
2C-C_HoxA_T3_F79	TAATTGGGAGTGATTTCCCTCCTGGAAGCCGGATGCCTGTTCCCCAGTCG
2C-C_HoxA_T3_F80b	TAATTGGGAGTGATTTCCCTTCGCAATAAATAATCGGCCCGCGGCAGCCG
2C-C_HoxA_T3_F81	TAATTGGGAGTGATTTCCCTGCCACGCGCGCTAACAGCATTATGTCCTCT
2C-C_HoxA_T3_F82	TAATTGGGAGTGATTTCCCTCCACATGACCGACAGCGGCCAATGGAAGGG
2C-C_HoxA_T3_F83	TAATTGGGAGTGATTTCCCTGATGTTAGTGCACACGCAAAAACCAGACAA
2C-C_HoxA_T3_F84	TAATTGGGAGTGATTTCCCTGGAGGGGAATGTTCCTGAGCTGGCTGCAGA
2C-C_HoxA_T3_F85	TAATTGGGAGTGATTTCCCTGAGGCAGAGGGAATCCAAGGCGACCCAGTC
2C-C_HoxA_T3_F86	TAATTGGGAGTGATTTCCCTAGAAAGGCTGCGCCGGGAGTCACGGGGCTA
2C-C_HoxA_T3_F87	TAATTGGGAGTGATTTCCCTGAACTCATAATTTTGACCTGTGATTTGTTG
2C-C_HoxA_T3_F88	TAATTGGGAGTGATTTCCCTGCGGACAGGAGAGGGATGGGGAGGATCCCAAG
2C-C_HoxA_T3_F89	TAATTGGGAGTGATTTCCCTCCCTAAAGGCCACCCAGCTGTGGAGGCTTC
2C-C_HoxA_T3_F90	TAATTGGGAGTGATTTCCCTCAAGACGCAAAGAGAAAAGACCAAGGCCCC
2C-C_HoxA_T3_F91	TAATTGGGAGTGATTTCCCTAGTGATCCCTAGCCTGGGTGCAAAGACAAA
2C-C_HoxA_T3_F93	TAATTGGGAGTGATTTCCCTACACCCTCTGGGCGGTCATCAAGTTCTGGG
2C-C_HoxA_T3_F94	TAATTGGGAGTGATTTCCCTCCATAAAGGCCGGGTCTGCGAACTGTCTGGAA
2C-C_HoxA_T3_F95	TAATTGGGAGTGATTTCCCTAGGCCTTGAGGTAACTATTGCAAAATATAC
2C-C_HoxA_T3_F96	TAATTGGGAGTGATTTCCCTTCCGAGTGGAGCGCGCATGAAGCCAGTTGG
2C-C_HoxA_T3_F96b	TAATTGGGAGTGATTTCCCTCCGCCGCTCACGGACAATCTAGTTGTACAA
2C-C_HoxA_T3_F97	TAATTGGGAGTGATTTCCCTTCTATCAACTGGAGGAGAACCACAAGCATA
2C-C_HoxA_T3_F98	TAATTGGGAGTGATTTCCCTCGTCCAGCAGAACAATAACGCGTAAATCACTCC
2C-C_HoxA_T3_F99	TAATTGGGAGTGATTTCCCTCAACCCTTAAATTCGCCTTTGCTACGAGGA
2C-C_HoxA_T3_F100	TAATTGGGAGTGATTTCCCTACAGAAAGCAGCGACTCCTAGAACAGGGGT

(continued)

Table 7
(continued)

Name	Sequence with adapter sequences (5′–3′)*
2C-C_HoxA_T3_F101	TAATTGGGAGTGATTTCCCTGTCTGCAAATGGGCTGGGCATGTCTGGATG
2C-C_HoxA_T3_F102	TAATTGGGAGTGATTTCCCTAAAGCTGCAGCGAATGTCCCCTAATCAGA
2C-C_HoxA_T3_F103	TAATTGGGAGTGATTTCCCTCCCAGAATTCCTACCACGCCCCCGGCGTCC
2C-C_HoxA_T3_F104	TAATTGGGAGTGATTTCCCTAACCTTACGGCACCAGGGCTATAGGGCCCA
2C-C_HoxA_T3_F105	TAATTGGGAGTGATTTCCCTAACTGAAGAAAATGAATCGAGAAAACCGGA
2C-C_HoxA_T3_F106	TAATTGGGAGTGATTTCCCTTGCTTACTTGGAAGATGGCCAGGCAGCTT
2C-C_HoxA_T3_F107	TAATTGGGAGTGATTTCCCTGCATTTGTCCGCCGAGTCGTAGAGGCAGTA
2C-C_HoxA_T3_F108	TAATTGGGAGTGATTTCCCTGCTCTCCGAGCATGACATTGTTGTGGGATA
2C-C_HoxA_T3_F109	TAATTGGGAGTGATTTCCCTCTAGTCCGTTGATCCAGTTAGGATCTTCTC
2C-C_HoxA_T3_F110	TAATTGGGAGTGATTTCCCTACGAACCTGAGTGCAGAAAAGCTTCAGAGC
2C-C_HoxA_T3_F111b	TAATTGGGAGTGATTTCCCTCTGGAGGGCACTCTCAATGCTTTCAAGACT
2C-C_HoxA_T3_F112	TAATTGGGAGTGATTTCCCTAGACATAGAGATGCACCTAGTAGGAAACCA
2C-C_HoxA_T3_F113	TAATTGGGAGTGATTTCCCTGCCCAGCCCGCTGCTATTGAGA
2C-C_HoxA_T3_F114	TAATTGGGAGTGATTTCCCTGAACCCAGACCAGGCTCCCTCTTGTTGAAG
2C-C_HoxA_T3_F115	TAATTGGGAGTGATTTCCCTCAGACACTGAGCAAATCCAAGTTTCATTAC
2C-C_HoxA_T3_F116	TAATTGGGAGTGATTTCCCTGGAGGTGGTCTGGGACTCTCTTGATTAAGA
2C-C_HoxA_T3_F117	TAATTGGGAGTGATTTCCCTACTGGAGTCCCGGCGCAACACATGGCTTTT
2C-C_HoxA_T3_F118	TAATTGGGAGTGATTTCCCTTGGTCGAAAGCCTGTGGCAGGACGCCGTTC
2C-C_HoxA_T3_F119	TAATTGGGAGTGATTTCCCTAGCTTACGTCTCCAAATTTCTACTTCACGGA
2C-C_HoxA_T3_F120	TAATTGGGAGTGATTTCCCTGGTGGCTTGTCCGATTTGCACGGTGACTTG
2C-C_HoxA_T3_F121	TAATTGGGAGTGATTTCCCTGAAGGCAAAATATGCTCCCCATCCTGCAGG
2C-C_HoxA_T3_F122	TAATTGGGAGTGATTTCCCTCTAGACAGACCTGCGGTTTTATAGCAGTTT
2C-C_HoxA_T3_F123	TAATTGGGAGTGATTTCCCTCAATGCGAGGCCTTGTTACCGAGGTTGTTG
2C-C_HoxA_T3_F124	TAATTGGGAGTGATTTCCCTACCCACTTGACAGACTGAATCTCACTCCCA
2C-C_HoxA_T3_F125	TAATTGGGAGTGATTTCCCTTAATCTTCTCTTCTCCCTTCAGTTTGGGA
2C-C_HoxA_T3_F126	TAATTGGGAGTGATTTCCCTAGCTGCTGTCGCCAACCCCCCAGACCAGAG
2C-C_HoxA_T3_F127	TAATTGGGAGTGATTTCCCTGTCCTAAAACCTAGGCATAAATCTCCCTCT
2C-C_HoxA_T3_F128	TAATTGGGAGTGATTTCCCTATAAAGAAGTTGTGAGTCCTCAGGAGAGGT
2C-C_HoxA_T3_F129	TAATTGGGAGTGATTTCCCTTAACAAAAATGTACAGCTTTAAAGCAGA
2C-C_HoxA_T3_F130	TAATTGGGAGTGATTTCCCTAGGCTTGTCAACGCGAGGTGGCGCCCTTGA
2C-C_HoxA_T3_F131	TAATTGGGAGTGATTTCCCTGATGATCCACAGAATTCACTTTATGTGAGA
2C-C_HoxA_T3_F132	TAATTGGGAGTGATTTCCCTGGGTTGACGTTTGACATTTAACGGGCTGGG
2C-C_HoxA_T3_F133	TAATTGGGAGTGATTTCCCTATAGGTCGTCATTTACCGGGCAGAGTGGAC

(continued)

Table 7
(continued)

Name	Sequence with adapter sequences (5′–3′)*
2C-C_HoxA_T3_F134	TAATTGGGAGTGATTTCCCTGGAGGAGCACGGAGGCTGTCATAGCCCGAG
2C-C_HoxA_T3_F135	TAATTGGGAGTGATTTCCCTCTGACCTGGTAACAACGCTTCCTCCTCCAG
2C-C_HoxA_T3_F136	TAATTGGGAGTGATTTCCCTCTGATCCTGGATTTGTCCTGACCAATGTAA
2C-C_HoxA_T3_F137	TAATTGGGAGTGATTTCCCTCCAGGCAGCCAACAAACTGACTTGCTGTGG
2C-C_HoxA_T3_F138	TAATTGGGAGTGATTTCCCTTCCCTGTCCGCTGGCCAACTTCAGCCCAGA
2C-C_HoxA_T3_F140	TAATTGGGAGTGATTTCCCTGTCGGACCAGGCTGGTGACATACTTCGCTG
2C-C_HoxA_T3_F141	TAATTGGGAGTGATTTCCCTTTGCTGCATCAGAATTATGGAGGAACACAA
2C-C_HoxA_T3_F142	TAATTGGGAGTGATTTCCCTCCAGAGTTAAATGCAGGCTCGACCTTCTGC
2C-C_HoxA_T3_F143	TAATTGGGAGTGATTTCCCTAGACTTGAGCATGACAGGGTGGGGGCCTC
2C-C_HoxA_T3_F144	TAATTGGGAGTGATTTCCCTGAGACGCTATTTCCACCTTGACAAGAAAGA
2C-C_HoxA_T3_F145	TAATTGGGAGTGATTTCCCTACCCCTAACTTGTAAATGAACATGCCAGCC
2C-C_HoxA_T3_F146	TAATTGGGAGTGATTTCCCTTCGTACAGTCGCAAACATTATTCCGTTCTT
2C-C_HoxA_T3_F147	TAATTGGGAGTGATTTCCCTTCTTGCTCCCATAGAACCGTGCCCTCACAG
2C-C_HoxA_T3_F148	TAATTGGGAGTGATTTCCCTAGTAAGGAGATCAAGACAGCCCTGGGAAGGAA
2C-C_HoxA_T3_F149	TAATTGGGAGTGATTTCCCTGTGAAGGAAACATGGAGGGTGACACAGCTA
2C-C_HoxA_T3_F150	TAATTGGGAGTGATTTCCCTTGACAAAGCACTGGCCATTTCTCCATGTTT
2C-C_HoxA_T3_F151	TAATTGGGAGTGATTTCCCTATAAAGGAATACTCTATATTCAGACGAGA
2C-C_HoxA_T3_F152	TAATTGGGAGTGATTTCCCTACTCCTGTAGGCATTGATGTGTGTCCCCTG
2C-C_HoxA_T3_F153	TAATTGGGAGTGATTTCCCTACAGCCTGGACTTCCTGAGCCACAACAAAT
2C-C_HoxA_T3_F153b	TAATTGGGAGTGATTTCCCTGATCCGAGCCAGCAAGGATGCTGGGACTCG
2C-C_HoxA_T3_F154	TAATTGGGAGTGATTTCCCTAAGCAGACTAAGTGGCACTCCTTGATCTCT
2C-C_HoxA_T3_F155b	TAATTGGGAGTGATTTCCCTCCCTAGGCAGGGACTTGGGAGAGGACCTTA
2C-C_HoxA_T3_F156	TAATTGGGAGTGATTTCCCTCAATCGAGCGAGGCCCACACCTGGCGCATC
2C-C_HoxA_T3_F157	TAATTGGGAGTGATTTCCCTACCTGGAGCCTCGACTACACAGCATCTTCT
2C-C_HoxA_T3_F158	TAATTGGGAGTGATTTCCCTTAGGGAAGGAACCAAAAGATGGACCCACAG
2C-C_HoxA_T3_F159	TAATTGGGAGTGATTTCCCTCCACGAACGGTTTCTTTAGTGATTATGTTG
2C-C_HoxA_T3_F160	TAATTGGGAGTGATTTCCCTGAGCTCATGTTGTTAAGTATACAAGGTCCA
2C-C_HoxA_1_R	TCAGAATGCAGCCTGTTGGACAAATGATGAATCACCGACTGCCCA TAGAGAGG
2C-C_HoxA_2_R	AGAAAATCTCCAGATACGAAGTTGTGATCTAGGATCACCGACTGCCCA TAGAGAGG
2C-C_HoxA_3_R	CACCCCAGAGTGTTTTCACATGGCCCTGGCATCACCGACTGCCCA TAGAGAGG
2C-C_HoxA_4_R	CTTGTGATATTTGAAAACCAGCATGGTCAGGTTGATCACCGACTGCCCA TAGAGAGG

(continued)

Table 7 (continued)

Name	Sequence with adapter sequences (5′–3′)*
2C-C_HoxA_5_R	GGAATCAGTGTCCAAGCTCTCTGGATATACATCACCGACTGCCCA TAGAGAGG
2C-C_HoxA_6_R	CTTTGCTTCCATAAATCCCCTAAGATTTTGAAATGATCACCGACTGCCCA TAGAGAGG
2C-C_HoxA_7_R	AGACTTAGCAGCCGTAGAGTTTCTTACTCAAATTATCACCGACTGCCCA TAGAGAGG
2C-C_HoxA_8_R	CTTAGTGTTCAGCTCCTGAACTTTTATTACCTGATCACCGACTGCCCA TAGAGAGG
2C-C_HoxA_9_R	TAATCAAACAACGTCCTTGCCTCAAAATAGATCACCGACTGCCCATAGAGAGG
2C-C_HoxA_10_R	CTTGATTCCCAGTCTTTGTAGCAATGTTCTGAGCATCACCGACTGCCCA TAGAGAGG
2C-C_HoxA_11_R	TTCACAGACGTGCTCAGCCATCAGGAAAGCATCACCGACTGCCCA TAGAGAGG
2C-C_HoxA_12_R	TTGAGAGCTCAGATAAACTGCTGGGACTCAATCACCGACTGCCCA TAGAGAGG
2C-C_HoxA_13_R	CACATTAAAGTATTTGGAAGCCAGCCATCTATCACCGACTGCCCATAGAGAGG
2C-C_HoxA_14_R	CCTTTACAGCCATGACGGCTATGAAATCAAATCACCGACTGCCCATAGAGAGG
2C-C_HoxA_15_R	AGTTTTGGCTTTTGAAGGGAGTTCTGTTTAATCACCGACTGCCCA TAGAGAGG
2C-C_HoxA_16_R	ACTTAGCACAGGAAGCCGGGTTTCTGAAGTATCACCGACTGCCCA TAGAGAGG
2C-C_HoxA_17_R	TCGTTTCCTGGCGGGGTGGCCGGAGAGATGATCACCGACTGCCCA TAGAGAGG
2C-C_HoxA_18_R	TTCCAGAGTAAACAGCGGGAGCGCACTGGGATCACCGACTGCCCA TAGAGAGG
2C-C_HoxA_19_R	TAAATCCCCGGCGCTCCGCCGTCTAGGTGAATCACCGACTGCCCA TAGAGAGG
2C-C_HoxA_20_R	TTCTGTAATTCGACTTGGAGGGGCGGATGAATCACCGACTGCCCA TAGAGAGG
2C-C_HoxA_21_R	GGCTGCTGAGGTCGTTAAATTGTTGTTTACATCACCGACTGCCCA TAGAGAGG
2C-C_HoxA_22_R	GGCGGGGAGAGAAGGAGCTCCTGTGGGAGAATCACCGACTGCCCA TAGAGAGG
2C-C_HoxA_23_R	TCCCCGTCCGAAAGCATTAAGTTAGAAGGCATCACCGACTGCCCA TAGAGAGG
2C-C_HoxA_24_R	AACGATCTATTATTAGCAATGTAAACAGGTATCACCGACTGCCCATAGAGAGG
2C-C_HoxA_25_R	AGTTCTGGCCCATTGTTGACAAGCAGTTGGATCACCGACTGCCCA TAGAGAGG
2C-C_HoxA_26_R	TTCGATAGAGATGGAGGGGACGGGGTCTATCACCGACTGCCCATAGAGAGG

(continued)

Table 7 (continued)

Name	Sequence with adapter sequences (5′–3′)*
2C-C_HoxA_28_R	GGGGGCGAGGGTGCCTGCGTGCCTCCTGATATCACCGACTGCCCA TAGAGAGG
2C-C_HoxA_29_R	CTTCGAGAGCCGCCTCCCGTTTTCCGCTTATCACCGACTGCCCATAGAGAGG
2C-C_HoxA_29b_R	CTCACCAAAAGCCTTGACAGCTTCTAGCCCATCACCGACTGCCCA TAGAGAGG
2C-C_HoxA_30_R	CTTCATCCTTAGGTGAATCTCTCCCCTAGCATCACCGACTGCCCATAGAGAGG
2C-C_HoxA_31_R	CAATTCTTTCCTCCTGACGCGATAACAGACATCACCGACTGCCCATAGAGAGG
2C-C_HoxA_32_R	AGACGTACACTTAGTCATCCTTGCACAGAGATCACCGACTGCCCA TAGAGAGG
2C-C_HoxA_33_R	GAGGACAGAGAGAGGAAGAGCGGCGTCAGGATCACCGACTGCCCA TAGAGAGG
2C-C_HoxA_34_R	TAAGTCTGAGTAGGGTCTTCCCAGAGGTAAATCACCGACTGCCCA TAGAGAGG
2C-C_HoxA_35_R	TCGCGTTGTTTCACGATCTTGATCGCACACATCACCGACTGCCCA TAGAGAGG
2C-C_HoxA_36_R	GCACAAGCCATAAAATCCTGCCAGAGTTTCATCACCGACTGCCCA TAGAGAGG
2C-C_HoxA_37_R	GTCAGCAGACCACCACCTTGGAGCTCATACATCACCGACTGCCCA TAGAGAGG
2C-C_HoxA_38_R	TTCCTAAGAAATAGAGCCCAGCAAGAGAGCCCCTATCACCGACTGCCCA TAGAGAGG
2C-C_HoxA_39_R	AGATGAAGGTCCAGCGTCCTGGCATGCAGAATCACCGACTGCCCA TAGAGAGG
2C-C_HoxA_40_R	GTTAAACACTTGGTTACTGCCGAGGCCGGCATCACCGACTGCCCA TAGAGAGG
2C-C_HoxA_41_R	ATTTCCCTCTTCCAGCTGCTCCGGCTGCTCATCACCGACTGCCCA TAGAGAGG
2C-C_HoxA_42_R	GTTTTGAGTTTAAGAGTTCTGTCTAAGCAAATCACCGACTGCCCATAGAGAGG
2C-C_HoxA_43_R	TGTGAGGACTCACTTATAGCAGCTGCAGGAATCACCGACTGCCCA TAGAGAGG
2C-C_HoxA_44_R	AATATTTGCAAGGCTGTTGGCAGATTAGATATCACCGACTGCCCATAGAGAGG
2C-C_HoxA_45_R	CACGCTTTTCCCGTAGGAAGAACCGATGATATCACCGACTGCCCA TAGAGAGG
2C-C_HoxA_46_R	GACTCTCCTGAAACCTCACTAATATGTGTTATCACCGACTGCCCATAGAGAGG
2C-C_HoxA_47_R	CCCATAAATCGTTCAGACTAATGAGTCACAATCACCGACTGCCCATAGAGAGG
2C-C_HoxA_48_R	CCATTCCTGGCAGAATATCTTCCTAAGAGGATCACCGACTGCCCATAGAGAGG
2C-C_HoxA_49_R	CATCCTGGGTATTAAGTACTCTACTATGCCATCACCGACTGCCCATAGAGAGG
2C-C_HoxA_50_R	ATTTGCAAGAAGATCCCTCTCTCCTCCACAGACATCACCGACTGCCCA TAGAGAGG

(continued)

Table 7
(continued)

Name	Sequence with adapter sequences (5′–3′)*
2C-C_HoxA_51_R	TCTCCCAGGTCTACTTTGCATGGGGTGAACATCACCGACTGCCCA TAGAGAGG
2C-C_HoxA_52_R	GTAGCCACATCACAATTTGTACAGTTCCACATCACCGACTGCCCATAGAGAGG
2C-C_HoxA_53_R	AGAGGCCGAGGCCGAATTGGAGGATCGCATATCACCGACTGCCCA TAGAGAGG
2C-C_HoxA_54_R	CTTTTCTTCAGCCCCAATGCCCAGCACCATTCTATCACCGACTGCCCA TAGAGAGG
2C-C_HoxA_55_R	AGGGGTACACCACGGGCTCCTTGCCCTTCAATCACCGACTGCCCA TAGAGAGG
2C-C_HoxA_56_R	CGAGCTCATGGTCATTAATTTGTGAAGTGCATCACCGACTGCCCA TAGAGAGG
2C-C_HoxA_57_R	GAAAGGAAGGGCGTTGGGACCGAGGGGCATATCACCGACTGCCCA TAGAGAGG
2C-C_HoxA_58_R	GATGGGAATGGATAGAAGCAACACCTCCACATCACCGACTGCCCA TAGAGAGG
2C-C_HoxA_59_R	AATCTTTAGCCATCCTCTCTAGACTGGAGGATCACCGACTGCCCA TAGAGAGG
2C-C_HoxA_60_R	AAAAGATGGGATGCCCTGAAATGTAGCAGAATCACCGACTGCCCA TAGAGAGG
2C-C_HoxA_61_R	TTGAAGGCTTCCCCACCACCTACACTAGACATCACCGACTGCCCA TAGAGAGG
2C-C_HoxA_62_R	TAGTTATTCACTTATAGCTTGTAGCATTTCATCACCGACTGCCCATAGAGAGG
2C-C_HoxA_63_R	GTCCATACCGTGATACACGCGTCACTCTGAATCACCGACTGCCCA TAGAGAGG
2C-C_HoxA_64_R	AGGAATCGCCAGCAGAGCACGCTTTAGTACAATCACCGACTGCCCA TAGAGAGG
2C-C_HoxA_65_R	GGGAACAAACGAGCCCCCGCACATTGCCGCATCACCGACTGCCCA TAGAGAGG
2C-C_HoxA_66_R	AAGAAAGGCGGAAATCTCCTACTGACAAAGATCACCGACTGCCCA TAGAGAGG
2C-C_HoxA_67_R	TTAAAGTTAGGCCTGGGGATGCGGCGCGGCATCACCGACTGCCCA TAGAGAGG
2C-C_HoxA_68_R	GTCTGTGCTATGGATGATTTTACGATCTAAATCACCGACTGCCCATAGAGAGG
2C-C_HoxA_69_R	CACACCGCTTGGAGTCACAGTTTTCATCACATCACCGACTGCCCA TAGAGAGG
2C-C_HoxA_70_R	CTTGTCCCCCTGGCGGACTTTGGAAGACATCACCGACTGCCCATAGAGAGG
2C-C_HoxA_71_R	CCATTTCCCCCACTATTTGTGAGCGCAGGGATCACCGACTGCCCA TAGAGAGG
2C-C_HoxA_72_R	TGTACTTGGTTCCCTCCTACGTAGGCACCCATCACCGACTGCCCA TAGAGAGG

(continued)

Table 7
(continued)

Name	Sequence with adapter sequences (5′–3′)*
2C-C_HoxA_73_R	TTTGAAAATACAGATATCACCTTCGGGGAAATCACCGACTGCCCATAGAGAGG
2C-C_HoxA_74_R	ACGCATTCAGGGCGCACCCCAGAACTCCGGATCACCGACTGCCCA TAGAGAGG
2C-C_HoxA_75_R	TTCAGAGTGTGTACACTACTTACATGGTTCCACATCACCGACTGCCCA TAGAGAGG
2C-C_HoxA_76_R	GGCTCGTGTACTTCCGGTCGGCGCCTTCGTATCACCGACTGCCCA TAGAGAGG
2C-C_HoxA_77_R	CGCCAAAAGCGACAGCAGCAAATCGCACCAGATCACCGACTGCCCA TAGAGAGG
2C-C_HoxA_78_R	ACTGGGCTCCTAAGACAGCCGAGGGACCTGATCACCGACTGCCCA TAGAGAGG
2C-C_HoxA_79_R	ATCCGTCCTGGAAAGGGTTTACTTTGCATAATCACCGACTGCCCATAGAGAGG
2C-C_HoxA_80b_R	CAGTCAGGAAGGCGGCGGACCTAGGATGCAATCACCGACTGCCCA TAGAGAGG
2C-C_HoxA_81_R	GCACCGAAACTTCCCACCAAGTAACACCCAATCACCGACTGCCCA TAGAGAGG
2C-C_HoxA_82_R	CCGAACAACTCATAAAGTTGTATTGCAAAGATCACCGACTGCCCATAGAGAGG
2C-C_HoxA_83_R	ACGTCTCTACACGGATAAAGGCACATATACATCACCGACTGCCCATAGAGAGG
2C-C_HoxA_84_R	TCTGTGGGTTAGCTTCTGCTTAGCAGGACTGATCACCGACTGCCCA TAGAGAGG
2C-C_HoxA_85_R	TCTGCGGCCGCTCAGTCCACAAAAGTTGGGATCACCGACTGCCCA TAGAGAGG
2C-C_HoxA_86_R	CCGGCTCGCAACAGCCTGGCTCCGCTCTTCATCACCGACTGCCCA TAGAGAGG
2C-C_HoxA_87_R	TCCGGCAGCTTTCAGTGTCGGTTTTACGAGATCACCGACTGCCCA TAGAGAGG
2C-C_HoxA_88_R	CTTGGTCCAGGGCTCACTAGCAGGAGTCGGATCACCGACTGCCCA TAGAGAGG
2C-C_HoxA_89_R	ACCTATCCAAGTACCAGCTCACATGGAGCTATCACCGACTGCCCATAGAGAGG
2C-C_HoxA_90_R	CGCCGCCGCCGTCTGTCTAGACTCAAGCGAATCACCGACTGCCCA TAGAGAGG
2C-C_HoxA_91_R	CTCGTACCTGTCGTGTACAAATGAGATTGGATCACCGACTGCCCA TAGAGAGG
2C-C_HoxA_93_R	GCTGCAGGCGTTCCTCCTGTGTCCGCCAGTATCACCGACTGCCCA TAGAGAGG
2C-C_HoxA_94_R	TTCGTCCCTTAATGAGTTTACAACTGTCCAGCCATCACCGACTGCCCA TAGAGAGG
2C-C_HoxA_95_R	AGTGTAAGTTCAGTCTGATGGAAACCCCAGATCACCGACTGCCCA TAGAGAGG

(continued)

Table 7
(continued)

Name	Sequence with adapter sequences (5′–3′)*
2C-C_HoxA_96_R	CTGCTGGGTTATCTGCGGGGAAGAGAAACAATCACCGACTGCCCA TAGAGAGG
2C-C_HoxA_96b_R	AAGGCTCTCTGGGCTGCACTGCTTTCGAAGATCACCGACTGCCCA TAGAGAGG
2C-C_HoxA_97_R	GTCAGTCAGGGACAAAGTGTGAGTGTCAAGATCACCGACTGCCCA TAGAGAGG
2C-C_HoxA_98_R	GCACGCTATTAATGGTCCGATGTTTTGCAGTCATCACCGACTGCCCA TAGAGAGG
2C-C_HoxA_99_R	CCCCACGGAGGAGCTGGCCAGGAGGGAGCGATCACCGACTGCCCA TAGAGAGG
2C-C_HoxA_100_R	AATCAAATTCACGTGTGGATACTGTGCCTGATCACCGACTGCCCATAGAGAGG
2C-C_HoxA_101_R	CCTTTTCCACGTTTTATGCCTGAGAAGACAATCACCGACTGCCCATAGAGAGG
2C-C_HoxA_102_R	TCTCCTAGTCAGCCGTTAGCGACAGGCGAAATCACCGACTGCCCA TAGAGAGG
2C-C_HoxA_103_R	GCGAAGGAGCAGCCAACCTAACCCTACCTGATCACCGACTGCCCA TAGAGAGG
2C-C_HoxA_104_R	GACAAATTACGACCGTCTAGGTAATATTTAATCACCGACTGCCCATAGAGAGG
2C-C_HoxA_105_R	TCCGGGAGCTCACAGCCAACTTTAATTTTTCCTGATCACCGACTGCCCA TAGAGAGG
2C-C_HoxA_106_R	GGCTGTCTCACCCCAGGGCCCTGATTGCCCATCACCGACTGCCCA TAGAGAGG
2C-C_HoxA_107_R	GGAGCTCTCTTCTTTGATGTTCTGCGCGAAATCACCGACTGCCCA TAGAGAGG
2C-C_HoxA_108_R	ATTTGGCGAAGGGAGCAGATAGCCCTTTCTATCACCGACTGCCCA TAGAGAGG
2C-C_HoxA_109_R	CTCTGGTGTTGTGCTCACCCGGCTTGCCTTATCACCGACTGCCCA TAGAGAGG
2C-C_HoxA_110_R	ACCGCAGCCAGTGGCCCTATCTTTAGCTTCATCACCGACTGCCCA TAGAGAGG
2C-C_HoxA_111b_R	CTGAATTATCCAGCGGTGGAGTTGGGCTGCATCACCGACTGCCCA TAGAGAGG
2C-C_HoxA_112_R	GTGACCTCCAGGACTTCTGCCCCTGGGAATATCACCGACTGCCCA TAGAGAGG
2C-C_HoxA_113_R	TCTCATTTTTACATCTAAGAAATCGCTGCATCACCGACTGCCCATAGAGAGG
2C-C_HoxA_114_R	CCTCCACGGGCCCCCGGACGGTCCCACTGCATCACCGACTGCCCA TAGAGAGG
2C-C_HoxA_115_R	GTGTGGTGGAGAGTTTGCCAAACTGCCACAATCACCGACTGCCCA TAGAGAGG
2C-C_HoxA_116_R	TCTCAGTGGTTAAGATTCCTAATAATCATATCACCGACTGCCCATAGAGAGG
2C-C_HoxA_117_R	ATAAAAATCTCCGGATTACCTCGCTATCAAATCACCGACTGCCCATAGAGAGG

(continued)

Table 7
(continued)

Name	Sequence with adapter sequences (5′–3′)*
2C-C_HoxA_118_R	CTGCCCACGGTGCTATAGAAATTGGACGAGATCACCGACTGCCCA TAGAGAGG
2C-C_HoxA_119_R	TCCGCTTCAAAGAGGCAGCTGCAGTGGAGAATCATCACCGACTGCCCA TAGAGAGG
2C-C_HoxA_120_R	ATTACACTCTCTCATTCATGGTCACTTCCGATCACCGACTGCCCATAGAGAGG
2C-C_HoxA_121_R	ATTGAAGGAGCAATGTTTGGAGGAAGCGAAATCACCGACTGCCCA TAGAGAGG
2C-C_HoxA_122_R	TGGCAGTCAACTTCAGCTTGTGCCTGAGCAATCACCGACTGCCCA TAGAGAGG
2C-C_HoxA_123_R	TGCTGGGGGCGTTTCCAGCACAGTCATTCAATCACCGACTGCCCA TAGAGAGG
2C-C_HoxA_124_R	GATGGCTACAGTTAAGTTTCTCTCTTGGTGATCACCGACTGCCCA TAGAGAGG
2C-C_HoxA_125_R	TCCACTAAAGTAAATGATAACTAGATTGCTTAATCACCGACTGCCCA TAGAGAGG
2C-C_HoxA_126_R	GCTCAGAGCTAACGCTAAGCCCCTTAAGGCATCACCGACTGCCCA TAGAGAGG
2C-C_HoxA_127_R	GCCTTTTGGATAACGCTATATCTTTGCTTAATCACCGACTGCCCATAGAGAGG
2C-C_HoxA_128_R	TGGAGAGGTTTCCGGCAGCCACTTTTGTAAATCACCGACTGCCCA TAGAGAGG
2C-C_HoxA_129_R	TCTGCAGAAATCTGATGCAGAGCAAGCCACATCACCGACTGCCCA TAGAGAGG
2C-C_HoxA_130_R	TCTACTAATCCAGCTAAGGCCAATTCATGAATCACCGACTGCCCATAGAGAGG
2C-C_HoxA_131_R	TCTCCCGAGAGTTTCCATCCCAACATGATCACCGACTGCCCATAGAGAGG
2C-C_HoxA_132_R	CTGATGGGGTTGGCGGAAGAACTGGCAGTCATCACCGACTGCCCA TAGAGAGG
2C-C_HoxA_133_R	TTCCAGAGGTGGGGAGGCTGCGCCTGCTCTATCACCGACTGCCCA TAGAGAGG
2C-C_HoxA_134_R	CCGCATGGAGAAGACCCCAGTGGCGCTGTTATCACCGACTGCCCA TAGAGAGG
2C-C_HoxA_135_R	AGACTTTGAGATAGAGCGAGCGATCCCTGTATCACCGACTGCCCA TAGAGAGG
2C-C_HoxA_136_R	GTGTCGCCCAATAAAACCTTCTATGACCCCATCACCGACTGCCCATAGAGAGG
2C-C_HoxA_137_R	CAAAAGAAAGTTTGGAGGACGATTGAGACGATCACCGACTGCCCA TAGAGAGG
2C-C_HoxA_138_R	TCTCAGGACCTACCAACCCTTTCCCATCACCGACTGCCCATAGAGAGG
2C-C_HoxA_140_R	GCGCATCTCCTTACTCACTCTTACTTTTCCATCACCGACTGCCCATAGAGAGG
2C-C_HoxA_141_R	AAAACATACCTCAGTCCTTAGTGTGTCCTAATCACCGACTGCCCATAGAGAGG

(continued)

Table 7
(continued)

Name	Sequence with adapter sequences (5′–3′)*
2C-C_HoxA_142_R	CCATAGAGAGTCCAAACGCATGAATTCCTGATCACCGACTGCCCA TAGAGAGG
2C-C_HoxA_143_R	CTACAGAAGCCCTGGAAACTCTATCGGTAGATCACCGACTGCCCA TAGAGAGG
2C-C_HoxA_144_R	TCTAGCTGTCCACCGAAAGCACAGGGATCACCGACTGCCCATAGAGAGG
2C-C_HoxA_145_R	CCAGCCATTCACAAAAGAGAATGCGTTGTCATCACCGACTGCCCA TAGAGAGG
2C-C_HoxA_146_R	ACTGTAAACGGCCCCGGCCACCTTTACGAGATCACCGACTGCCCA TAGAGAGG
2C-C_HoxA_147_R	CAATCCAGTGTTTGATTCTCCCTATAGAAAATCACCGACTGCCCATAGAGAGG
2C-C_HoxA_148_R	TTCATGATCTGCCCTGCTGGAGAGGTTCAAAGAATCACCGACTGCCCA TAGAGAGG
2C-C_HoxA_149_R	CATAAATGATTGCACCAAAGGCTGAGATAAATCACCGACTGCCCATAGAGAGG
2C-C_HoxA_150_R	AGTGGGGTTGATAGGAAGACTTCCCTGGGTATCACCGACTGCCCA TAGAGAGG
2C-C_HoxA_151_R	TCTGTGTGCTCACAGGCAAACAGGTCTAAGATCACCGACTGCCCA TAGAGAGG
2C-C_HoxA_152_R	TTGGAGTCATAGAATACCATGGTGGGGGCAATCACCGACTGCCCA TAGAGAGG
2C-C_HoxA_153_R	GGGGCCTACTAGTGTGTTGGTGGCATGATTATCACCGACTGCCCA TAGAGAGG
2C-C_HoxA_153b_R	CCGCGACTTCCGCCTCTGGCGATCGGCAGTATCACCGACTGCCCA TAGAGAGG
2C-C_HoxA_154_R	CACTCCTACTCGGCTTTCATCTCACCCAGAATCACCGACTGCCCA TAGAGAGG
2C-C_HoxA_155b_R	TGAGTCAAACCTCTATGAACCCCAACCTTTATCACCGACTGCCCATAGAGAGG
2C-C_HoxA_156_R	ACTGCCGAGCCATTAGCTGCGGGTTTCCTTATCACCGACTGCCCA TAGAGAGG
2C-C_HoxA_157_R	GGGTCTGGCGTCTGCCAGCACCTGATCTCTATCACCGACTGCCCA TAGAGAGG
2C-C_HoxA_158_R	AGTCTTCCAAGCCACTGTTTGTCACTGGTAATCACCGACTGCCCA TAGAGAGG
2C-C_HoxA_159_R	ACTCCACCTTCCTAGTTTCAAACCCTGCACATCACCGACTGCCCA TAGAGAGG
2C-C_HoxA_160_R	AATCACTTGGAGGTGACCTACGGGAAAGTCATCACCGACTGCCCA TAGAGAGG

*Forward 2C-ChIP library primers are with the complementary T3 sequence (T3c; 5′-TAATTGGGAGTGA TTTCCCT-3′). Reverse 2C-ChIP library primers are with the complementary "P1-key" adapter sequence (5′-ATCA CCGACTGCCCATAGAGAGG-3′) used in the PGM™ system. Note that Reverse 2C-ChIP primers must be 5′-phosphorylated

quantitatively amplify 2C-ChIP input libraries produced with 320 primers and ~6000 cells.

1. Set up PCR reactions according to Table 8. Here, as when controlling for the quality of 2C-ChIP libraries (Subheading 3.3), we recommend preparing reactions in a specific order to minimize background issues. Samples should be prepared on ice by adding "master mix 1" first to the tubes (A), followed by the 2C-ChIP library (B), then the barcoded Forward 2C-ChIP amplification primer (C), and lastly the "master mix 2" (D). Do not forget to include a PCR "water control" without a 2C-ChIP library to control for reagent specificity.
2. Amplify the samples using the PCR program in Table 9.

Table 8
PCR amplification of 2C-ChIP libraries for PGM™ sequencing

A. master mix 1	*per* reaction
10X PCR HIFI buffer	10.0 μL
50 mM Mg_2SO_4	8.0 μL
25 mM dNTP mix	0.8 μL
20 μM P1-key (Reverse primer)	2.0 μL
volume master mix 1	**20.8** μL
B. 2C-ChIP library	14.0 μL ***added individually***
C. 20 μM A-key (Forward Primer)*	2.0 μL ***added individually***
D. master mix 2	
NEB Taq DNA polymerase	0.8 μL
water	61.4 μL
volume master mix 2	**62.2** μL
total reaction volume	100.0 μL

**see* Note 17

Table 9
2C-ChIP sequencing PCR amplification pro3gram

No. of PCR cycles	Denature	Anneal	Extend
1	95 °C, 5 min		
16*	95 °C, 30 s	60 °C, 30 s	72 °C, 15 s
1	95 °C, 30 s	60 °C, 30 s	72 °C, 3 min

**see* Note 18

3.6 Purifying 2C-ChIP Libraries Before Deep Sequencing

Free primers are easily removed with Sera-Mag SpeedBeads™ (Thermo Fisher Scientific; *see* **Note 19**). Below, we provide experimental conditions that maximize recovery and virtually eliminate any unincorporated 2C-ChIP library amplification primers from samples.

1. Using a P100 micropipette, transfer/measure the amplified 2C-ChIP library to a new microtube, and record this volume for the next steps.
2. Add 0.8× volume of Sera-Mag beads to the 2C-ChIP library and mix by resuspending ten times with the micropipette.

   This Sera-Mag bead concentration mostly captures DNA products above ~150 bp. The supernatant will contain smaller DNA products, but fragments up to ~180 bp will remain.
3. Incubate at room temperature for 5 min, and then place the tube on a magnetic bar for 2 min.
4. Transfer the supernatant containing the 2C-ChIP library to a new microfuge tube. Do not discard the Sera-Mag beads until the end of the purification.
5. Extract the amplified 2C-ChIP products of ~160 bp from the supernatant by adding 0.6× volume of Sera-Mag beads (this volume is calculated from the original recorded sample volume), and mix by resuspending ten times with a micropipette (*see* **Note 20**).
6. Incubate at room temperature for 5 min, and then place the tube on a magnetic bar for 2 min.
7. Remove the supernatant containing the unwanted primers, and keep in a separate tube until end of purification.
8. Wash the Sera-Mag beads with 500 μL of 80% ethanol while on the magnetic bar (do not mix by pipetting up and down). Incubate for 30 s before carefully removing the ethanol wash with a micropipette.
9. Repeat the ethanol wash.
10. Quick spin in a microcentrifuge to collect the remaining ethanol at the bottom of the tube and return to the magnetic bar for 2 min. Remove any residual ethanol wash with a P10 micropipette.
11. Elute the 2C-ChIP amplified library products from the beads by adding 23 μL of 0.1× TE buffer. Resuspend with the micropipette to mix and incubate for 2 min at room temperature.
12. Transfer the 2C-ChIP supernatant to a new microfuge tube and SpeedVac for 2 min to evaporate any residual ethanol. Figure 4 shows an example of 2C-ChIP library before and after purification on Sera-Mag SpeedBeads™ (Thermo Fisher Scientific; *see* **Note 21**).

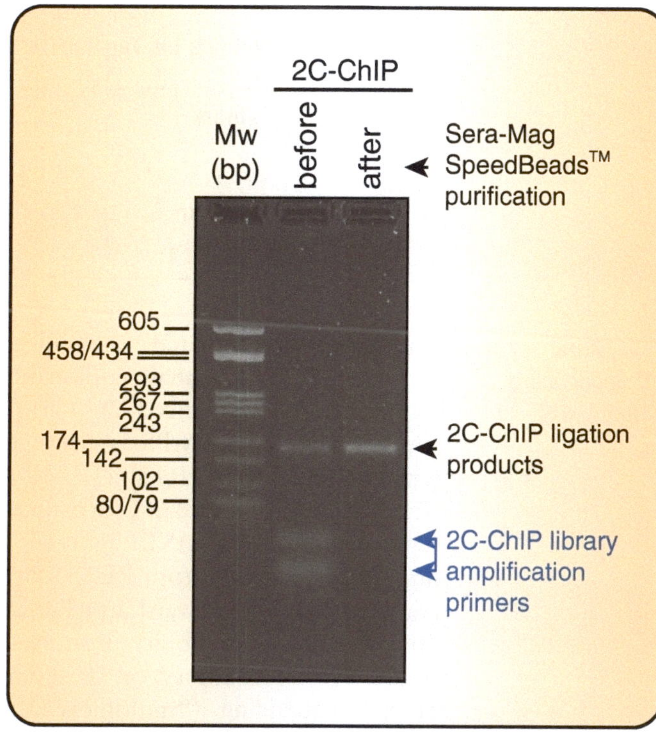

Fig. 4 Before and after purification of an input 2C-ChIP library using Sera-Mag SpeedBeads™ to remove unincorporated 2C-ChIP amplification primers. The input 2C-ChIP library is as described in Fig. 3 and was resolved on a 2.5% agarose gel as described in Subheading 3.3

3.7 Quantifying and Pooling Barcoded 2C-ChIP Libraries for Deep Sequencing

For multiplexed sequencing of 2C-ChIP libraries, the pooling of libraries will depend on the sequencing platform used. When using the PGM™ system, 2C-ChIP libraries should be pooled at individual concentrations of 9 pM. We quantify 2C-ChIP libraries by TaqMan™ using the Ion Library TaqMan™ Quantitation Kit (Thermo Fisher Scientific) according to the manufacturer's recommendations. For each quantitation, we recommend setting up a titration curve from at least two sets of tenfold titrations (ranging from 1/10 to 1/10,000) of the *E. coli* DH10B library control provided in the kit. To fall within the linear range of the TaqMan™ titration curve, 2C-ChIP samples must be diluted according to their expected concentration. In general, input 2C-ChIP libraries require more dilution than ChIP samples. Also, 2C-ChIP libraries originating from histone marks are generally more concentrated than libraries from non-histone proteins whether they bind chromatin directly or not. Table 10 summarizes dilution factors that worked well for 2C-ChIP libraries generated as described in sections above.

Table 10
2C-ChIP library dilutions for TaqManTM quantitation

Type of 2C-ChIP library	Dilution
Input	1:1000
α-H3K4me3, α-H3K27me3, α-Suz12	1:100
α-Ash2L, α-UTX, α-CTCF	1:10

3.8 Processing 2C-ChIP Sequencing Data with LAMPS

The Ligation-mediated Amplified, Multiplexed Primer-pair Sequence (LAMPS) analysis pipeline [3] is a computational tool for processing sequenced 2C-ChIP library reads (Fig. 5). The source code for LAMPS is made publicly available at https://github.com/BlanchetteLab/LAMPS. LAMPS maps sequences (in either FASTQ or BAM format) to previously defined genomic regions using the BLAST alignment algorithm [8]. Sequences are first mapped to a custom BLAST database containing all possible combinations of Forward and Reverse 2C-ChIP primers (Fig. 5; "possible 2C-ChIP library product list"), which may consist of a single 2C-ChIP library or of several barcoded samples pooled before sequencing ("multiplexed"). Sequences, which do not uniquely map to this list (Fig. 5; "unmappable reads"), are remapped once again with BLAST to the list of Forward and Reverse 2C-ChIP primers used in the experiment. A quality control (QC) report is generated at this step to inform on the nature of short reads. Sequences that map to expected primer pairs ("expected mapped reads") correspond to ligation products between adjacent Forward and Reverse 2C-ChIP library primers. Those corresponding to "unexpected mapped reads" are background ligation products between non-adjacent primers or between either two Forward or Reverse primers. The QC report generated at this step informs on the specificity of 2C-ChIP reactions.

The "expected mapped reads" are next normalized based on read count (in reads per million [RPM]) and the corresponding original "DNA density" of each sample (*see* **Note 22**). Normalization based on DNA density includes the library dilution made for sequencing at 9 pM ("TaqMan™ dilution"), the input or ChIP dilution made before generating 2C-ChIP libraries ("ChIP dilution"), and the fact that the input signal only represents a percentage of what was used for ChIP ("input percentage"). The resulting normalized read counts per sample are provided in bedGraph format for easy integration with most genomic browsers. QC reports are also provided at each normalization step by LAMPS, which helps identify problematic primers and other error source in the protocol (e.g., PCR artifacts, human error). More information about the LAMPS analysis pipeline is found in [3].

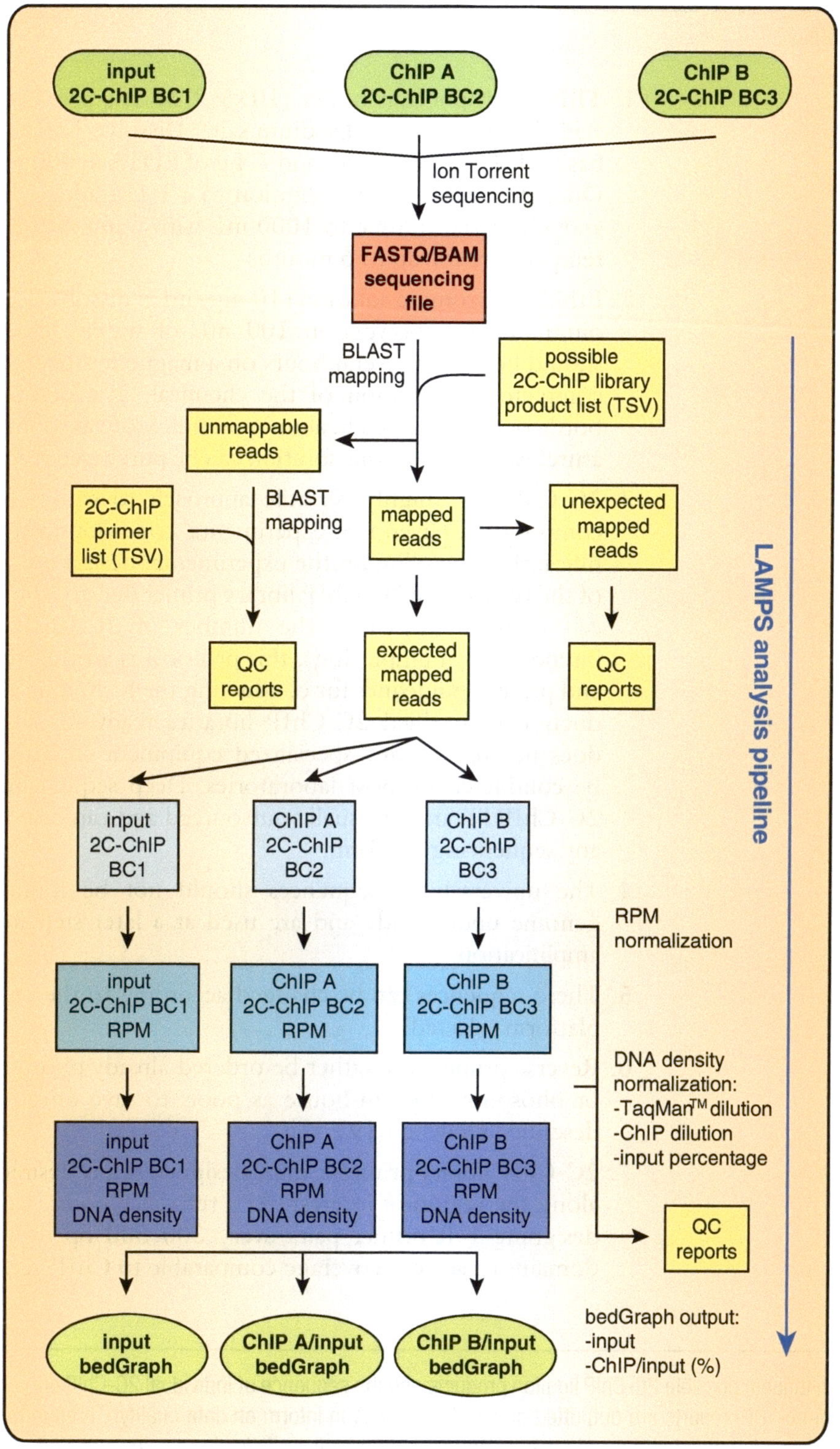

Fig. 5 Flowchart of the 2C-ChIP sequencing data processing procedure. 2C-ChIP sequencing data can be processed with the LAMPS analysis pipeline [3]. This pipeline accepts either FASTQ or BAM files (user must install "SAMtools" for BAM files) and will map sequence reads against a user-provided TSV file containing all

4 Notes

1. TBE electrophoresis buffer (10×): 1 M Tris base, 1 M boric acid, 20 mM EDTA (disodium salt). Dissolve 121.1 g of Tris base, 61.8 g of boric acid, and 7.4 g of EDTA in 800 mL water. Once dissolved, transfer solution to a 1 L graduated cylinder and adjust the volume to 1000 mL with water. Store at room temperature for up to 6 months.
2. Ethidium bromide solution (10 mg/mL): dissolve 1 g of ethidium bromide powder in 100 mL of water. The solution should be stirred several hours on a magnetic stirrer to ensure complete resuspension of the chemical. Transfer to a dark bottle or one wrapped in aluminum foil. Store at room temperature. Alternatively, this solution can be purchased ready-made.
3. 2C-ChIP is a highly versatile approach to study chromatin composition. Given that experimental conditions can vary significantly depending on the experimental design (i.e., the size of the region, the 2C-ChIP library primer density, the available cell number for ChIP, the number of 2C-ChIP libraries intended for multiplexing), this protocol is written to explain and present guidelines for conducting the technique. The production of purified 2C-ChIP libraries ready for sequencing does not require any specialized equipment such that it can be conducted in most laboratories. Deep sequencing of the 2C-ChIP libraries is usually outsourced and can be adapted to any sequencing platform.
4. The universal tail sequences should not be found in the genome under study and are used at a later step for library amplification.
5. These sequences can be changed according to the sequencing platform selected.
6. Reverse primers can either be ordered already phosphorylated or phosphorylated in-house as pools to save on cost as was described elsewhere [9].
7. 2C-ChIP library primers can be designed at any desired density along the region of interest. As a reference point, we suggest designing F-R primer pairs every 300–500 bp over defined domains to achieve coverage comparable to ChIP-seq profiles.

Fig. 5 (continued) possible 2C-ChIP ligation products and the sequence of individual 2C-ChIP primers. Several quality control (QC) reports are outputted during the analysis to inform on data quality. Through this analysis pipeline, distinctly barcoded datasets are separated and normalized following a user-provided configuration file (TSV) that will specify the normalization factors and the input track to be used in ChIP ratio calculations. LAMPS processes sequencing output into individual bedGraph files

Alternatively, regulatory DNA elements can be probed specifically by tilling F-R pairs over these target regions. The 2C-ChIP primer homology regions can be designed with any free PCR primer design program such as "Primer-BLAST," "Primer3," or other accessibles here (https://toptipbio.com/free-primer-design-programs/). For simplicity, we typically select all primers complementary to the antisense strand of the selected genome reference. When analyzing mammalian genomes, we design primers to bear a GC content ranging from 40% to 60%, with homology lengths varying between 22 and 34 nucleotides. Primers must be unique based on BLAST results and have limited possibility of self-annealing and/or hairpin formation when analyzed by the "Oligo Calc" tool available at (http://biotools.nubic.northwestern.edu/OligoCalc.html).

8. 2C-ChIP is PCR-based, and therefore we recommend taking the same necessary precautions as one would when conducting PCR reactions. For instance, all reaction solutions should be prepared with ultrapure water (e.g., Milli-Q® water with a resistivity of 18 MΩ cm at 25 °C). If possible, separate between Pre- and Post-PCR steps (using different bench space).
9. We recommend dispensing the primers into aliquots to avoid freeze–thaw cycles, which can render them inactive.
10. When a large number of primers (>100 primers) are used, we recommend creating high concentration pools by mixing equal volumes of stock primers in 1× TE. When preparing pools, we recommend keeping Forward and Reverse 2C-ChIP library primers separate from each other as higher background has been observed when these primers are mixed at higher concentrations.
11. Example: If your design consists of 80 Forward and 80 Reverse 2C-ChIP library primers, you can create two separate pools (one "Forward" and one "Reverse" primer pool) at 1 μM concentration in 1× TE simply by mixing 10 μL of each Forward or Reverse 2C-ChIP primer.
12. Example: If starting from two pools (one Forward, one Reverse) each at 1 μM concentration, we recommend first separately diluting 10 μL into 15 μL of water to reach 400 nM dilutions. Each diluted sample should then be serially diluted (tenfold dilutions) three times by transferring 10 μL of the diluted samples into 90 μL of water (40 nM, 4 nM, 400 pM dilutions). Finally, equal volume of the 400 pM Forward and Reverse library primer dilutions should be mixed immediately before use (200 pM "diluted primer mix").
13. *The input or ChIP gDNA volume per reaction will depend on the amount of cells and the antibody used. For the input,

we generally find that purified DNA samples from 100,000 cells per μL must be diluted one hundred times (1/100) and that 5.8 μL of that dilution gives signals within the quantitative linear range of 2C-ChIP. For the ChIP material, we find that purified samples yielding over 5% of the input as quantified by ChIP-qPCR of a positive control region should be analyzed at a 1/20 dilution using the maximum volume allowed. We highly recommend quantifying the amount of genomic DNA in samples to make sure that no more than 1–2 ng of template DNA is used in 2C-ChIP reactions. Too much genomic DNA can prevent quantitative detection of DNA as we have reported previously [2].

14. ∗∗∗Individual final primer concentrations of 34 pM.
15. ∗Any barcoded Forward 2C-ChIP library amplification primer from Table 1 can be used.
16. ∗We limit the 2C-ChIP PCR extension time to 15 s to enrich the amplification of the small ~160 bp 2C-ChIP products.
17. ∗Each 2C-ChIP library should be amplified with a different preselected barcoded Forward 2C-ChIP amplification primer.
18. ∗2C-ChIP libraries can be amplified on a real-time quantitative PCR machine to identify the lowest possible cycle number to produce libraries.
19. We found that unincorporated 2C-ChIP amplification primers (Figs. 3 and 4) can cause quantitation and sequencing artifacts, and therefore we recommend removing them prior deep sequencing.
20. The PEG concentration in the 2C-ChIP library sample will now be at 1.4× the original sample volume, which captures products above 100 bp. Supernatant will contain DNA products between 1 and ~ 130 bps. The resulting supernatant will therefore contain the unincorporated primers.
21. Do not SpeedVac the beads as they might dry and prevent elution of the 2C-ChIP products.
22. The LAMPS data processing pipeline can account for any desired number of user-provided normalization factors. The values of at least three dilution steps must be kept on record while generating the libraries to obtain quantitative 2C-ChIP/ChIP "percentage of input" bedGraph files from LAMPS:
    (a) The amount of starting material: the amount of chromatin in the input sample is generally 1/10th of what was used to produce ChIP samples.

        For ease, set input as 1× and ChIP samples as 10× multipliers.

    (b) The amount of cells or genomic DNA (if quantitating with the Quant-iT™ PicoGreen™ dsDNA Assay Kit) used at the 2C-ChIP library primer-annealing step. While input samples will likely be diluted, some ChIP samples such as those from non-histone binding proteins might not be diluted at all. Thus, the dilution factor will either be based on volume or genomic DNA amounts.

(c) The amount of purified 2C-ChIP products generated in individual purified and amplified 2C-ChIP reactions. As described in Subheading 3.6, TaqMan™ should be used to quantify the barcoded 2C-ChIP libraries before sequencing. Each library should then be diluted and pooled at an individual concentration of 9 pM before sequencing.

Acknowledgments

We thank members of our laboratories for important feedback on this protocol. This work was supported by the Canadian Institutes of Health Research (CIHR; MOP-142451 to J.D.) and the Natural Sciences and Engineering Research Council of Canada (NSERC; Discovery grants to M.B. and J.D.). X.Q.D.W. was supported by a scholarship from the Fonds de Recherche Québec Santé (FRQS) and by the CIHR.

References

1. Johnson DS, Mortazavi A, Myers RM, Wold B (2007) Genome-wide mapping of in vivo protein-DNA interactions. Science 316 (5830):1497–1502. https://doi.org/10.1126/science.1141319
2. Wang XQD, Cameron CJF, Paquette D, Segal D, Warsaba R, Blanchette M, Dostie J (2019) 2C-ChIP: measuring chromatin immunoprecipitation signal from defined genomic regions with deep sequencing. BMC Genomics 20(1):162. https://doi.org/10.1186/s12864-019-5532-5
3. Cameron CJF, Wang XQD, Dostie J, Blanchette M (2020) LAMPS: an analysis pipeline for sequence-specific, ligation-mediated amplification reads. BMC Research Notes, in press. https://doi.org/10.1186/s13104-020-05106-1
4. Mukhopadhyay A, Deplancke B, Walhout AJ, Tissenbaum HA (2008) Chromatin immunoprecipitation (ChIP) coupled to detection by quantitative real-time PCR to study transcription factor binding to DNA in *Caenorhabditis elegans*. Nat Protoc 3(4):698–709. https://doi.org/10.1038/nprot.2008.38
5. Taneyhill LA, Adams MS (2008) Investigating regulatory factors and their DNA binding affinities through real time quantitative PCR (RT-QPCR) and chromatin immunoprecipitation (ChIP) assays. Methods Cell Biol 87:367–389. https://doi.org/10.1016/S0091-679X(08)00219-7
6. Kim TH, Dekker J (2018) ChIP-quantitative polymerase chain reaction (ChIP-qPCR). Cold Spring Harb Protoc 2018(5):pdb prot082628. https://doi.org/10.1101/pdb.prot082628
7. Addgene (2018) How to run an agarose gel. https://www.addgene.org/protocols/gel-electrophoresis/
8. Altschul SF, Gish W, Miller W, Myers EW, Lipman DJ (1990) Basic local alignment search tool. J Mol Biol 215(3):403–410. https://doi.org/10.1016/S0022-2836(05)80360-2
9. van Berkum NL, Dekker J (2009) Determining spatial chromatin organization of large genomic regions using 5C technology. Methods Mol Biol 567:189–213. https://doi.org/10.1007/978-1-60327-414-2_13

Chapter 9

Single-Cell DamID to Capture Contacts Between DNA and the Nuclear Lamina in Individual Mammalian Cells

Kim L. de Luca and Jop Kind

Abstract

The organization of DNA within the eukaryotic nucleus is important for cellular processes such as regulation of gene expression and repair of DNA damage. To comprehend cell-to-cell variation within a complex system, systematic analysis of individual cells is necessary. While many tools exist to capture DNA conformation and chromatin context, these methods generally require large populations of cells for sufficient output. Here we describe single-cell DamID, a technique to capture contacts between DNA and a given protein of interest. By fusing the bacterial methyltransferase Dam to nuclear lamina protein lamin B1, genomic regions in contact with the nuclear periphery can be mapped. Single-cell DamID generates contact maps with sufficient throughput and resolution to reliably identify patterns of similarity as well as variation in nuclear organization of interphase chromosomes.

Key words DamID, Single-cell genomics, Chromatin, Nuclear lamina, Lamin B1

1 Introduction

A long-standing question in biology is how the same genome can give rise to all different cell types and their corresponding functions in the organism. It has become clear that gene transcription occurs in a cell type-specific manner, after which mRNA translation and downstream processes eventually culminate in functional proteins. Still, the regulatory events underlying differences in gene expression and cell fate commitment remain poorly understood. DNA is packaged within chromatin, which, in turn, is coiled into higher-order structures and resides in nuclear compartments with more permissive or restrictive features (reviewed in [1]). Densely packed heterochromatin is typically located at the nuclear lamina (NL), while less compacted, transcriptionally active, euchromatin resides in the nuclear interior (reviewed in [2]). Withal, heterogeneity in cellular function is partly the result of variability in spatial genome organization.

Beatrice Bodega and Chiara Lanzuolo (eds.), *Capturing Chromosome Conformation: Methods and Protocols*, Methods in Molecular Biology, vol. 2157, https://doi.org/10.1007/978-1-0716-0664-3_9,

There are multiple ways of experimentally capturing nuclear organization and chromatin context, e.g., chromatin immunoprecipitation followed by next-generation sequencing (ChIP-seq) [3]; DNA adenine methyltransferase identification (DamID) [4, 5]; assay for transposase-accessible chromatin using sequencing (ATAC-seq) [6]; chromosome conformation capture (3C) [7] and its derivatives 4C [8], 5C [9], and Hi-C [10]; and chromatin interaction analysis by paired-end tag sequencing (ChIA-PET) [11]. These methods generally require large populations of cells for sufficient output. If single-cell adaptations exist [12–14], they lack the resolution to conclusively identify variation in chromosomal organization across cells. More recently, genome architecture mapping (GAM) [15] and super-resolution chromatin tracing [16] have interrogated chromatin organization of single cells. However, these techniques are experimentally challenging and largely unavailable to those without designated hardware and prior training. Here we describe the method for performing a modified version of DamID that is suited to single cells [17]. DamID is easily downscaled, with the capacity to process hundreds of cells per day while retaining high genomic coverage at a resolution of ~10 kb. As a result, single-cell DamID (scDamID) robustly assesses cellular heterogeneity with easy implementation and common lab equipment.

DamID involves the tethering of *Escherichia coli* DNA adenine methyltransferase (Dam) to a protein of interest (POI) and subsequent identification of Dam-methylated DNA sequences. The NL is a meshwork of proteins lining the inner nuclear membrane (INM), providing structure to the nucleus and serving as a scaffold for the genome. By tethering Dam to NL component lamin B1 (Dam-LMNB1) and expressing the fusion protein in vivo, the genome that comes in close proximity to the NL can be recognized, amplified, and sequenced. Untethered Dam is standardly included in the experimental design, as a proxy for accessible chromatin that is methylated by a freely diffusing enzyme. Dam deposits a methyl group on N^6 of the adenine within a GATC motif ($G^{m6}ATC$) on both strands of the DNA. These $G^{m6}ATCs$ are cleaved with methylation-sensitive restriction enzyme DpnI, leaving blunt ends to which a universal adapter is ligated. Polymerase chain reaction (PCR) primers hybridize to this adapter in order to specifically amplify methylated genomic fragments. Expanding on the previously published protocol, we include here a multiplexing strategy to pool hundreds of cells in one sequencing library, by adding a cell-specific barcode to the PCR primer.

This chapter describes the single-cell DamID workflow from plasmid choice to sample preparation and data processing. Population-based DamID has been described extensively with regards to the Dam-POI [5, 18] and computational analysis [19]; therefore, we devote particular attention to the novelties of the single-cell approach. We reflect on all aspects of the experimental setup and execution in Subheading 4.

2 Materials

2.1 Hardware

1. Benchtop centrifuge with tube and plate rotors.
2. Nucleic acid spectrophotometer such as NanoDrop (Thermo Scientific).
3. Conventional gel electrophoresis equipment.
4. Real-time thermal cycler with 96-well plate format.
5. Thermal cycler with 96-well plate format.
6. Fluorometer such as Qubit (Invitrogen).
7. Automated electrophoresis system such as Bioanalyzer or TapeStation (Agilent).
8. (Access to a facility providing) fluorescence-activated cell sorter.
9. (Access to a facility providing) Illumina sequencer.
10. Optional: UV PCR workstation.
11. Optional: liquid-handling robot such as Nanodrop II (BioNex).

2.2 Plasmids

See the van Steensel lab website for elaborate explanations on plasmids for DamID in mammalian cells (http://research.nki.nl/vansteensellab/Mammalian_plasmids.htm). For conventional transfection, we have used the pPTuner IRES2 plasmid (Clontech PT4040-5) in which to clone Dam-POI (*see* ref. 20).

For lentiviral transduction, we have used the pCCL.sin.cPPT.hPGK.ΔLNGFR.WPRE lentiviral plasmid [21] in which to clone Dam-POI (*see* ref. 17).

2.3 Cell Culture

1. Cell culture dishes and plasticware.
2. Cell type-specific culture medium.
3. Transfection or transduction reagents.
4. Antibiotics for selection after transfection or transduction.
5. Dissociation reagent.
6. FACS tubes (strainer cap is recommended).
7. Cell-permeable fluorescent DNA stain to evaluate DNA content in live cells during FACS, such as Hoechst.
8. Cell-impermeable fluorescent DNA stain to discriminate live and apoptotic cells during FACS, such as DAPI or PI.
9. Optional: chemical induction agent such as Shield1 or indole-3-acetic acid (IAA; auxin-class hormone).

2.4 MboI-qPCR Assay

1. qPCR 96-well plates.
2. RT plate sealers.
3. Genomic DNA isolation kit or equivalent separate reagents.

Table 1
Primer sequences for MboI-qPCR

Name	Sequence
LAD1_for	CATTGGCTTCTTTGGTGCCAGGT
LAD1_rev	ACGGTGGAGGCAGTCAAAAGGC
LAD2_for	ACAGCAGGAAGTACTTGAGATCC
LAD2_rev	ATTAATCTGGCCCGGAGAGT
LAD3_for	AGCTTATATCAAATAAATCCCTGAAA
LAD3_rev	TGTGCATGACAAATATAAAACCAA
iLAD1_for	GAAGGTTCCCCCACAGAAAT
iLAD1_rev	CTGAGGCAAAGACAGGGAAG
iLAD2_for	ACAGCAGGAAGTACTTGAGATCC
iLAD2_rev	ATTAATCTGGCCCGGAGAGT
UBE2B_for	ACTCAGGGGTGGATTGTTGA
UBE2B_rev	GCCAGAGATTTCAGGGAAAG

4. 1% agarose gel including DNA stain.
5. Gel loading dye.
6. 1 Kb+ DNA ladder.
7. MboI enzyme plus buffer.
8. 10 μM primers flanking GATCs (*see* Table 1 for sequences).
9. Real-time PCR mix including dye.

2.5 Single-Cell DamID and Next-Generation Sequencing

1. PCR 96-well plates.
2. Lysis buffer: 10 mM Tris acetate pH 7.5, 10 mM magnesium acetate, 50 mM potassium acetate, 0.67% Tween-20, 0.67% IGEPAL CA-630, freshly added 0.67 mg/mL Proteinase K.
3. DpnI enzyme plus buffer.
4. T4 DNA ligase plus buffer.
5. 50 μM DamID double-stranded adapter. Dissolve Adapter_top and Adapter_bottom to 100 μM in annealing buffer and then mix equal volumes of both oligonucleotides in a tightly closed tube. Place tube in a container with water of ~94 °C and let cool to room temperature to allow slow annealing of adapters.

   Adapter_top

   5′ CTAATACGACTCACTATAGGGCAGCGTGGTCG CGGCCGAGGA 3′

   Adapter_bottom

   5′ TCCTCGGCCGCG 3′
6. Annealing buffer: 100 mM potassium acetate, 30 mM HEPES, pH 7.5.

7. 25 μM barcoded primers.
   5′ NNNNNNBARCODGTGGTCGCGGCCGAGGATC 3′
8. PCR mix (recommended: including gel loading dye).
9. SPRI beads or spin column purification kit.
10. Illumina library preparation kit or reagents.
11. Qubit DNA HS reagents and tubes.

3 Methods

A proper DamID experimental design includes Dam-LMNB1 as well as untethered Dam. In scDamID, clonal cell lines have to be established (for both constructs separately) to avoid intercellular variation due to differences in, e.g., protein expression level and inducibility. Initial experimental steps are performed with multiple clones per construct to select for clones with desired signal-to-noise methylation levels.

3.1 Choosing the DamID Plasmid

Dependent on the preferred mode of delivery for your cells, select a vector for the Dam constructs. Including an antibiotic resistance cassette to select for successful plasmid integration is recommended. For temporal regulation of protein expression, consider including a degron system or inducible promoter. We routinely use the ProteoTuner system (degron-tagged Dam construct is degraded by default and stabilized upon addition of Shield1) or the auxin-inducible degron (AID; degron-tagged Dam construct is stable by default and degraded upon addition of auxin).

3.2 Creating a Clonal Cell Line that Stably Expresses the Dam-Fusion Constructs

1. Introduce the Dam plasmids into cells by your method of choice. We have derived cell lines by conventional liposome transfection and lentiviral transduction.
2. Select successfully transfected cells by antibiotic resistance and/or fluorescence-activated cell sorting (FACS; if the vector includes expression of a fluorescent protein).
3. Once all cells in the culture population are resistant against the antibiotic or contain another selectable marker, establish clones by FACS or limiting dilution followed by colony picking. *See* **Note 1**.
4. Testing clones by MboI-qPCR assay. This is an optional, though recommended, step to evaluate levels of adenine methylation in the different clones before proceeding to scDamID. Purified gDNA is digested with restriction enzyme MboI, which specifically cuts unmethylated (but not hemimethylated or fully methylated) GATC sequences. Quantitative PCR (qPCR) is performed on digested and undigested DNA, with primers flanking GATC sequences in lamina-associated

domains (LADs) and inter-LADs (iLADs). Percentage of methylation is calculated by $1/2^{(Ct\ (digested)\ -\ Ct\ (undigested))} \times 100\%$. Comparing the percentages of methylation in LADs and iLADs gives an estimation of NL-specific Dam methylation. Typically, the Dam-LMNB1 clones with highest LAD/iLAD ratios are selected for subsequent experiments.

5. Grow cells under appropriate culture conditions. Dependent on Dam fusion, add or remove chemical induction agent 12 h before cell collection. *See* **Note 2**.
6. Isolate genomic DNA from cells using the Wizard Genomic DNA Purification Kit or another method of choice.
7. Measure concentration of purified gDNA using a spectrophotometer.
8. Check integrity of gDNA by running ~500 ng of DNA on a 1% agarose gel. Include a 1 Kb+ DNA ladder. Intact gDNA runs as a single, tight band larger than 10 Kb in size.
9. For each clone, prepare two digestion reactions with 1 μg gDNA each: one containing 5 units (U) MboI and one undigested control.
10. Incubate reactions at 37 °C for 4 h to digest unmethylated DNA.
11. Incubate reactions at 65 °C for 20 min to heat inactivate MboI. *See* **Note 3**. Dilute digestion reactions 1/5 with nuclease-free water (NFW).
12. Assemble qPCR reactions on ice, protected from light as much as possible. Each reaction of 10-μL final volume contains 4 μL of DNA, 0.5 μL of 10 μM forward primer, 0.5 μL of 10 μM reverse primer, and 5 μL of 2× qPCR mix (including PCR buffer, polymerase, and dye). We recommend performing each reaction in triplicates. *See* Table 1 for primer sequences. Select (at least) two primer sets per region, i.e., two for LADs and two for iLADs. Do include reactions with positive control primers that have performed robustly in your hands, as well as negative control reactions without DNA template.
13. Run the assembled reactions in a qPCR thermocycler using the program described in Table 2.
14. Calculate the percentage of methylation per primer set as follows. Take the average of replicate Ct values. Calculate the differences between digested and undigested samples (ΔCt). Then calculate $1\ /\ (2^{\Delta Ct}) \times 100\%$ for the percentage of methylation.
15. Compare the methylation levels in LADs versus iLADs. A good Dam-LMNB1 clone typically has a high (>3) LAD/iLAD ratio, in addition to a high (>40) LAD methylation

Table 2
Thermal cycling program for MboI-qPCR

Cycle	Denature	Anneal	Extend
1	95 °C for 3 min		
2–40	95 °C for 5 s	60 °C for 10 s	72 °C for 10 s. Acquire at end of step.
Optional	Melt profile analysis		

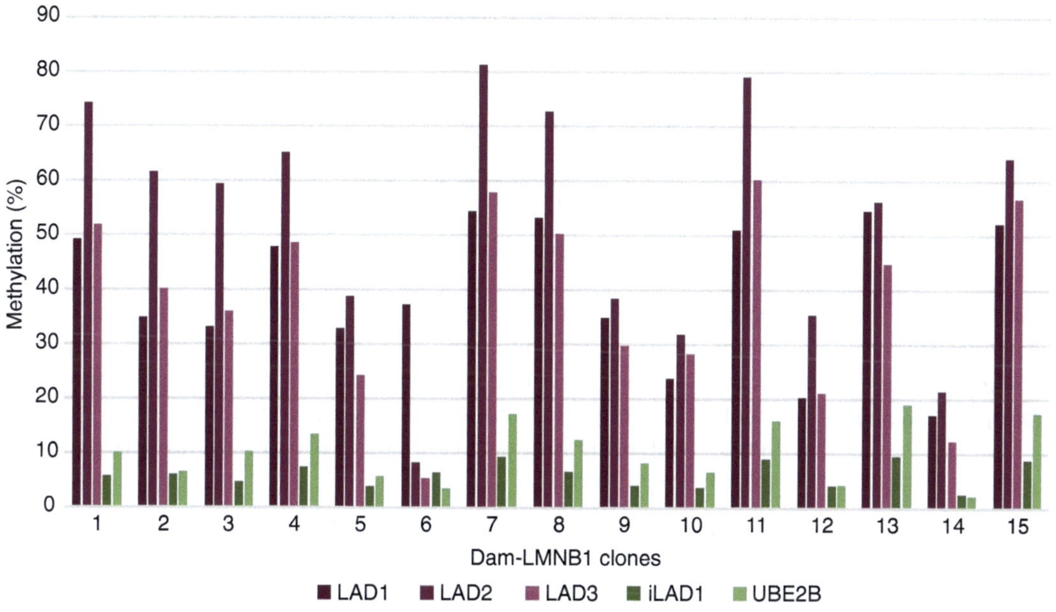

Fig. 1 Methylation levels across 15 Dam-LMNB1 clones, assayed by MboI-qPCR. Primers flank GATC sequences in LADs (purple) and iLADs (green). UBE2B primers flank a GATC in the *UBE2B* gene promoter

percentage. A good Dam clone typically has a low (<1) LAD/iLAD ratio and a low (<20) LAD methylation percentage. *See* Fig. 1 for an example MboI-qPCR result.

16. Proceed to scDamID with selected clones, (at least) two per construct. *See* **Note 4**.

3.3 Obtaining Single Cells for DamID

scDamID involves depositing single cells into reaction reservoirs (typically tubes, strips, or plates of 200-μL volume capacity), followed by additive reactions without cleaning the sample in between, keeping material loss at a minimum.

All subsequent steps are performed at room temperature, unless otherwise specified. If available, using a multistep pipette or liquid-handling robot will highly decrease hands-on time and increase throughput. A decontaminated working environment, such as a UV PCR workstation, is advised.

1. Grow cells under appropriate culture conditions. Dependent on Dam-fusion construct, add or remove chemical induction agent 12 h before cell collection.
2. Prior to cell collection, prepare an appropriate amount of lysis buffer including freshly added Proteinase K and keep on ice until dispensation. Per sample well, 3 μL of lysis buffer is needed.
3. Dispense 3 μL of lysis buffer per well in an appropriate number of 96-well plates. Keep plates on ice until cell collection.
4. If needed, make a single-cell suspension of the target sample (e.g., by trypsin dissociation).
5. Collect live single cells in the prepared 96-well plates by FACS. *See* **Notes 5–9**.
6. As soon as collection per plate is complete, seal plate and centrifuge at 1000 × *g* for 1 min.
7. Continue with Subheading 3.4 or freeze plates at −20 °C.

3.4 DamID: Amplification of Dam-Marked Genomic Fragments

Genomic DNA that has been methylated at GATC motifs is specifically digested, leaving blunt ends to which a universal adapter is ligated. Using barcoded primers that hybridize to this adapter, the methylated fragments are specifically enriched for by PCR. To avoid cross-contamination, take care not to touch samples between wells or cause spillovers.

1. Incubate plates at 42 °C for 4 h to lyse cells and digest all proteins.
2. Incubate plates at 80 °C for 20 min to heat inactivate Proteinase K.
3. Assemble DpnI digestion mix and add 7 μL of mix to each well.

   DpnI digestion mix per well: 6.9 μL of 1× buffer and 0.1 μL of DpnI enzyme.
4. Incubate plates at 37 °C for 8 h to digest methylated DNA.
5. Incubate plates at 80 °C for 20 min to heat inactivate DpnI and then cool on ice.
6. Assemble adapter ligation mix and add 10 μL of mix to each well.

   Adapter ligation mix per well: 9.55 μL of 1× T4 ligase buffer, 0.2 μL of 50 μM adapter, and 0.25 μL of T4 ligase.
7. Incubate plates at 16 °C for 12–16 h.
8. Incubate plates at 65 °C for 10 min to heat inactivate T4 ligase and then cool on ice. *See* **Note 10**.
9. Add 2 μL of 25 μM cell-specific, barcoded primer to each well. Take care that each barcoded primer corresponds to a single sample. *See* **Notes 11** and **12**.

Table 3
Thermal cycling program for DamID PCR

Cycle	Denature	Anneal	Extend
Heated lid on			
1			68 °C for 10 min
2	94 °C for 3 min	65 °C for 5 min	68 °C for 15 min
3–6	94 °C for 1 min	65 °C for 1 min	68 °C for 10 min
7–32	94 °C for 1 min	65 °C for 1 min	68 °C for 2 min

10. On ice, assemble PCR mix and add 28 μL of mix to each well.
    PCR mix per well: 28 μL PCR buffer including polymerase (e.g., 10 μL of 5× PCR buffer, 0.5 μL of polymerase, and 17.5 μL of NFW).
11. Run the assembled reactions in a thermocycler using the program described in Table 3. When testing an uncharacterized cell line, evaluate the PCR product by agarose gel electrophoresis after 20 cycles, and adjust cycles as necessary. *See* **Note 13**.
12. Run 4 μL of the PCR product on 1% agarose gel to check control samples, smear intensity, and distribution of fragment size. Include a 1 Kb+ DNA ladder. Run more or fewer cycles of PCR if necessary. *See* Fig. 2 for an example DamID PCR result.

3.5 Preparation of Illumina Sequencing Libraries

The amplified product is multiplexed by pooling together all samples with different barcodes. The pools are subsequently cleaned, concentrated, and further processed into libraries for deep sequencing. We have used proprietary kits as well as separate reagents to perform library preparation (end repair and phosphorylation, dA-tailing, adapter ligation, and PCR).

1. Evaluate PCR product and estimate the relative concentration of the different samples.
2. Pool barcoded samples together according to their estimated concentration; the aim is to generate a mixture with equal numbers of molecules across the samples. *See* **Notes 11** and **12**.
3. Purify the multiplexed material with SPRI beads (2:1 volume beads/samples) or spin columns and elute in an appropriate volume of resuspension buffer; 30 μL of MQ is usual practice in our lab.
4. Prepare libraries for next-generation sequencing with method of choice, preferably starting with 250–500 ng DNA.
5. Measure the concentration of each library with a Qubit fluorometer, per manufacturer's instructions.

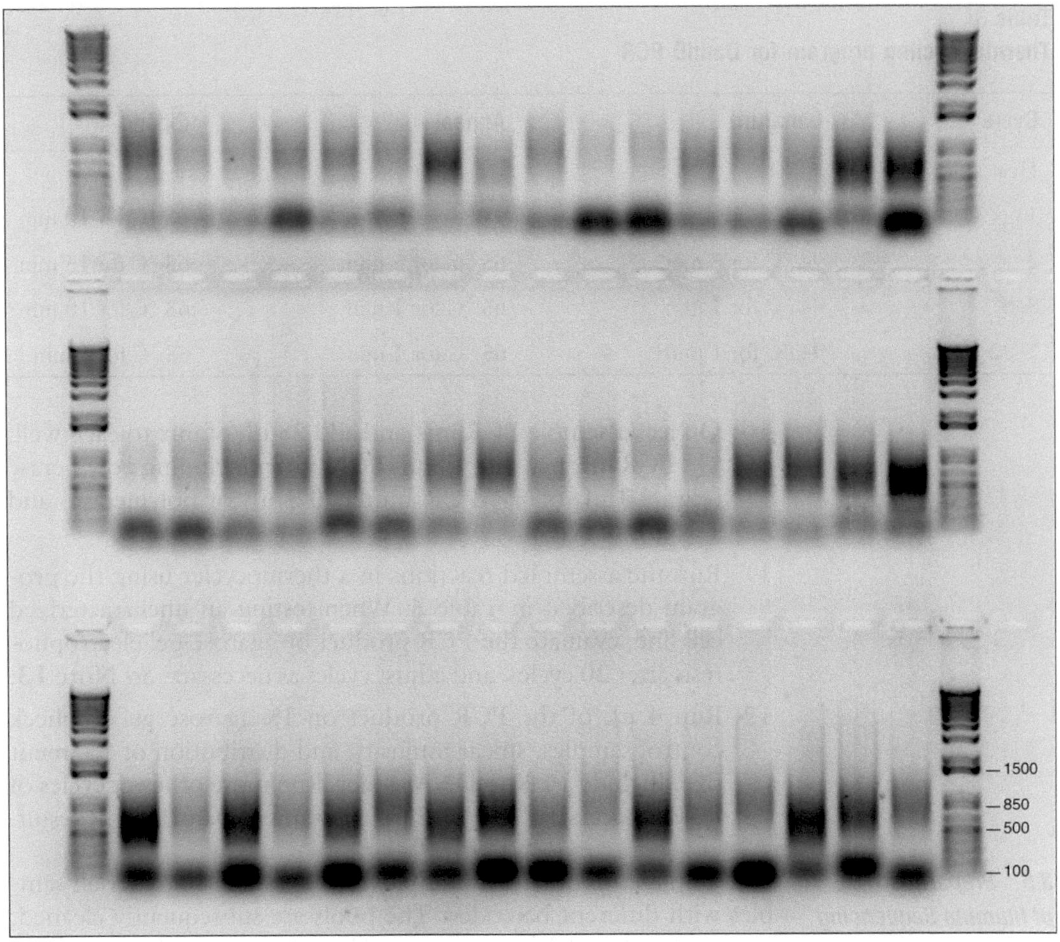

Fig. 2 DamID PCR product of 48 single cells (with different barcoded primers), loaded on 1% agarose gel

6. Evaluate the fragment distribution of each library with an automated electrophoresis system such as Agilent Bioanalyzer or TapeStation.
7. For Dam-LMNB1 in mammalian cells, sequence single end to a depth of approximately 500 K raw reads per single-cell sample.

3.6 Raw Data Processing and Visualization

Raw reads are demultiplexed by their library-specific index and by their sample-specific DamID barcode, after which the DamID primer sequence is removed and sequences are aligned to the reference genome. Reads per GATC are counted, summed across sequencing lanes, aggregated in genomic segments, and optionally smoothened for visualization.

4 Notes

1. We recommend seeding an initial number of ~200 clones, as survival rate may be poor. Depending on the cell line, we have experienced survival anywhere between 5–90%. Subsequent testing of clones by MboI-qPCR and conventional DamID has resulted in 10–60% of clones screened as appropriate for scDamID.
2. Depending on the construct and degron or induction system, timing of the protein stabilization may need to be optimized. We have good experiences with inductions between 4 and 24 h.
3. Optional: check MboI digestion by running an aliquot of the digestion reaction on a 1% agarose gel.
4. After having selected clones, we recommend performing conventional population-based DamID followed by next-generation sequencing. This information serves as a comparison between bulk DamID and other methods such as ChIP-seq or Hi-C, if available, but most importantly as a comparison for the scDamID data.
5. Collecting cells by FACS is by far the most straightforward and high-throughput method. If FACS is unavailable, pipetting by mouth with a microcapillary may be an option.
6. If sorting live cells, it is recommended to exclude apoptotic cells by briefly incubating cell suspension with a DNA dye (e.g., DAPI or PI, depending on the cytometer's laser setup).
7. Dam methylation accumulates in G1 and G2. During DNA replication, the mark remains on only one DNA strand (hemimethylation). We recommend collecting cells at the G1/S or G2/M transition.
8. To sort on the cell cycle phase, incubate cells with Hoechst, a cell membrane-permeable DNA dye. Live incubation of the cell suspension with 1 μg/mL Hoechst at 37 °C for 30 min before sorting has generated acceptable DNA content profiles in our hands. Alternatively, the FUCCI reporter can be used to discriminate cell cycle phases.
9. During cell collection, it is recommended to include multiple low-cell samples (between 10 and 100 cells) and empty wells (0 cells).
10. In conventional DamID, a digestion step with MboI is included to destroy and thereby avoid PCR amplification of fragments with unmethylated GATCs. We do not include this MboI digestion in scDamID, but it is not advised against per se.

11. We recommend pooling multiple experimental conditions in the same library to avoid batch effects (or, at the very least, enable batch correction). Therefore, if the number of samples exceeds the number of available barcodes, take care to add barcodes to your samples such that multiple conditions can be pooled together. While one library per condition does facilitate future resequencing of particular samples, it is best practice not to pool in that manner until after you have established potential differences between experimental conditions.
12. We recommend pooling and processing 100-cell (or higher) samples separately from the single-cell samples.
13. The number of DamID PCR cycles depends on various aspects, most significantly the expression level of Dam-POI in vivo. We recommend starting with 20 cycles. An ultimately time-saving but somewhat laborious approach is to pause the PCR program (at 4 °C) after 20 cycles, remove 4 μL from each PCR reaction, and store cold in, e.g., a 384-well plate, and repeat this process after every subsequent two cycles. Loading the successive material on 1% agarose gel should give a good indication of when enough product has been generated. Keep in mind that, when pooling many samples in one sequencing run, little material is required per sample. Therefore, avoid overamplification since it will generate many PCR duplicates and starkly reduce complexity of the sample. *See* Fig. 3 for an example gel of DamID PCR products after successive cycling.

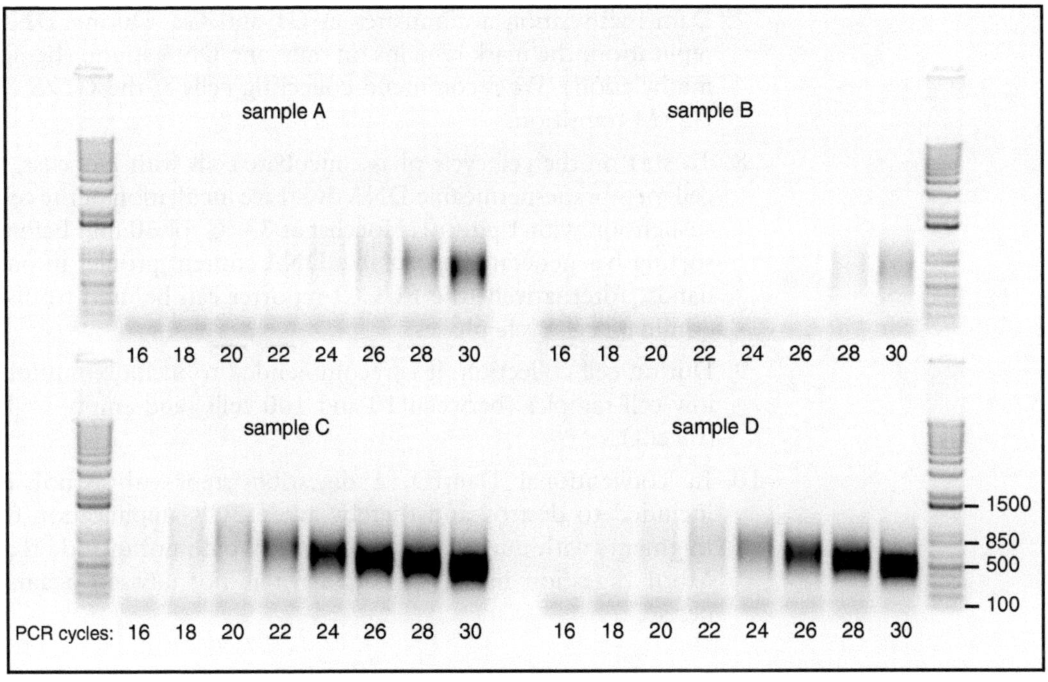

Fig. 3 DamID PCR products of four different samples after successive cycling, loaded on 1% agarose gel

References

1. Bickmore WA, van Steensel B (2013) Genome architecture: domain organization of interphase chromosomes. Cell 152(6):1270–1284. https://doi.org/10.1016/j.cell.2013.02.001
2. Zuleger N, Robson MI, Schirmer EC (2011) The nuclear envelope as a chromatin organizer. Nucleus 2(5):339–349. https://doi.org/10.4161/nucl.2.5.17846
3. Johnson DS, Mortazavi A, Myers RM, Wold B (2007) Genome-wide mapping of in vivo protein-DNA interactions. Science 316 (5830):1497–1502. https://doi.org/10.1126/science.1141319
4. van Steensel B, Henikoff S (2000) Identification of in vivo DNA targets of chromatin proteins using tethered dam methyltransferase. Nat Biotechnol 18(4):424–428. https://doi.org/10.1038/74487
5. Vogel MJ, Peric-Hupkes D, van Steensel B (2007) Detection of in vivo protein-DNA interactions using DamID in mammalian cells. Nat Protoc 2(6):1467–1478. https://doi.org/10.1038/nprot.2007.148
6. Buenrostro JD, Giresi PG, Zaba LC, Chang HY, Greenleaf WJ (2013) Transposition of native chromatin for fast and sensitive epigenomic profiling of open chromatin, DNA-binding proteins and nucleosome position. Nat Methods 10(12):1213–1218. https://doi.org/10.1038/nmeth.2688
7. Dekker J, Rippe K, Dekker M, Kleckner N (2002) Capturing chromosome conformation. Science 295(5558):1306–1311. https://doi.org/10.1126/science.1067799
8. Simonis M, Klous P, Splinter E, Moshkin Y, Willemsen R, de Wit E, van Steensel B, de Laat W (2006) Nuclear organization of active and inactive chromatin domains uncovered by chromosome conformation capture-on-chip (4C). Nat Genet 38(11):1348–1354. https://doi.org/10.1038/ng1896
9. Dostie J, Richmond TA, Arnaout RA, Selzer RR, Lee WL, Honan TA, Rubio ED, Krumm A, Lamb J, Nusbaum C, Green RD, Dekker J (2006) Chromosome Conformation Capture Carbon Copy (5C): a massively parallel solution for mapping interactions between genomic elements. Genome Res 16 (10):1299–1309. https://doi.org/10.1101/gr.5571506
10. Lieberman-Aiden E, van Berkum NL, Williams L, Imakaev M, Ragoczy T, Telling A, Amit I, Lajoie BR, Sabo PJ, Dorschner MO, Sandstrom R, Bernstein B, Bender MA, Groudine M, Gnirke A, Stamatoyannopoulos J, Mirny LA, Lander ES, Dekker J (2009) Comprehensive mapping of long-range interactions reveals folding principles of the human genome. Science 326 (5950):289–293. https://doi.org/10.1126/science.1181369
11. Fullwood MJ, Liu MH, Pan YF, Liu J, Xu H, Mohamed YB, Orlov YL, Velkov S, Ho A, Mei PH, Chew EG, Huang PY, Welboren WJ, Han Y, Ooi HS, Ariyaratne PN, Vega VB, Luo Y, Tan PY, Choy PY, Wansa KD, Zhao B, Lim KS, Leow SC, Yow JS, Joseph R, Li H, Desai KV, Thomsen JS, Lee YK, Karuturi RK, Herve T, Bourque G, Stunnenberg HG, Ruan X, Cacheux-Rataboul V, Sung WK, Liu ET, Wei CL, Cheung E, Ruan Y (2009) An oestrogen-receptor-alpha-bound human chromatin interactome. Nature 462(7269):58–64. https://doi.org/10.1038/nature08497
12. Nagano T, Lubling Y, Stevens TJ, Schoenfelder S, Yaffe E, Dean W, Laue ED, Tanay A, Fraser P (2013) Single-cell Hi-C reveals cell-to-cell variability in chromosome structure. Nature 502(7469):59–64. https://doi.org/10.1038/nature12593
13. Buenrostro JD, Wu B, Litzenburger UM, Ruff D, Gonzales ML, Snyder MP, Chang HY, Greenleaf WJ (2015) Single-cell chromatin accessibility reveals principles of regulatory variation. Nature 523(7561):486–490. https://doi.org/10.1038/nature14590
14. Rotem A, Ram O, Shoresh N, Sperling RA, Goren A, Weitz DA, Bernstein BE (2015) Single-cell ChIP-seq reveals cell subpopulations defined by chromatin state. Nat Biotechnol 33(11):1165–1172. https://doi.org/10.1038/nbt.3383
15. Beagrie RA, Scialdone A, Schueler M, Kraemer DC, Chotalia M, Xie SQ, Barbieri M, de Santiago I, Lavitas LM, Branco MR, Fraser J, Dostie J, Game L, Dillon N, Edwards PA, Nicodemi M, Pombo A (2017) Complex multi-enhancer contacts captured by genome architecture mapping. Nature 543 (7646):519–524. https://doi.org/10.1038/nature21411
16. Bintu B, Mateo LJ, Su JH, Sinnott-Armstrong NA, Parker M, Kinrot S, Yamaya K, Boettiger AN, Zhuang X (2018) Super-resolution chromatin tracing reveals domains and cooperative interactions in single cells. Science 362(6413). https://doi.org/10.1126/science.aau1783
17. Kind J, Pagie L, de Vries SS, Nahidiazar L, Dey SS, Bienko M, Zhan Y, Lajoie B, de Graaf CA,

Amendola M, Fudenberg G, Imakaev M, Mirny LA, Jalink K, Dekker J, van Oudenaarden A, van Steensel B (2015) Genome-wide maps of nuclear lamina interactions in single human cells. Cell 163 (1):134–147. https://doi.org/10.1016/j.cell.2015.08.040

18. Robson MI, Schirmer EC (2016) The application of DamID to identify peripheral gene sequences in differentiated and primary cells. Methods Mol Biol 1411:359–386. https://doi.org/10.1007/978-1-4939-3530-7_23
19. Gomez-Saldivar G, Meister P, Askjaer P (2016) DamID analysis of nuclear organization in Caenorhabditis elegans. Methods Mol Biol 1411:341–358. https://doi.org/10.1007/978-1-4939-3530-7_22
20. Kind J, Pagie L, Ortabozkoyun H, Boyle S, de Vries SS, Janssen H, Amendola M, Nolen LD, Bickmore WA, van Steensel B (2013) Single-cell dynamics of genome-nuclear lamina interactions. Cell 153(1):178–192. https://doi.org/10.1016/j.cell.2013.02.028
21. Amendola M, Venneri MA, Biffi A, Vigna E, Naldini L (2005) Coordinate dual-gene transgenesis by lentiviral vectors carrying synthetic bidirectional promoters. Nat Biotechnol 23 (1):108–116. https://doi.org/10.1038/nbt1049

Chapter 10

Formaldehyde-Mediated Snapshot of Nuclear Architecture

Federica Lucini, Andrea Bianchi, and Chiara Lanzuolo

Abstract

Genome architecture and function are strictly related to nuclear structures, which contact chromatin at specific regions, regulating its compaction and three-dimensional higher-order structure, therefore contributing to specialized gene expression programs. Recently, growing evidence uncovers a dynamic role of nuclear structures in the plasticity of transcriptional programs. When the cellular microenvironment changes, external cues are transmitted to the nucleus through complex signalling cascades, finally resulting in a genome reorganization that allows the adjustment of the cell to a new condition. This process can be very rapid, especially in cells whose function is to contain sudden threats to the organism. Some examples are stem cells that switch from a quiescent to an activated state to replace damaged tissues or immune cells that, with a similar dynamic, identify and eliminate pathogens.

Experimental treatments often require the isolation of cells from their physiological environment, exposing them to possible sudden changes in their nuclear architecture. Here we propose an early cross-linking on primary cells, a fixing method that can help to minimize the risk of nuclear structure alteration during the isolation process. We also bring some examples of downstream studies on early-fixed cells.

Key words Nuclear architecture, Chromatin conformation, Cross-linking, Chromatin immunoprecipitation, FISH

1 Introduction

Our body is made up of tens of trillions of cells. Although all their nuclei contain the same genome, each of them can conduct different functions by using only a small part of it [1, 2]. In the last years, increasing technological advances have enabled us to finely characterize and classify hundreds of distinct cell types, confirming that the genetic information is much more than just a linear sequence of nucleotides [3, 4]. DNA-protein and DNA-RNA interplay, together with chromatin three-dimensional organization, are essential determinants for the identity and function of a cell in every moment of its life [5–10]. Responsible for the plasticity of genomic structure and function is the epigenome: an ensemble of countless mechanisms constantly at work in our cells to provide a

Beatrice Bodega and Chiara Lanzuolo (eds.), *Capturing Chromosome Conformation: Methods and Protocols*,
Methods in Molecular Biology, vol. 2157, https://doi.org/10.1007/978-1-0716-0664-3_10,

multiple layered control on the genetic information [11]. On the one hand, epigenetic mechanisms are responsible for the organized compaction of the linear genomic sequence into the small nuclear volume, preventing its damage and reducing errors during replication and cell division [12, 13]. On the other hand, they drive genome function by regulating gene expression [14]. Both chromatin conformation and its highly organized spatial positioning inside the nucleus are principal determinants of genome function [8]. They are established thanks to the interaction of distinct portions of chromatin with specific nuclear structures, which serve as compartmentalizing agents for the segregation of activities inside the nucleus [15–17], separating compact, transcriptionally inactive microenvironments from open and transcriptionally active ones [18]. The best known repressive microenvironment is certainly the nuclear lamina (NL): a dense fibrillary network underlying the inner nuclear membrane and interacting with nearly 40% of the genome [19]. Principal components of the NL are lamin proteins, which directly contact chromatin at regions called lamina-associated domains (LADs), characterized by low gene density, low expression levels, and the H3K9me2/3 signature [19]. Genomic regions positioned in close proximity to the nucleolus are also associated with low gene density and low transcriptional levels. They are called nucleolus-associated domains (NADs) and they encompass centromeric and pericentromeric satellite repeats and subtelomeric regions [20]. Deep sequencing data revealed a partial overlap between NADs and LADs, suggesting a possible turnover of silent regions between the nucleolus and the nuclear periphery [21]. Interestingly, recent findings also suggested that lamin proteins can directly crosstalk with epigenetic regulators to ensure the correct transcriptional regulation at the gene level [22–27]. In line with these findings, alteration of NL integrity is often associated with loss of LAD conformation, heterochromatin protein dispersion, and aberrant gene expression [28–30]. On the other hand, an example of active chromatin environment is represented by the nuclear pore complexes (NPCs) [31]: multiprotein channels that connect the nucleo- to the cytoplasm. Besides mediating molecular exchange, these structures have been shown to dynamically interact with the genome at active sites and to regulated multiple cellular processes [32]. For instance, NPCs have been demonstrated to have a role in the maintenance of gene loop interactions between the promoter and the end of some genes in yeast, allowing a sort of "memory" of recent transcription which is useful for rapid reinduction upon environmental condition changes [33].

Since cells need to constantly optimize their transcriptional state according to the stimuli they receive from the environment, genome architecture needs to be plastic. Epigenetic regulators must be sensitive and responsive to external cues, remodeling chromatin to achieve the expression of an optimal set of genes in a short time [34]. In this fundamental biological communication, lamins and nuclear envelope transmembrane proteins (NETs) have been recognized as key transductor elements [35].

In some cells, such as those belonging to the stem cell niches, the sensitivity to environmental changes and the production of a fast adaptative response is vital for the survival of the organism. Muscle stem cells (MuSCs), for instance, are predominantly quiescent in homeostatic conditions but they need to rapidly activate upon injury to reconstitute the damaged muscle [36]. In principle, these sudden changes in their cellular state could influence downstream molecular biological studies. In fact, recent reports showed that standard MuSC extraction protocols determine the activation of a MuSC subpopulation in a very short time [37]. Single-cell RNA-sequencing data revealed that after dissociation and FACS isolation, this subpopulation expresses high levels of immediate early genes (IEGs)—including Fos, Jun, Socs3, and heat shock proteins—which are not detected in quiescent MuSCs in vivo by single-molecule RNA fluorescence in situ hybridization (sm-FISH). In addition, the transcriptomic profile of ex vivo MuSCs features some genes commonly induced in vivo after muscle injury [37, 38]. More studies, using different approaches, converge to this view, underlining how in vivo MuSCs compared to ex vivo MuSCs overexpress the genes involved in RNA metabolism, fructose metabolism, and quiescence [39]. Even though the altered ex vivo MuSC subpopulation can be initially relatively small, its proportion increases with the time of dissociation and can generate, as what happens for all contaminants, divergent or confusing results. These changes in MuSC transcriptome profile could be dependent on mechanical stress that, as it has also been demonstrated for other stem cells, can directly trigger epigenetic modifications [40, 41]. Altogether, these studies suggest that MuSC extraction protocol could have an impact on chromatin architecture, sustaining the expression of early genes involved in cell activation [39, 41].

Another example of plastic, highly environment-sensitive cells is represented by the immune system. Leukocytes continuously patrol the organism, navigating blood and lymphatic vessels and homing to tissues. When they meet specific pathogen-associated signals, they rapidly activate to generate the most suitable response to eliminate the threat [42]. The activation of resting T lymphocytes, for example, involves engagement of the T cell receptor (TCR) by antigen-presenting cells (APCs). Following this contact, the expression of lamin A in the lymphocyte rapidly increases, allowing a more efficient immunological synapse between the T cell and the APC and promoting, through interaction with the linker of the nucleoskeleton and cytoskeleton (LINC) complex, the polymerization of F-actin, critical for functional immunological synapse formation and optimal cell activation [43, 44]. Unfortunately, the extreme susceptibility to external stimuli hampers the

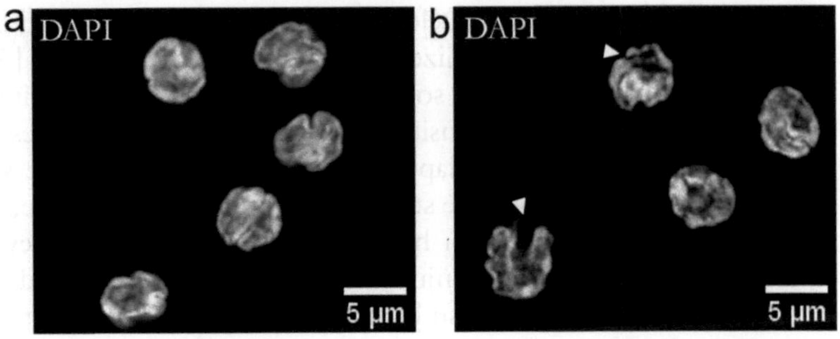

Fig. 1 Confocal microscope images of T lymphocyte nuclei fixed before (**a**) or after (**b**) sorting, counterstained with DAPI. Nuclear ruptures are indicated by arrowheads

study of these cells in physiological conditions. Density gradient centrifugation, magnetic enrichment, FACS, and every other manipulation implicated in their isolation could potentially modify their nuclear architecture in unpredictable ways. There is therefore a growing need for ways of rapidly blocking epigenetic responses and nuclear structure modifications [45] to study cells in a condition as similar as possible to their physiological state.

Here we propose a method for the early cross-linking of cells as a solution for the preservation of their physiological nuclear architecture from modification induced by electrical and other environmental stressing stimuli (Fig. 1). We further present some downstream analysis performed on these cells.

2 Materials

2.1 Reagents Penicillin-Streptomycin Solution

1. DPBS without calcium and magnesium.
2. HBSS with calcium and magnesium.
3. MACS separation buffer (Miltenyi Biotec).
4. Ficoll-Paque™ Plus.
5. RPMI 1640 with stable L-glutamine.
6. DMEM high glucose without L-glutamine, with sodium pyruvate.
7. Fetal bovine serum (FBS).
8. Bovine serum albumin lyophilized powder (BSA).
9. 96%–100% ethanol.
10. Nuclease-free water.
11. Paraformaldehyde (PFA) powder.
12. 36.5–38% formaldehyde in H_2O.
13. Formamide, pH 7.0.

14. Phenol/chloroform/isoamyl alcohol 25:24:1 solution.
15. 0.5 M EDTA.
16. 0.5 M EGTA.
17. 5 M NaCl.
18. 1 M $MgCl_2$.
19. 50 mM $CaCl_2$.
20. 0.1 M sodium citrate, pH 6.0.
21. 1 M Hepes-KOH, pH 7.5.
22. 2.5 M glycine.
23. 3 M sodium acetate, pH 5.2.
24. 1 M HCl.
25. 1 M Tris-HCl, pH 7.8.
26. 1 M Tris-HCl, pH 8.0.
27. 1 M Tris-HCl pH 8.1.
28. 1 M Tris-HCl, pH 8.6.
29. 2% NaN_3.
30. RNAseA (DNase and protease-free).
31. Dispase II.
32. Collagenase A.
33. DNase I.
34. Proteinase K.
35. 0.1 M phenylmethanesulfonyl fluoride (PMSF).
36. Glycogen.
37. Triton X-100.
38. Sodium dodecyl sulfate (SDS).
39. Agarose powder.
40. TAE 10×.
41. DNA HS kit for Bioanalyzer.
42. Poly-L-lysine hydrobromide.
43. Mounting medium.
44. Kit for depletion of non-$CD4^+$ cells.
45. Anti-CD31 (PECAM-1) monoclonal antibody, eFluor 450.
46. Anti-CD45 monoclonal antibody, eFluor450.
47. Anti-TER-119 monoclonal antibody, eFluor 450.
48. Anti-Ly-6A/E (Sca-1) monoclonal antibody, FITC.
49. Anti-alpha7-integrin monoclonal antibody, APC.
50. Alexa Fluor® 488 Donkey Anti-Goat IgG.

51. Alexa Fluor® 488 Donkey Anti-Rabbit IgG.
52. Alexa Fluor® 647 Donkey Anti-Rabbit IgG.
53. Anti-CD4 monoclonal antibody, VioGreen.
54. Anti- CD25 monoclonal antibody, VioBright FITC.
55. Anti-CD45RO monoclonal antibody, APC.
56. Anti-CD127 monoclonal antibody, PE.
57. Anti-CD183 (CXCR3) monoclonal antibody, PE-Cy™5.
58. Anti-H3K27me3 polyclonal antibody.
59. Anti-H3K4me3 polyclonal antibody.
60. Protein G beads.
61. DNA library preparation kit.
62. AMPure beads.
63. Anti-lamin A/C polyclonal antibody.
64. Anti-Ring1b monoclonal antibody.
65. Fluorescent probe for chromosome 18.
66. CHORI BAC CH29-101F16 for Pparγ locus.
67. BAC extraction kit.
68. dNTPs (C-G-A).
69. dTTP.
70. Cy3 dUTP.
71. β-Mercaptoethanol.
72. DNA polymerase I.
73. Dextran sulfate.
74. Mouse Cot-1 DNA.
75. Unlabelled salmon sperm DNA.
76. Glycerol.
77. Pepsin.
78. Dry ice.
79. DAPI.
80. RNase cocktail (RNase A 500 U/ml and RNase T1 20,000 U/ml).
81. Tween-20.
82. ChIP control primers.

2.2 Solutions

1. Ethanol solutions (v/v): 100% and 80% ethanol in nuclease-free water.
2. $HBSS^+$: HBSS with calcium and magnesium supplemented with 1% Pen/Strep and 0.2% BSA.

3. Muscle digestion solution: 2.4 U/mL Dispase II (resuspended in DPBS), 2 μg/μL collagenase A, 400 μM $CaCl_2$, 5 mM $MgCl_2$, 10 μg/mL DNase I (*see* **Note 1**).
4. Formaldehyde solution: 50 mM Hepes-KOH pH 7.5, 100 mM NaCl, 1 mM EDTA, 0.5 mM EGTA, 11% formaldehyde.
5. Lysis buffer: 50 mM Tris-HCl pH 8.1, 100 mM NaCl, 5 mM EDTA pH 8.0, 0.2% NaN_3, 0.5% SDS.
6. Equilibration buffer: 100 mM Tris-HCl pH 8.6, 100 mM NaCl, 5 mM EDTA pH 8.0, 0.2% NaN_3, 5% Triton X-100.
7. Washing solution high salt: 20 mM Tris-HCl pH 8, 500 mM NaCl, 2 mM EDTA, 1% Triton X-100, 0.1% SDS.
8. Washing solution low salt: 20 mM Tris-HCl pH 8, 150 mM NaCl, 2 mM EDTA, 1% Triton X-100, 0.1% SDS.
9. TE: 10 mM Tris-HCl pH 8, 1 mM EDTA.
10. TE NaCl solution: 10 mM Tris-HCl pH 8, 1 mM EDTA, 50 mM NaCl.
11. Elution solution: 50 mM Tris-HCl pH 8, 10 mM EDTA, 1% SDS.
12. Murine muscle staining mix: anti-CD31 Pacific Blue, anti-CD45 Pacific Blue, anti-CD119 Pacific Blue, anti-SCA FITC, anti- α7 integrin APC in HBSS+ to a final volume of 50 μL per sample.
13. Human lymphocyte staining mix: anti-CD4 VioGreen, anti-CD45RO APC, anti-CD127 PE, anti-CD25 FITC, anti-CXCR3 PE-Cy5.
14. SSC 20× solution: 3 M NaCl, 300 mM sodium citrate, pH 7.
15. BAC probe staining solution: 0.02 mM dNTPs (C-G-A), 0.01 mM dTTP, 0.01 mM Cy3-dUTP, 50 mM Tris-HCl pH 7.8, 5 mM $MgCl_2$, 10 mM β-mercaptoethanol, 10 ng/μL BSA, 0.05 U/μL DNA polymerase I, 0.004–0.007 U/μL DNase I, 1–3 μg BAC DNA, DNase-RNase free water to final volume of 50 μL.

2.3 Equipment and Supplies

1. Tilting shaker.
2. 15- and 50-mL conical tubes.
3. 1.5-mL microcentrifuge tubes.
4. 6-well plates.
5. 24-well plates.
6. Refrigerated swing-bucket rotor centrifuge (20,000 × *g* max.).
7. Refrigerated microcentrifuge (21,130 × *g* max.).
8. Microscope glass slides (26 × 76 mm).

9. 6–12-mm diameter round glass coverslips.
10. Branson Digital Sonifier 250 with 102-C converter, SEE-1 sonifier sound enclosure, and 3-mm diameter tapered microtip.
11. Parafilm.
12. Jewelers forceps, Dumont No. 5.
13. Cold room.
14. Tube rotator.
15. Thermomixer.
16. Nanodrop.
17. 2100 Bioanalyzer System.
18. Sequencing platform.
19. Laser scanning confocal microscope.
20. High numerical aperture (NA) oil-immersion apochromat objectives.
21. High-resolution and high-sensitivity CCD camera.
22. Software for image collection and Z-stack analyses.
23. Biological safety cabinet.
24. Fume hood.
25. 5-mL polypropylene round-bottom tubes for FACS.
26. Sterile cell strainers of 40- and 70 μm mesh size.
27. Fluorescence-activated cell sorting (FACS) system.
28. Silicone.
29. 75 °C heated plate.
30. Humidified chamber.
31. Sterile syringe filters of 0.2 μm pore size.
32. Microbiological incubator (temperature working range 37 to 60 °C).
33. Cell counting chamber.
34. Widefield fluorescence microscope.
35. Ultra low temperature freezer (–80 °C).
36. 1-mL syringe with 0.50 × 16 mm needle.
37. Electrophoresis equipment.
38. Magnetic rack.
39. Thermal cycler and quantitative PCR instrument.

3 Methods

3.1 Human T Lymphocytes

3.1.1 Isolation and Cross-Linking of Cells

1. After collection of Ficoll-Paque-separated human peripheral blood mononuclear cells (PBMCs), resuspend them in sterile DPBS (*see* **Note 2**).
2. Centrifuge at 500 × *g* for 8 min at room temperature and discard the supernatant.
3. Resuspend the pellet in DPBS and count the cells (*see* **Note 3**).
4. Centrifuge at 200 × *g* for 10 min at room temperature and discard the supernatant.
5. Resuspend the pellet using 40 μL of MACS separation buffer per 10^7 cells.
6. Add 10 μL of $CD4^+$ T cell biotin-antibody cocktail per 10^7 cells and incubate for 5 min at 4 °C (biotinylated antibodies bind to all non-$CD4^+$ cells).
7. Add 30 μL of MACS buffer per 10^7 cells.
8. Add 20 μL of $CD4^+$ T cell microbead cocktail per 10^7 cells and incubate for 10 min at 4 °C (microbeads bind to biotinylated antibodies).
9. Proceed to magnetic separation of bead-bound (non-$CD4^+$) and bead-unbound ($CD4^+$) cells using LS columns, either manually or with autoMACS® Pro Separator (*see* **Note 4**).
10. Collect unlabeled cells and count them.
11. Resuspend the cells in RPMI supplemented with 10% FBS (*see* **Note 5**).
12. Under a fume hood, add formaldehyde solution in a 1:10 ratio to the cell suspension (i.e., for 9 mL of cell suspension, add 1 mL of formaldehyde solution).
13. Incubate on a tilting shaker at room temperature for 9 min.
14. Add glycine to a final concentration of 125 mM (*see* **Note 6**).
15. Incubate on a tilting shaker at room temperature for 5 min.
16. Centrifuge the suspension at 1500 × *g* for 15 min at 4 °C and discard the supernatant.
17. Resuspend the pellet in cold DPBS filling the whole tube (*see* **Note 7**).
18. Centrifuge at 1500 × *g* for 15 min at 4 °C and discard the supernatant.
19. Resuspend the pellet in cold MACS buffer and transfer the suspension to a sterile FACS tube (*see* **Note 8**).
20. Incubate at 4 °C until staining.

3.1.2 Fluorescence-Activated Cell Sorting

1. Count the cells.
2. Centrifuge at 1000 × *g* for 10 min at 4 °C and discard the supernatant.
3. Add the human lymphocyte staining mix.
4. Incubate at 37 °C in the dark for 20 min.
5. Wash in 4 mL of MACS buffer and discard the supernatant.
6. Resuspend the cells in about 300 μL of MACS buffer (*see* **Note 9**).
7. Filter the suspension (*see* **Note 10**).
8. Sort the suspension, collecting in RPMI supplemented with 10% FBS:
   (a) T naïve $CD4^{+}CD45RO^{-}$.
   (b) Thl $CD4^{+}CD45RO^{+}CXCR3^{+}$.
   (c) Treg $CD4^{+}CD25^{+}CD127^{-}$.

3.1.3 Immuno-fluorescence (IF) (See **Note 11**)

1. Rinse a 9-mm round glass coverslip in 100% ethanol and let it air-dry.
2. Coat coverslip with poly-L-lysine solution (*see* **Notes 12** and **13**).
3. Count cells obtained from FACS and bring the suspension to the desired concentration (*see* **Note 14**).
4. Deposit a drop of cell suspension on the poly-L-lysine-coated coverslip (*see* **Notes 15** and **16**).
5. Let the drop sit on the coverslip for 30 min at room temperature, then carefully remove the liquid avoiding cell aspiration (*see* **Note 17**).
6. Under a fume hood, incubate the coverslip with 4% (w/v) PFA for 7 min at room temperature (*see* **Notes 18** and **19**).
7. Remove PFA and rapidly wash twice in 0.05% Triton/DPBS.
8. Wash in 0.05% Triton/DPBS for 5 min at room temperature on a tilting shaker—repeat three times.
9. Incubate coverslip in 0.5% Triton/DPBS for 10 min at room temperature.
10. Wash in 0.05% Triton/DPBS for 5 min at room temperature on a tilting shaker—repeat twice.
11. Wash in DPBS for 5 min at room temperature on a tilting shaker.
12. Incubate in 20% glycerol/DPBS at room temperature overnight.
13. Dry the back of the coverslip and put it on dry ice for about 30 s.
14. Let it gradually thaw at room temperature.

15. Soak it in 20% glycerol/DPBS.
16. Repeat **steps 13–15** four times.
17. Wash in 0.5% Triton/PBS for 5 min on a tilting shaker.
18. Wash in 0.05% Triton/PBS for 5 min at room temperature on a tilting shaker—repeat twice.
19. Incubate coverslip in 0.1 M HCl for 15 min at room temperature.
20. Wash in DPBS for 5 min at room temperature on a tilting shaker—repeat three times.
21. Incubate the coverslip in 4% (w/v) BSA at room temperature for 30 min.
22. Dilute primary antibody in 4% BSA and incubate with cells in a humidified chamber overnight at 4 °C (*see* **Note 20**).
23. Wash in 0.05% Triton/DPBS for 5 min at room temperature on a tilting shaker—repeat three times.
24. Dilute secondary antibodies in 1% BSA and incubate with cells in a humidified chamber for 1 h at room temperature in the dark.
25. Wash in 0.05% Triton/DPBS for 5 min at room temperature on a tilting shaker in the dark—repeat three times.
26. Incubate with 0.1% DAPI/DPBS for 5 min at room temperature in the dark.
27. Wash in DPBS for 5 min at room temperature on a tilting shaker in the dark—repeat twice.
28. Mount the coverslip on a microscope slide.
29. Acquire images through a confocal microscopy system (Fig. 2).

3.1.4 Chromosome Painting

1. Follow **steps 1–18** of Subheading 3.1.3.
2. Incubate coverslip in 2.5 μL RNase cocktail/150 μL DPBS for 20 min at 37 °C (*see* **Note 21**).
3. Rinse in DPBS.
4. Incubate coverslip in 0.1 M HCl for 15 min at RT.
5. Wash in PBS for 5 min at room temperature on a tilting shaker—repeat twice.
6. Rinse in 2× SSC.
7. Incubate in 50% formamide pH 7.0/2× SSC overnight at room temperature (*see* **Note 22**).
8. Denature the fluorescent probes for the desired chromosome in a thermomixer at 70 °C, 300 rpm for 5 min.
9. Accurately dry the coverslip (*see* **Note 23**).
10. Put a drop of probe on a microscope slide.

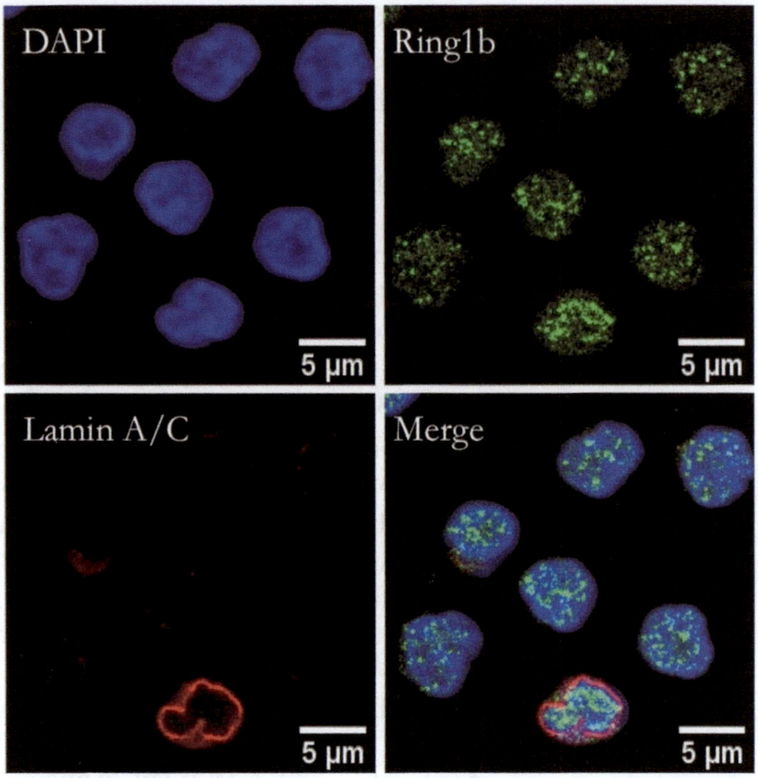

Fig. 2 Confocal microscope images of T lymphocyte nuclei stained for Ring1b subunit of polycomb group (PcG) proteins and lamin A/C as indicated

11. Turn the coverslip with the cells facing down on the drop of probe.
12. Seal the borders of the coverslip with silicone and let it dry in the dark.
13. Put the microscope slide on a heated plate at 75 °C for 4 min (*see* **Note 24**).
14. Transfer the slide to a humidified chamber and incubate at 37 °C overnight in the dark.
15. Remove silicone and detach the coverslip from the microscope slide.
16. Transfer the coverslip to a 6-well plate.
17. Wash in 2× SSC at 37 °C, 90 rpm for 5 min in a shaking incubator in the dark—repeat three times.
18. Wash in 0.1× SSC at 60 °C, 90 rpm for 5 min in a shaking incubator in the dark—repeat three times.
19. Wash in 0.2% Tween/4× SSC at 37 °C, 90 rpm for 5 min in a shaking incubator in the dark—repeat twice.
20. Wash in DPBS at RT for 5 min on a tilting shaker in the dark.

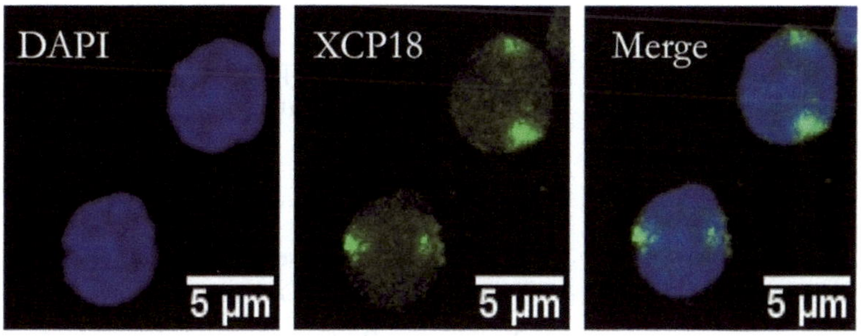

Fig. 3 Maximum intensity projection of a confocal z stack of fixed and sorted T lymphocyte nuclei stained for chromosome 18

21. Incubate with 0.2% DAPI/DPBS 5 min at RT in the dark.
22. Wash in DPBS for 5 min at room temperature on a tilting shaker in the dark—repeat twice.
23. Mount the coverslip on a microscope slide.
24. Acquire images through a confocal microscopy system (Fig. 3).

3.2 Murine Muscle Stem Cells (MuSCs)

3.2.1 Isolation and Cross-Linking of Cells

1. Sacrifice mice, extract muscles, and digest them as described in [46].
2. Filter the digestion suspension with a 70-μm strainer (*see* **Note 25**).
3. Centrifuge at 300 × *g* for 10 min at 4 °C and discard the supernatant.
4. Resuspend the pellet in 10 mL of HBSS$^+$.
5. Filter the solution with a 40-μm strainer (*see* **Note 25**).
6. Centrifuge at 300 × *g* for 10 min at 4 °C and discard the supernatant.
7. Count the cells.
8. Resuspend the cells in DMEM supplemented with 10% FBS (*see* **Note 5**).
9. Follow **steps 12–18** of Subheading 3.1.1.
10. Resuspend the pellet in cold HBSS$^+$ and transfer the suspension to a sterile FACS tube (*see* **Note 8**).
11. Incubate at 4 °C until staining.

3.2.2 Fluorescence-Activated Cell Sorting

1. Centrifuge the samples at 1000 × *g* for 10 min at 4 °C and reduce the supernatant to 300 μL.
2. Add the murine muscle staining mix.
3. Incubate on ice, in the dark for 30 min.
4. Wash in 4 mL of HBSS$^+$ and discard the supernatant.
5. Resuspend the cells in about 300 μL of HBSS$^+$ (*see* **Note 9**).

6. Filter the cells (*see* **Note 10**).
7. Sort the suspension, collecting $CD31^{-}CD45^{-}CD119^{-}$ SCA^{-}alpha7-integrin^{+} cells in DMEM supplemented with 10% FBS.

3.2.3 Fluorescence In Situ Hybridization (FISH)

1. Prepare BAC DNA following the guidelines of your kit (we used PhasePrep™ BAC DNA Kit Sigma-Aldrich), and then quantify DNA with nanodrop.
2. Prepare BAC probe staining solution and incubate at 16 °C for 1–2 h in the dark (*see* **Note 26**).
3. When the length of the fluorescent probe reaches about 50 bp, precipitate the DNA fragments incubating the reaction at −80 °C for 1–2 h in 3 volumes of 100% ethanol and 1/10 volume of 3 M sodium acetate pH 5.2.
4. Centrifuge at 15,000 × *g* at 4 °C for 30 min and discard the supernatant.
5. Wash the pellet with 1 mL of 70% ethanol and centrifuge again at 15,000 × *g* at 4 °C for 5 min.
6. Discard the supernatant and dry the pellet.
7. Resuspend the pellet in 50 μL of DNase-RNase free water and quantify DNA.
8. Mix 100–300 ng of labelled probe with 3.5 ng of mouse Cot-1 DNA and 20 μg of unlabelled salmon sperm DNA.
9. Precipitate as in **steps 3–6**.
10. Resuspend the pellet in 6 μL of 50% formamide/2× SSC/10% dextran sulfate (*see* **Note 27**).
11. Keep at 40 °C, 300 rpm in the dark until hybridization. Alternatively, the probe can be stored at −20 °C in the dark.
12. Prepare a coverslip with cells and treat it as described in **steps 1–7** of Subheading 3.1.4.
13. Equilibrate cells in 2× SSC for 2 min.
14. Incubate coverslip in DPBS for 3 min and then in 0.01 M HCl/0.0025% pepsin in DNase-RNase free water for 1 or 2 min (*see* **Note 28**).
15. Inactivate pepsin by washing twice for 5 min in $MgCl_2$/DPBS in agitation.
16. Wash three times for 2 min in DPBS with gentle shaking.
17. Fix the cells for 1 min in 1% PFA.
18. Wash once for 5 min in DPBS and twice for 5 min in 2× SSC.
19. Incubate the cells again in 50% formamide pH 7.0/2× SSC for at least 30 min.

Fig. 4 Widefield microscope images of FISH experiment on fixed and sorted murine muscle stem cells (MuSCs) with probes spanning Pparγ locus

20. For the hybridization, follow **steps 9–15** of Subheading 3.1.4 using 6 μL of the previously prepared probe (*see* **step 11**).
21. Transfer the coverslip in a 6-well plate containing 5 mL of 37 °C preheated 2× SSC.
22. Wash 3 × 5 min in 6-well plate with 2× SSC preheated at 37 °C shaking at 90 rpm in the dark.
23. Wash 3 × 5 min in 6-well plate with 0.1% SSC preheated at 60 °C shaking at 90 rpm in the dark.
24. Rinse briefly in 4× SSC/0.2% Tween preheated at 37 °C.
25. Wash 3 × 3 min at 4× SSC/0.2% Tween preheated at 37 °C in the dark.
26. Rinse in DPBS.
27. Transfer the coverslip in 24-well plate and cross-link with 2% formaldehyde/PBS for 10 min at room temperature in the dark.
28. Wash in DPBS for 5 min at room temperature on a tilting shaker in the dark—repeat twice.
29. Incubate with 0.2% DAPI/DPBS for 5 min at room temperature in the dark.
30. Wash in DPBS for 5 min at room temperature on a tilting shaker in the dark—repeat twice.
31. Mount the coverslip on a microscope slide.
32. A confocal or widefield fluorescent microscopy system (Fig. 4).

3.2.4 Chromatin Immunoprecipitation (ChIP)

1. Centrifuge cells obtained from FACS (Subheading 3.2.2) at 1500 × *g* for 15 min at 4 °C and discard the supernatant.
2. Resuspend the pellet in lysis buffer (100 μL/10^6 cells) and store it at −80 °C until sonication.
3. Thaw samples on ice and collect the number of cells needed in a single 1.5-mL tube (*see* **Note 29**).
4. Resuspend the cells in fresh lysis buffer and incubate in mild agitation at 4 °C for 3 h, passing suspension through a 0.50 × 16 mm syringe needle every hour.

5. Add equilibration buffer in a ratio 1:2 to lysis buffer (*see* **Note 30**).
6. Sonicate the cells in a cold water bath using Branson Digital Sonifier to shear chromatin to 200-bp fragments.
7. Check the fragmentation of chromatin by de-cross-linking a small aliquot of the suspension and running it on the Bioanalyzer (*see* **Note 31**).
8. When the chromatin is sonicated to 200-bp fragments (Fig. 5a), centrifuge at 21,000 × *g* for 30 min at 4 °C and transfer the supernatant to a new 1.5-mL tube.
9. Take 2% of the total volume of the sample and keep it at 4 °C to use it as input chromatin.
10. Incubate the remaining fragmented chromatin with 1 mM PMSF and 2/4 μg of the antibody of interest on a rotating wheel at 4 °C overnight.
11. Add 40 μL of protein G beads to the sample and incubate for 2 h on the rotating wheel at 4 °C.
12. Put the sample on a magnet and discard the supernatant.
13. Wash the beads with 0.6 mL of low-salt solution incubating on the rotating wheel at 4 °C for 5 min, then put the sample on the magnet, and discard the supernatant.
14. Wash the beads with 0.6 mL of high-salt solution incubating on the rotating wheel at 4 °C for 5 min, then put the sample on the magnet, and discard the supernatant.
15. Wash the beads with 0.6 mL of low-salt solution incubating on the rotating wheel at 4 °C for 5 min, then put the sample on the magnet, and discard the supernatant.
16. Wash the beads with 1 mL of TE NaCl solution incubating on the rotating wheel at 4 °C for 5 min, then put the sample on the magnet, and discard the supernatant.
17. Reverse the cross-linking of chromatin by incubating the beads in 150 μL of elution buffer at 65 °C overnight.
18. De-cross-link also input chromatin in 150 μL of elution buffer at 65 °C overnight.
19. Dilute the samples with one volume of TE.
20. Add RNase A to a final concentration of 0.2 μg/mL and incubate at 37 °C for 2 h.
21. Add then Proteinase K to a final concentration of 0.2 μg/mL and incubate at 55 °C for 2 h.
22. Isolate DNA through standard phenol/chloroform extraction (*see* **Note 32**).
23. Resuspend the precipitated DNA in 31 μL of 10 mM Tris-HCl pH 8.0.

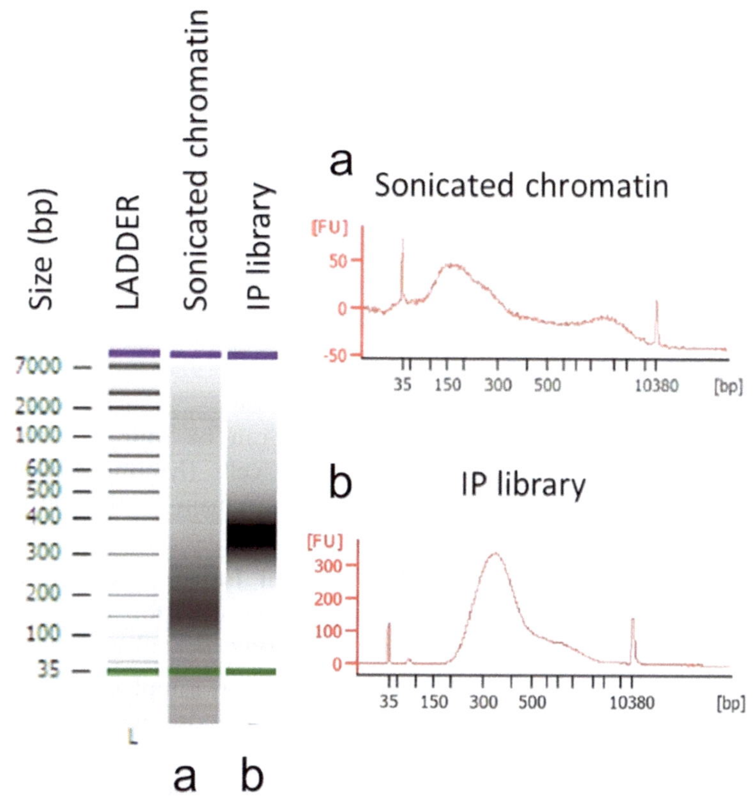

Fig. 5 Bioanalyzer run of (**a**) sonicated chromatin used for immunoprecipitation and (**b**) DNA library of the immunoprecipitated sample

24. Quantify IP and input.
25. Validate the IP by real-time PCR amplification with primer pairs spanning a previously described binding region (positive control) and a flanking unbound region (negative control).
26. Prepare DNA libraries of the samples (*see* **Note 33**).
27. Quantify and check the quality of the libraries using the Bioanalyzer (Fig. 5b).
28. Multiplex libraries with distinct adapter indexes and, after cluster generation on FlowCell, sequence (Fig. 6).

4 Notes

1. Make the solution fresh before muscle harvesting. Filter it with 0.2-μm filter.
2. Ficoll-Paque solution contains a highly branched, high-molecular-weight polymer of sucrose. Mixing this solution with blood allows stratification of blood cells on the basis of their density: erythrocytes sediment on the bottom and

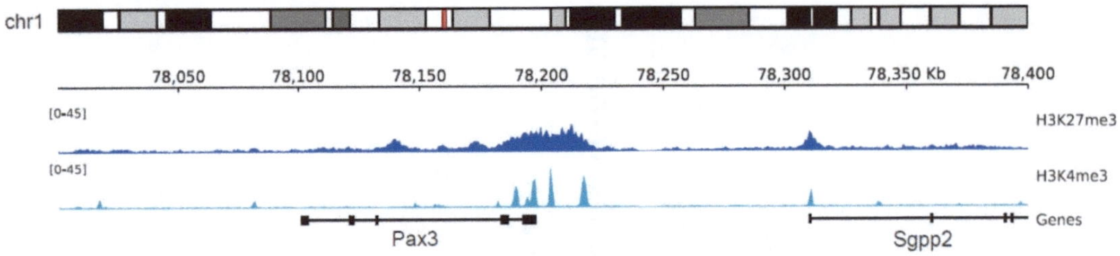

Fig. 6 ChIP-seq on fixed and sorted murine muscle stem cells (MuSCs): tracks of H3K27me3 (blue) and H3K4me3 (cyan) marks plotted as the linear ratio IP/input signal of the effective genome size normalized coverage on Pax3 and Sgpp2 loci

granulocytes can be found in an intermediate layer while mononuclear cells (lymphocytes and monocytes), together with platelets, remain on the top of Ficoll-Paque and can be easily collected.

3. CD4+ T Cell Isolation Kit from Miltenyi Biotec requires counting of the cells for addition of optimized volumes of buffers, antibody cocktail and microbead suspension.
4. If processing many samples, automation is suggested. Manual columns are single use, while autoMACS® Pro Separator columns can perform up to one hundred separations. Furthermore, manual separation is time consuming and error-prone.
5. To avoid clumping of the cells during cross-linking, do not exceed the concentration of 0.5×10^6 cells/mL.
6. Glycine quenches formaldehyde by absorbing its reactivity and exhausting its cross-linking activity.
7. FBS addition, in a percentage between 10 and 20%, is recommended to help the sedimentation of the cross-linked cell pellet at the bottom of the tube.
8. We recommend using polypropylene rather than polystyrene tubes to minimize cell binding to the plastic surface.
9. The optimal resuspension volume depends on the number of cells in the sample and on cytometer settings. We normally use a 70-μm nozzle to sort lymphocytes and MuSCs with an optimal flow rate of about 10,000–15,000 events/s.
10. Cell suspension filtering through a 50–70-μm nylon mesh just prior to sorting is needed to remove cell clumps and reduce the probability of plugging of the instrument during sorting.
11. The immunofluorescence protocol of Subheading 3.1.3 can be used also for murine muscle stem cells.
12. Poly-L-lysine coating confers a net positive charge to the surface of the glass, helping the attachment of the cells.

13. Put poly-L-lysine solution on the coverslips and let it sit for a couple of minutes, then discard the solution, rinse with water, and let the coverslips dry. Repeat twice.
14. The optimal concentration of cell suspension depends on cell dimension. For lymphocytes and MuSCs (around 5–10 μm of diameter), we recommend 0.6–1 × 10^6 cells/mL.
15. The optimal volume of the drop depends on the dimension of the coverslip used. We recommend 0.3 mL for 9-mm round coverslip and 0.15 mL for 6-mm round coverslips.
16. We recommend to always use freshly coated coverslips, to avoid poly-L-lysine degradation and poor attachment of cells.
17. During this time, cells will deposit on the coverslip and contact the poly-L-lysine coating. However, they are not stably attached until cross-linked through formaldehyde treatment. Therefore, the removal of the drop and the addition of PFA solution should be very gentle, to avoid cell displacement or clumping.
18. A 4% PFA should be prepared fresh by heat-mediated dissolution of powder in DPBS.
19. The short PFA treatment at this step is due to the fact that cells have already been cross-linked. This treatment is needed to fix the cells on the coated coverslip.
20. For double staining, we suggest a separated incubation in a humidified chamber for the two primary antibodies: the first at 4 °C overnight and the second at room temperature for 1–3 h.
21. Digestion of RNA ensures minimal aspecific binding of DNA probes to RNA.
22. Cells can be kept in this solution at 4 °C for up to 1 month.
23. Residual liquid on the coverslip could dilute the probe and decrease the binding efficiency.
24. The optimal denaturation temperature depends on cell and probe properties. Higher denaturation temperatures could increase specific binding.
25. Before pipetting the suspension, wet the 10-mL pipette and the cell strainer with $HBSS^+$ to avoid cell adhesion on plastic surfaces and on the strainer membrane. After passing the suspension through the strainer, wash the falcon which previously contained the suspension with additional 10 mL of $HBSS^+$ and pass them through the strainer to recover the maximum amount of cells.
26. After 1 h of reaction, check the length of the fluorescent probe on a 2% (w/v) agarose gel. If the probe length is around 50 bp, you can stop the reaction.

27. First resuspend in 3 μL of 100% formamide, avoiding to touch the pellet with the pipette tip. Incubate in a thermomixer at 300 rpm, 40 °C overnight in the dark. The day after, add 3 μL of 4× SSC/20% dextran sulfate.
28. Pepsin digests proteins, increasing the DNA accessibility for the probe.
29. For histone modification analysis in MuSCs, we recommend about 1–1.5 × 10^6 cells/IP.
30. This will bring the solution to the composition of IP buffer: 100 mM NaCl, 33 mM Tris-HCl pH 8.1, 33 mM Tris-HCl pH 8.6, 0.3% SDS, 1.7% Triton X-100, 5 mM EDTA.
31. Take a small volume of sonicated suspension and bring it to 100 μL with TE buffer, incubate overnight in a thermomixer at 65 °C, and then follow **steps 19–22** of Subheading 3.2.4. Resuspend in a small volume of TE and run on a Bioanalyzer.
32. We recommend to extract twice the aqueous phase, by replacing the first one with an equal volume of TE.
33. DNA libraries can be prepared with commercially available kits. We used NEBNext® Ultra™ II DNA Library Prep Kit for Illumina® with NEBNext® Multiplex Oligos for Illumina®.

Acknowledgments

The authors declare no competing financial interests. This work was supported by grants from the Italian Minister of Health n. GR-2013-02355413, My First AIRC Grant (MFAG), n. 18535, AFM-Telethon n. 21030, and Cariplo Foundation n. 2017-0649.

References

1. Calero-Nieto FJ, Ng FS, Wilson NK, Hannah R, Moignard V, Leal-Cervantes AI, Jimenez-Madrid I, Diamanti E, Wernisch L, Gottgens B (2014) Key regulators control distinct transcriptional programmes in blood progenitor and mast cells. EMBO J 33 (11):1212–1226. https://doi.org/10.1002/embj.201386825
2. Schoenfelder S, Mifsud B, Senner CE, Todd CD, Chrysanthou S, Darbo E, Hemberger M, Branco MR (2018) Divergent wiring of repressive and active chromatin interactions between mouse embryonic and trophoblast lineages. Nat Commun 9(1):4189. https://doi.org/10.1038/s41467-018-06666-4
3. Kim K, Lee K, Bang H, Kim JY, Choi JK (2016) Intersection of genetics and epigenetics in monozygotic twin genomes. Methods 102:50–56. https://doi.org/10.1016/j.ymeth.2015.10.020
4. Xiang Z, Yang Y, Chang C, Lu Q (2017) The epigenetic mechanism for discordance of autoimmunity in monozygotic twins. J Autoimmun 83:43–50. https://doi.org/10.1016/j.jaut.2017.04.003
5. Li G, Zhu P (2015) Structure and organization of chromatin fiber in the nucleus. FEBS Lett 589(20 Pt A):2893–2904. https://doi.org/10.1016/j.febslet.2015.04.023
6. Liang Y, Franks TM, Marchetto MC, Gage FH, Hetzer MW (2013) Dynamic association of NUP98 with the human genome. PLoS Genet 9(2):e1003308. https://doi.org/10.1371/journal.pgen.1003308
7. Luo H, Xi Y, Li W, Li J, Li Y, Dong S, Peng L, Liu Y, Yu W (2017) Cell identity bookmarking

through heterogeneous chromatin landscape maintenance during the cell cycle. Hum Mol Genet 26(21):4231–4243. https://doi.org/10.1093/hmg/ddx312

8. Shachar S, Misteli T (2017) Causes and consequences of nuclear gene positioning. J Cell Sci 130(9):1501–1508. https://doi.org/10.1242/jcs.199786
9. Simon JA, Tamkun JW (2002) Programming off and on states in chromatin: mechanisms of Polycomb and trithorax group complexes. Curr Opin Genet Dev 12(2):210–218
10. Wang KC, Yang YW, Liu B, Sanyal A, Corces-Zimmerman R, Chen Y, Lajoie BR, Protacio A, Flynn RA, Gupta RA, Wysocka J, Lei M, Dekker J, Helms JA, Chang HY (2011) A long noncoding RNA maintains active chromatin to coordinate homeotic gene expression. Nature 472(7341):120–124. https://doi.org/10.1038/nature09819
11. D'Urso A, Brickner JH (2014) Mechanisms of epigenetic memory. Trends Genet 30(6):230–236. https://doi.org/10.1016/j.tig.2014.04.004
12. Piskadlo E, Tavares A, Oliveira RA (2017) Metaphase chromosome structure is dynamically maintained by condensin I-directed DNA (de)catenation. eLife 6:e26120. https://doi.org/10.7554/eLife.26120
13. Takata H, Hanafusa T, Mori T, Shimura M, Iida Y, Ishikawa K, Yoshikawa K, Yoshikawa Y, Maeshima K (2013) Chromatin compaction protects genomic DNA from radiation damage. PLoS One 8(10):e75622. https://doi.org/10.1371/journal.pone.0075622
14. Boettiger AN, Bintu B, Moffitt JR, Wang S, Beliveau BJ, Fudenberg G, Imakaev M, Mirny LA, Wu CT, Zhuang X (2016) Super-resolution imaging reveals distinct chromatin folding for different epigenetic states. Nature 529(7586):418–422. https://doi.org/10.1038/nature16496
15. Chubb JR, Boyle S, Perry P, Bickmore WA (2002) Chromatin motion is constrained by association with nuclear compartments in human cells. Curr Biol 12(6):439–445. https://doi.org/10.1016/s0960-9822(02)00695-4
16. Osborne CS, Chakalova L, Brown KE, Carter D, Horton A, Debrand E, Goyenechea B, Mitchell JA, Lopes S, Reik W, Fraser P (2004) Active genes dynamically colocalize to shared sites of ongoing transcription. Nat Genet 36(10):1065–1071. https://doi.org/10.1038/ng1423
17. Spector DL (2003) The dynamics of chromosome organization and gene regulation. Annu Rev Biochem 72:573–608. https://doi.org/10.1146/annurev.biochem.72.121801.161724
18. Gilbert N, Boyle S, Fiegler H, Woodfine K, Carter NP, Bickmore WA (2004) Chromatin architecture of the human genome: gene-rich domains are enriched in open chromatin fibers. Cell 118(5):555–566. https://doi.org/10.1016/j.cell.2004.08.011
19. Guelen L, Pagie L, Brasset E, Meuleman W, Faza MB, Talhout W, Eussen BH, de Klein A, Wessels L, de Laat W, van Steensel B (2008) Domain organization of human chromosomes revealed by mapping of nuclear lamina interactions. Nature 453(7197):948–951. https://doi.org/10.1038/nature06947
20. Bersaglieri C, Santoro R (2019) Genome organization in and around the nucleolus. Cell 8(6). https://doi.org/10.3390/cells8060579
21. van Koningsbruggen S, Gierlinski M, Schofield P, Martin D, Barton GJ, Ariyurek Y, den Dunnen JT, Lamond AI (2010) High-resolution whole-genome sequencing reveals that specific chromatin domains from most human chromosomes associate with nucleoli. Mol Biol Cell 21(21):3735–3748. https://doi.org/10.1091/mbc.E10-06-0508
22. Briand N, Collas P (2018) Laminopathy-causing lamin A mutations reconfigure lamina-associated domains and local spatial chromatin conformation. Nucleus 9(1):216–226. https://doi.org/10.1080/19491034.2018.1449498
23. Cesarini E, Mozzetta C, Marullo F, Gregoretti F, Gargiulo A, Columbaro M, Cortesi A, Antonelli L, Di Pelino S, Squarzoni S, Palacios D, Zippo A, Bodega B, Oliva G, Lanzuolo C (2015) Lamin A/C sustains PcG protein architecture, maintaining transcriptional repression at target genes. J Cell Biol 211(3):533–551. https://doi.org/10.1083/jcb.201504035
24. Marullo F, Cesarini E, Antonelli L, Gregoretti F, Oliva G, Lanzuolo C (2016) Nucleoplasmic Lamin A/C and Polycomb group of proteins: an evolutionarily conserved interplay. Nucleus 7(2):103–111. https://doi.org/10.1080/19491034.2016.1157675
25. Oldenburg A, Briand N, Sorensen AL, Cahyani I, Shah A, Moskaug JO, Collas P (2017) A lipodystrophy-causing lamin A mutant alters conformation and epigenetic regulation of the anti-adipogenic MIR335 locus. J

Cell Biol 216(9):2731–2743. https://doi.org/10.1083/jcb.201701043

26. Zheng X, Hu J, Yue S, Kristiani L, Kim M, Sauria M, Taylor J, Kim Y, Zheng Y (2018) Lamins organize the global three-dimensional genome from the nuclear periphery. Mol Cell 71(5):802–815.e807. https://doi.org/10.1016/j.molcel.2018.05.017
27. Bianchi A, Mozzetta C, Pegoli G, Lucini F, Valsoni S, Rosti V, Petrini C, Cortesi A, Gregoretti F, Antonelli L, Oliva G, De Bardi M, Rizzi R, Bodega B, Pasini D, Ferrari F, Bearzi C, Lanzuolo C (2020) Dysfunctional polycomb transcriptional repression contributes to lamin A/C–dependent muscular dystrophy. J Clin Invest 130(5):2408–2421 https://doi.org/10.1172/JCI128161
28. Bianchi A, Manti PG, Lucini F, Lanzuolo C (2018) Mechanotransduction, nuclear architecture and epigenetics in Emery Dreifuss Muscular Dystrophy: tous pour un, un pour tous. Nucleus 9(1):276–290. https://doi.org/10.1080/19491034.2018.1460044
29. Serebryannyy L, Misteli T (2018) Protein sequestration at the nuclear periphery as a potential regulatory mechanism in premature aging. J Cell Biol 217(1):21–37. https://doi.org/10.1083/jcb.201706061
30. Lanzuolo C (2012) Epigenetic Alterations in Muscular Disorders. Comp Funct Genomics 2012:1-12. https://doi.org/10.1155/2012/256892
31. Fiserova J, Efenberkova M, Sieger T, Maninova M, Uhlirova J, Hozak P (2017) Chromatin organization at the nuclear periphery as revealed by image analysis of structured illumination microscopy data. J Cell Sci 130 (12):2066–2077. https://doi.org/10.1242/jcs.198424
32. Raices M, D'Angelo MA (2012) Nuclear pore complex composition: a new regulator of tissue-specific and developmental functions. Nat Rev Mol Cell Biol 13(11):687–699. https://doi.org/10.1038/nrm3461
33. Tan-Wong SM, Wijayatilake HD, Proudfoot NJ (2009) Gene loops function to maintain transcriptional memory through interaction with the nuclear pore complex. Genes Dev 23 (22):2610–2624. https://doi.org/10.1101/gad.1823209
34. Gialitakis M, Arampatzi P, Makatounakis T, Papamatheakis J (2010) Gamma interferon-dependent transcriptional memory via relocalization of a gene locus to PML nuclear bodies. Mol Cell Biol 30(8):2046–2056. https://doi.org/10.1128/MCB.00906-09
35. Robson MI, de Las Heras JI, Czapiewski R, Sivakumar A, Kerr ARW, Schirmer EC (2017) Constrained release of lamina-associated enhancers and genes from the nuclear envelope during T-cell activation facilitates their association in chromosome compartments. Genome Res 27(7):1126–1138. https://doi.org/10.1101/gr.212308.116
36. Cheung TH, Quach NL, Charville GW, Liu L, Park L, Edalati A, Yoo B, Hoang P, Rando TA (2012) Maintenance of muscle stem-cell quiescence by microRNA-489. Nature 482 (7386):524–528. https://doi.org/10.1038/nature10834
37. van den Brink SC, Sage F, Vertesy A, Spanjaard B, Peterson-Maduro J, Baron CS, Robin C, van Oudenaarden A (2017) Single-cell sequencing reveals dissociation-induced gene expression in tissue subpopulations. Nat Methods 14(10):935–936. https://doi.org/10.1038/nmeth.4437
38. Warren GL, Summan M, Gao X, Chapman R, Hulderman T, Simeonova PP (2007) Mechanisms of skeletal muscle injury and repair revealed by gene expression studies in mouse models. J Physiol 582(Pt 2):825–841. https://doi.org/10.1113/jphysiol.2007.132373
39. van Velthoven CTJ, de Morree A, Egner IM, Brett JO, Rando TA (2017) Transcriptional profiling of quiescent muscle stem cells in vivo. Cell Rep 21(7):1994–2004. https://doi.org/10.1016/j.celrep.2017.10.037
40. Le HQ, Ghatak S, Yeung CY, Tellkamp F, Gunschmann C, Dieterich C, Yeroslaviz A, Habermann B, Pombo A, Niessen CM, Wickstrom SA (2016) Mechanical regulation of transcription controls Polycomb-mediated gene silencing during lineage commitment. Nat Cell Biol 18(8):864–875. https://doi.org/10.1038/ncb3387
41. Machado L, Esteves de Lima J, Fabre O, Proux C, Legendre R, Szegedi A, Varet H, Ingerslev LR, Barres R, Relaix F, Mourikis P (2017) In situ fixation redefines quiescence and early activation of skeletal muscle stem cells. Cell Rep 21(7):1982–1993. https://doi.org/10.1016/j.celrep.2017.10.080
42. Sundberg-Kovamees M, Grunewald J, Wahlstrom J (2016) Immune cell activation and cytokine release after stimulation of whole blood with pneumococcal C-polysaccharide and capsular polysaccharides. Int J Infect Dis 52:1–8. https://doi.org/10.1016/j.ijid.2016.07.004
43. Gonzalez-Granado JM, Silvestre-Roig C, Rocha-Perugini V, Trigueros-Motos L,

Cibrian D, Morlino G, Blanco-Berrocal M, Osorio FG, Freije JMP, Lopez-Otin C, Sanchez-Madrid F, Andres V (2014) Nuclear envelope lamin-A couples actin dynamics with immunological synapse architecture and T cell activation. Sci Signal 7(322):ra37. https://doi.org/10.1126/scisignal.2004872

44. Tsopoulidis N, Kaw S, Laketa V, Kutscheidt S, Baarlink C, Stolp B, Grosse R, Fackler OT (2019) T cell receptor-triggered nuclear actin network formation drives CD4(+) T cell effector functions. Sci Immunol 4(31). https://doi.org/10.1126/sciimmunol.aav1987

45. Croft JA, Bridger JM, Boyle S, Perry P, Teague P, Bickmore WA (1999) Differences in the localization and morphology of chromosomes in the human nucleus. J Cell Biol 145 (6):1119–1131. https://doi.org/10.1083/jcb.145.6.1119

46. Mozzetta C (2016) Isolation and culture of muscle stem cells. Methods Mol Biol 1480:311–322. https://doi.org/10.1007/978-1-4939-6380-5_27

Chapter 11

Visualization of Chromatin Dynamics by Live Cell Microscopy Using CRISPR/Cas9 Gene Editing and ANCHOR Labeling

Ezio T. Fok, Stephanie Fanucchi, Kerstin Bystricky, and Musa M. Mhlanga

Abstract

The organization of the eukaryotic nucleus facilitates functional chromatin contacts which regulate gene transcription. Despite this being extensively studied through population-based chromatin contact mapping and microscopic observations in single cells, the spatiotemporal dynamics of chromatin behavior have largely remained elusive. The current methods to label and observe specific endogenous genomic loci in living cells have been challenging to implement and too invasive to biological processes. In this protocol, we describe the use of a recently developed DNA labelling strategy (ANCHOR) with CRISPR/Cas9 gene editing, to discreetly label genes for live cell imaging to study chromatin dynamics. Our approach improves on some of the fundamental shortfalls associated with current labelling strategies and has the potential for multiplexed observations.

Key words DNA visualization, Chromatin dynamics, Live cell microscopy, CRISPR/Cas9, ParS/ParB, Single cell analysis

1 Introduction

The eukaryotic nucleus is a dense cellular compartment that houses the genetic code for life. Chromosomal DNA is highly condensed to fit into this space through the volumetric compaction of the chromatin fiber by the extensive formation of loops. Insights into the organization of the nucleus over the last two decades have revealed that this compaction is nonrandom, but rather intricately arranged to facilitate function. Genome-wide mapping of the chromatin loops by high-throughput chromosomal conformation capture (3C)-based techniques, such as HiC, have revealed that the chromatin fiber is segmented into discrete regions of enriched chromosomal contacts called topologically associated domains (TADs) [1–3]. With this organization, long-range chromosomal contacts that occur in *cis* are enriched through the formation of

Beatrice Bodega and Chiara Lanzuolo (eds.), *Capturing Chromosome Conformation: Methods and Protocols*, Methods in Molecular Biology, vol. 2157, https://doi.org/10.1007/978-1-0716-0664-3_11,

these TADs and *trans* interactions are likely to be facilitated by the proximal positioning of TADs [4]. Importantly, chromosomal contact is strongly correlated with the transcriptional activity of the engaged DNA elements, hinting to an important role of nuclear architecture in regulating gene transcription [5–8].

Perhaps the best studied example demonstrating this is the functional looping of enhancer elements to promoters to activate gene transcription [9, 10]. The transcription of enhancer regions produces a class of long noncoding RNAs called enhancer RNAs (eRNAs). Functionally, these eRNAs mediate the formation of chromosomal loops that bring the enhancer region and its associated transcription factors into close spatial proximity to its cognate promoter. Remarkably, the ablation of eRNAs by RNA interference results in the loss of these looping events and the abrogation of gene transcription [10]. Through this spatial control of transcriptional elements, gene expression can be tightly regulated to allow for precise spatiotemporal and tissue-specific control. This is particularly apparent during lineage commitment, where different cell types adopt distinct DNA packaging configurations that lead to the nuclear repositioning of enhancer elements for their alternate usage in specific lineage-determinant gene activation [11–14]. Furthermore, chromosomal looping can bring multiple distal genes and regulators into close three-dimensional proximity to form multigene complexes that allow for co-regulated transcription [6, 15–17]. Strikingly, discrete perturbations of loops engaged in a multigene complex disrupt gene transcription, revealing a causal role of chromosomal contacts in the regulation of gene expression [18].

Despite the wealth of information provided by comprehensive genome-wide contact maps, the inherent caveat to these techniques is that they represent an average across a cell population, shrouding our ability to observe more nuanced chromosomal looping events, such as those that occur infrequently and/or dynamically. Owing to the development of high-resolution imaging approaches, such as fluorescent in situ hybridization (FISH), specific DNA loci and RNA species can be labelled to observe nuclear organization and gene transcription in single cells [19]. Such single-cell imaging approaches, in conjunction with polymer modelling, have been used to deconvolve population-based chromatin interaction datasets to reveal a repertoire of possible chromatin conformations within TADs [20]. This underscores the heterogeneity of chromatin conformations between cells, which can partially be attributed to dynamic events. Indeed, such dynamic events have been observed with the synthetic transcriptional activation of a single locus [21], the stimulus-induced assembly of long-range multigene complexes [17], and the repositioning of enhancer elements during development [13]. Recent insights into these processes have revealed that these dynamic events are coordinated through liquid phase transitions where clusters of chromatin-associated transcriptional machinery interact to form phase separated biomolecular condensates. Within these membraneless compartments, genetic elements are brought into close spatial proximity and gene transcription can take place [22]. However, single-cell FISH-based imaging approaches only provide static snapshots of these events and cannot provide insights into the spatiotemporal dynamics of chromatin behavior in the nucleus. Therefore, the application of real-time live cell microscopy is integral to studying this area of chromatin biology.

The visualization of specific endogenous genomic loci in living cells remains difficult. This has been largely due to the insufficiencies of techniques to achieve site-specific labelling of genomic loci that is discrete enough to preserve the native characteristics and function of the chromatin, simple to construct and implement, and able to produce long-lived fluorescent foci that can be clearly distinguished from background noise. The attainment of these characteristics is a balancing act that is difficult to achieve. The classical methods to label and observe discrete loci in living cells involve the insertion of large arrays of bacterial-derived operator sequence repeats into the genome. These repeat arrays recruit cognate repressor proteins fused to fluorescent markers to produce a visible focus at the insertion site [23–25]. This type of labelling regime has proven to be very powerful, providing many insights into the dynamics of chromatin during numerous cellular processes [26, 27]. However, targeted insertion of these arrays has been limited by their large size and the inefficiencies of classical recombineering techniques, thus limiting their general application to yeast. Furthermore, the repetitive nature of these sequences can also introduce genomic instability. As a result of these difficulties, it has been desirable to avoid the direct modification of the genome. To this end, designer DNA binding proteins have been adapted to direct fluorescent markers to specific loci, negating the need to alter the genomic sequence itself. One of the first examples of this was the use of the now legacy transcription activator-like effector (TALE) technologies as fusions with fluorescent proteins [28]. However, due to the cumbersome nature of producing site-specific TALEs, the use of this technology has been limited to labelling repetitive regions of the genome [25]. With the advent of CRISPR/dCas9 technology, this kind of labelling strategy has been unencumbered through the development of methods to easily generate guide RNA libraries for the tiling of dCas9-GFP fusion molecules along any specific genomic sequence [29].

All of the abovementioned labelling strategies rely on the function of fluorescent molecule multiplicity and require a sufficient number of fluorescent molecules to be loaded onto a locus to produce a discernible focus. As a result, attaining a strong signal over background noise can be challenging. Furthermore, the binding kinetics of these proteins have shown that their interaction with DNA is strong and can potentially alter the behavior of the chromatin and lead to the formation of heterochromatin, both of which can influence gene transcription [30].

The ANCHOR DNA labelling system can circumvent these technical drawbacks by being less invasive to biological processes. ANCHOR was derived from components of the ParABS DNA partitioning machinery found in prokaryotes [31]. It is composed of two parts, a short (~1 kb) *ParS* DNA seed sequence (ANCH) and a protein component consisting of a fluorescently tagged ParB protein (OR). Fluorescently labelled ParB protein dimers are able to recognize and specifically bind to the *ParS* seed sequence, initiating a nucleation event that promotes subsequent ParB recruitment through protein-protein interactions and the nonspecific binding and spreading of the ParB protein to adjacent DNA (Fig. 1). This oligomerization of fluorescently labelled ParB to the *ParS* locus amplifies the local fluorescent signal to culminate in the formation of a visible focus [32]. Importantly, the interaction of the ParB protein with DNA is relatively weak and highly

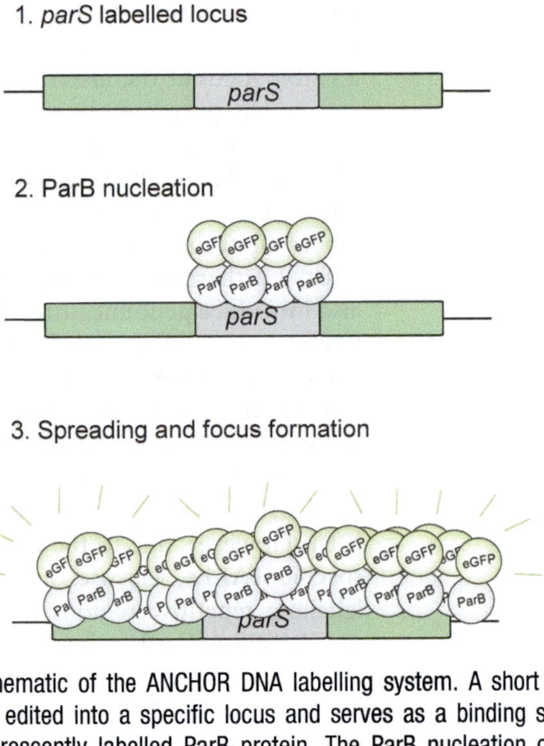

Fig. 1 A schematic of the ANCHOR DNA labelling system. A short *ParS* DNA sequence is edited into a specific locus and serves as a binding site for the cognate fluorescently labelled ParB protein. The ParB nucleation of this site subsequently leads to an oligomerization event, where multiple copies of the fluorescently labelled ParB protein are recruited to the site by weak protein-protein and protein-DNA interactions. This results in the formation of a nucleoprotein complex that spreads along the DNA to form a discernible fluorescent focus

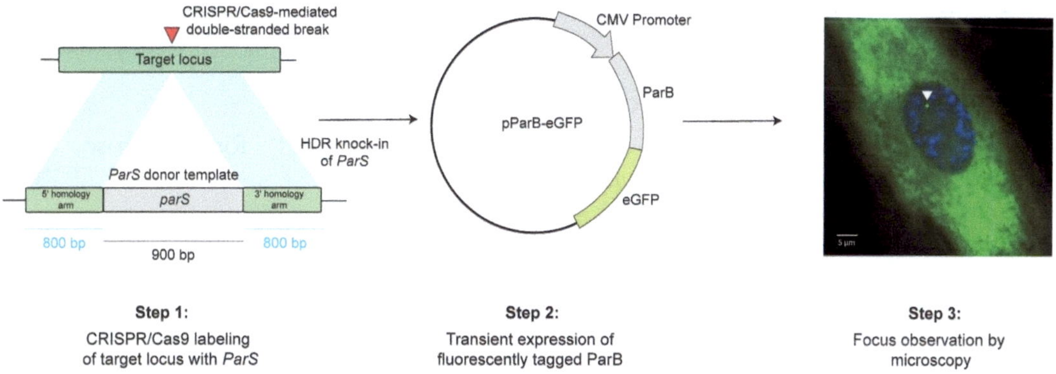

Fig. 2 An overview of applying the ANCHOR system with CRISPR/Cas9 gene editing for the visualization of specific genomic loci. CRISPR/Cas9 gene editing is used to insert the *ParS* sequence into the locus of interest by HDR. Repaired cells are subsequently transfected to transiently express the fluorescently labelled ParB protein. The genomic locus of interest can be observed by fluorescent microscopy in fixed or live cells. Arrow points to a *ParS*-ParB labelled locus

dynamic, creating a metastable nucleoprotein complex that has been shown to be innocuous to gene transcription [32], exonuclease activity [33], and virus infection and replication [31]. To date, four distinct ANCHOR variants have been developed and can be used in combinations for multiplex gene labelling.

Here we describe a protocol that we use to overcome the limitations of current DNA labelling strategies for live cell imaging. By applying precise CRISPR/Cas9 gene editing in combination with the ANCHOR DNA labelling system, we have been able to label specific genomic loci in a discrete and minimally invasive way. We outline our entire workflow in generating a discreetly *ParS*-labelled cell for live imaging in several sub-protocols (Fig. 2). Briefly, we begin by describing the design and cloning procedures to generate locus-specific CRISPR/Cas9 guide RNAs (gRNAs) and the targeting *ParS* donor template. We then go on to methods that we use to test these reagents for their cleavage and insertion activity. Penultimately, we use fixed-cell imaging as a way to test the functionality of the ANCHOR system in the cellular model of interest prior to finally carrying out live cell imaging experiments. Using this workflow, we applied the ANCHOR system with CRISPR/Cas9 gene editing to label endogenous genes in mammalian cells. In this protocol, we demonstrate this approach by real-time imaging of the *SAMD4A* locus. It has been documented that TNFα stimulation causes *SAMD4A* to engage in a multigene complex consisting of both *cis* and *trans* long-range interactions with *TNFAIP2* and *SLC6A5*, respectively. Furthermore, we show that specific combinations of *ParS*-ParB variants do not engage in cross-talk, potentially allowing for multiplexed labelling.

2 Materials

2.1 Equipment

1. Thermocycler.
2. Gel electrophoresis equipment for agarose and polyacrylamide gels.
3. 37 °C shaking incubator.
4. CO_2 incubator and laminar flow hood.
5. Spinning disk confocal microscope equipped with a mercury lamp, the appropriate filter, 100× objective, live cell imaging chamber, and CCD camera.

2.2 Reagents

2.2.1 CRISPR gRNA Cloning and Testing

1. gRNA oligonucleotides designed to target the locus of interest (standard desalting).

   gRNA F:

   5' - CACC**(G)**NNNNNNNNNNNNNNNNNNNN -3'

   gRNA R:

   5' - AAACNNNNNNNNNNNNNNNNNNNN**(C)** -3'

2. pX330 sequencing primer (standard desalting).
   5′-GAGGGCCTATTTCCCATGATTCC-3′
3. T4 PNK.
4. pX330 CRISPR/Cas9 plasmid (Addgene #42230).
5. *Bbs*I restriction enzyme and accompanying digestion buffer.
6. FastAP.
7. T4 ligase and T4 ligation buffer.
8. *Stbl3* chemically competent *E. coli* cells.
9. Standard bacterial culture reagents.
10. Maxiprep plasmid isolation kit.
11. Target locus-specific primers for T7E1 assay (standard desalting) (*see* Subheading 3.2).
12. NEB Q5 High-Fidelity Hot Start DNA Polymerase.
13. PCR column purification kit.
14. 10× DreamTaq buffer.
15. T7E1 enzyme and NEB buffer 2.
16. Polyacrylamide solution (1:30 bisacrylamide to acrylamide).
17. 10% (w/v) ammonium persulfate (APS).
18. Tetramethylethylenediamine (TEMED).
19. DNA loading dye and ladder.
20. TBE buffer.
21. Ethidium bromide (10 mg/ml).

2.2.2 ParS Donor Cloning

22. PCR primers for the amplification and overlap-extension PCR assembly of the *ParS* insert and the flanking homology arms (PAGE purified) (*see* Subheading 3.3).
23. NEB Q5 High-Fidelity Hot Start DNA Polymerase.
24. PCR column purification kit.
25. Unique restriction enzymes of choice and accompanying buffers (*see* Subheading 3.3).
26. pcDNA 3.1(+) plasmid backbone.
27. T4 DNA ligase and T4 ligation buffer.
28. *Stbl3* chemically competent *E. coli* cells.
29. Standard bacterial culture reagents.
30. Maxiprep plasmid isolation kit.
31. Agarose.
32. DNA loading dye and ladder.
33. TBE buffer.
34. Ethidium bromide (10 mg/ml).

2.2.3 Cell Culture and DNA Delivery

35. 24-well cell culture plate.
36. Appropriate cell culture medium and standard cell culture reagents.
37. 12-mm glass coverslips for imaging of fixed cells.
38. 35-mm glass bottom MatTek dishes for live cell imaging.

2.2.4 PCR Detection of Locus-Specific ParS Integration

39. PCR primers designed to detect site-specific *ParS* integration (*see* Subheading 3.5).
40. NEB Q5 High-Fidelity Hot Start DNA Polymerase.
41. QuickExtract genomic DNA extraction solution.
42. Agarose.
43. DNA loading dye and ladder.
44. TBE buffer.
45. Ethidium bromide (10 mg/ml).

2.2.5 Fixed-Cell Imaging

46. 1× PBS.
47. 37% formaldehyde.
48. DAPI counterstain. Dissolved in dimethylformamide to 5 mg/ml and stored in aliquots at −20 °C.
49. Equilibration buffer: 2× SSC, 0.4% (w/v) glucose.
50. Catalase from *A. niger*. Stored at 4 °C.

51. Glucose oxidase (type VII) from *A. niger* made up in water to 3.7 mg/ml. Stored at 4 °C.
52. Freshly made mounting medium: equilibration buffer, catalase, and glucose oxidase at a ratio of 100:1:1.

3 Methods

3.1 CRISPR gRNA Design and Cloning

1. We design CRISPR gRNAs using the publicly available CRISPOR design tool (http://crispor.tefor.net) (*see* **Notes 1** and **2**). We have adopted a gRNA cloning protocol developed by Feng Zhang, which can be accessed at this website: https://www.addgene.org/crispr/zhang/. Briefly, the targeting gRNA oligonucleotides are ordered with the appropriate sticky ends.
2. Annealed gRNA oligonucleotides are then cloned downstream of the U6 promoter on the pX330 plasmid by complementary *Bbs*I restriction sites (*see* **Note 3**).
3. The ligated plasmids are transformed into chemically competent *Stbl3 E. coli* and colonies are directly screened by Sanger sequencing. For this, we use the pX330 sequencing primer that binds to the U6 promoter so that the entire gRNA and the tracrRNA portion of the plasmid can be checked for sequence integrity.
4. Positive clones from a retained replicate master plate are inoculated into liquid culture for maxi plasmid preparation.

3.2 Assessing gRNA Activity

Because the design process is only predictive of gRNA function, we generally choose between 3 and 5 of the best gRNA candidates to test for DNA cleavage efficiency in cells using the T7E1 assay. Our protocol was adapted to improve the detection sensitivity from previously described methods [34].

1. Design and order primers that flank the cut site to produce an amplicon ranging from 200 to 300 bp.
2. Deliver the CRISPR/Cas9 plasmid into the cells by lipid transfection or electroporation (*see* **Note 4**).
3. Forty-eight hours post-transfection, aspirate the culture medium and add 250 μl of QuickExtract solution directly to the cells. Isolate the genomic DNA (gDNA) according to the manufacturer's protocol.
4. PCR amplify the CRISPR/Cas9 target region from the genome according to the Q5 High-Fidelity Hot Start Polymerase manufacturer's instructions.
5. Visualize the PCR product on a 1% agarose gel (*see* **Note 5**).
6. Purify the PCR product using a standard column-based purification kit.

7. Mix 600 ng of the purified amplicon with 3 μl of 10× DreamTaq buffer in a 30-μl reaction. Subject the mixture to a denaturation and reannealing cycle in a thermocycler (95 °C for 10 min, followed by a controlled ramp down at −0.2 °C/s to 25 °C) to form heteroduplexes.
8. Digest 150 ng of the heteroduplexes with 3U T7E1 endonuclease in a 10-μl reaction containing 1× NEB buffer 2.
9. Prepare an 8% polyacrylamide gel by polymerizing 3.2 ml of the polyacrylamide solution (1:30) with 200 μl APS and 10 μl TEMED in 8.8 ml of TBE buffer.
10. Separate the digested heteroduplex fragments on the set polyacrylamide gel for approximately 2–3 h at 60 V and 4 °C.
11. Stain the DNA by submerging the gel in 10 ml of TBE containing 8 μl of ethidium bromide for 15 min. Rinse the gel twice with distilled water.
12. Visualize using a UV illuminated gel imaging system. The efficiency of the CRISPR/Cas9-mediated cleavage can be calculated by using densitometry.

3.3 ParS Donor Design and Assembly

Once the most efficient gRNA has been determined, the precise position of the cut can be used to design the most efficient *ParS* donor template. We have tested single-stranded (ss) and double-stranded (ds) DNA donor templates with varying lengths of homology arms and found that a dsDNA donor consisting of two 800-bp homology arms that directly flank the cut site is the easiest to generate and the most efficient for the insertion of the *ParS* sequence. We generate our homology arms by PCR amplification directly from gDNA. Our assembly strategy for this construct makes use of a 3-way overlap extension PCR, which necessitates the use of longer primers to create PCR products that have overlapping complementary ends.

1. Design appropriate overlapping primers to amplify the components for the *ParS* donor construct (*see* **Note 6**).
2. *PCR round 1: Amplification of the individual components.* PCR amplify the 5′ and 3′ homology arms directly from gDNA harvested from the relevant cell type as described in Subheading 3.2. Amplify the *ParS* sequence from the ANCHOR plasmid collection (patent WO124047) (*see* **Note 8**).
3. Visualize the PCR products on a 1% agarose gel to confirm that the amplicons are of the correct size (*see* **Note 7**).
4. Column purify the PCR products.
5. *PCR round 2: 3-way overlap extension PCR assembly.* Mix each of the PCR fragments at an equimolar ratio to make up a total of 50 ng of template DNA for this reaction. Thermal cycling conditions and reaction setup were carried out according to Q5 High-Fidelity Hot Start Polymerase manufacturer's protocol.

6. Purify the final PCR product by gel extraction and clone into the pcDNA3.1(+) backbone by making use of the unique embedded restriction sites at the terminal ends of the assembled product.
7. Mass PCR amplify and pool for column purification of the assembled *ParS* donor from the pcDNA3.1(+) clone to generate sufficient amounts of linear dsDNA donor which can be used for transfection.

3.4 Cell Culture and DNA Delivery for Homology-Directed Repair (HDR)

Standard cell culture protocols were followed. Retinal pigment epithelial cells (RPE1) are a near diploid cell line and were used as a more tractable cellular system for the testing of reagents. Primary human umbilical vein endothelial cells (HUVECs) were reserved for live cell observations with the expectation that they were the closest to displaying "normal" dynamic chromatin behavior.

3.4.1 Transfection of RPE1 Cells

1. Seed 0.8×10^5 cells per well in a 24-well plate 24 h before transfection.
2. Prepare a DNA mixture consisting of 1 μg of pX330 CRISPR/Cas9 plasmid containing the most efficient gRNA and 2 μg of the linear *ParS* donor DNA.
3. Mix this with 2.5 μl of Lipofectamine 2000 and make up to 25 μl with serum-free medium. Allow the DNA to complex with the Lipofectamine 2000 for 15 min at room temperature before applying the mixture to the cells in a dropwise manner.
4. Allow for homology-directed repair to occur for 72 h.

3.4.2 Electroporation of HUVECS

1. This was performed exactly according to the Neon Transfection System protocol and seeded in a 24-well plate.

3.5 PCR Detection of ParS Insertion into the Target Locus

We recommend that the insertion of the *ParS* sequence into the correct genomic site is validated if this is the first time using newly made reagents. We have devised a PCR-based approach to detect the *ParS* sequence in the targeted locus by using a primer pair of which one binds to an endogenous site outside of the homology region of the *ParS* donor and the other directly to the *ParS* sequence itself (Fig. 3a).

1. Harvest gDNA 72 h post-transfection by using 250 μl of QuickExtract solution and following the manufacturer's instructions.
2. Set up a PCR according to the provided protocol for the Q5 High-Fidelity Hot Start Polymerase.
3. The presence of a correctly sized amplicon, as visualized on an agarose gel, is indicative of successful HDR-mediated integration of the *ParS* sequence. Remember to set a negative control condition where the *ParS* donor is excluded in the transfection.

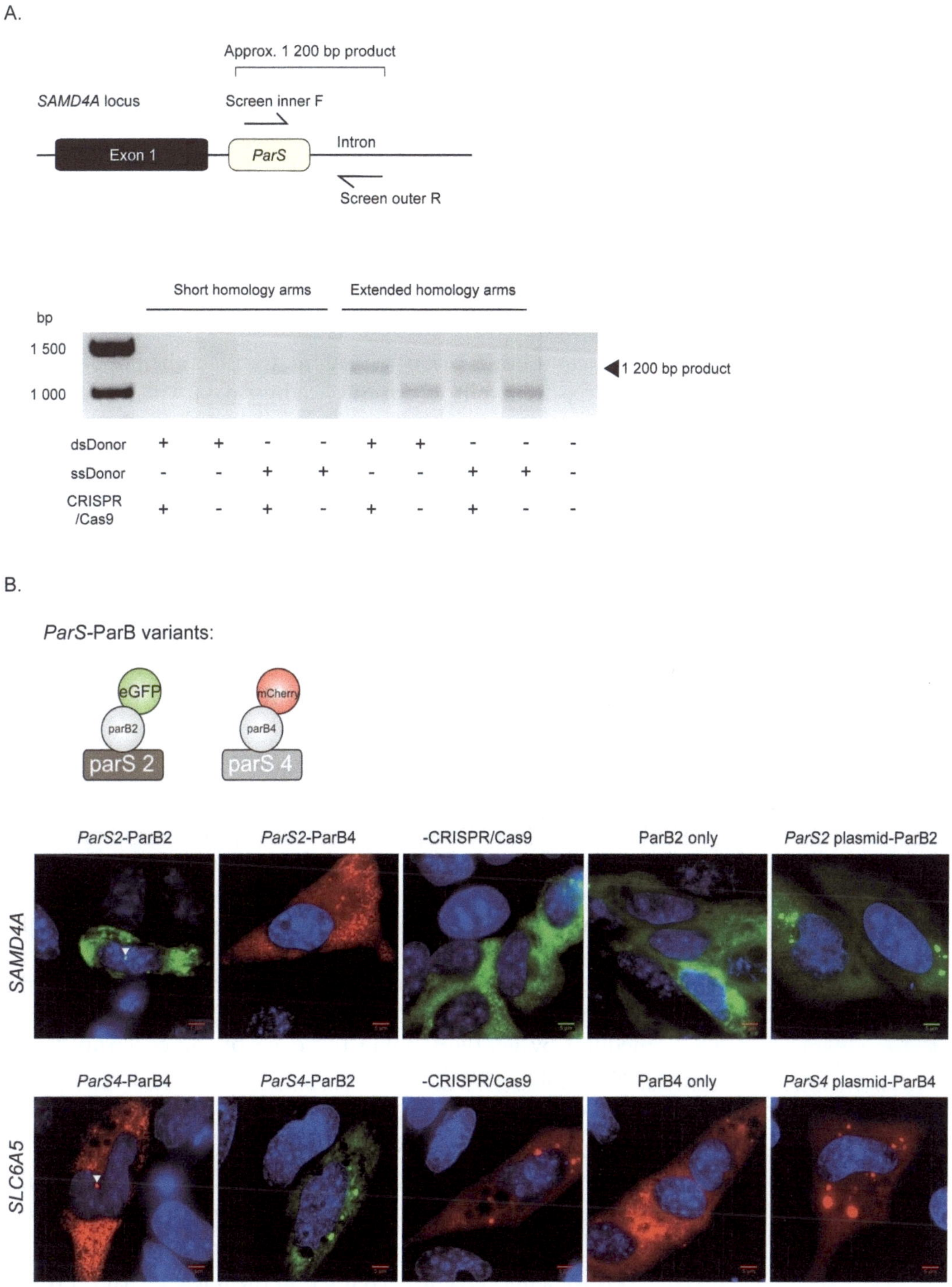

Fig. 3 (**a**) Successful site-specific integration of the *ParS* sequence can be checked using PCR. A dsDNA donor with extended homology arms (800 bp vs 25 bp) has been found to be the most efficient for HDR with CRISPR/Cas9. (**b**). The *ParS*2/4 and ParB2/4 variants exhibit minimal cross-talk, highlighting the possibility of multiplex labelling. Insertion of the *ParS* sequence produces distinct foci in the nucleus (shown by arrows), whereas expression of the fluorescently labelled ParB protein in the absence of *ParS* produces a uniformly dispersed fluorescent signal in the cytoplasm. Expression of the fluorescently labelled ParB protein in the presence of episomal *ParS* plasmid results in the formation of large fluorescent aggregates in the cytoplasm. Bar = 5 μm

3.6 Fixed-Cell Imaging for System Testing by Focus Detection

We recommend that this step is carried out before commencing with live cell imaging experiments to ensure that all the components of the ANCHOR system are functional in the cell. In this part of the protocol, we transfect the RPE1 cells, which we know have undergone successful HDR repair with the *ParS* donor, with a plasmid that expresses a fluorescently tagged ParB protein. The appearance of single or doublet fluorescent foci in the nucleus by imaging fixed preparations of these cells is indicative of a functional ANCHOR system (Fig. 3b).

1. Forty-eight hours prior to imaging, transfect the population of *ParS* repaired cells (from Subheading 3.4) with 3 μg of the plasmid encoding the fluorescently labelled ParB protein as described in Subheading 3.4 (*see* **Note 8**). Remember to seed these cells on 12-mm glass coverslips in a 24-well plate so that they can be imaged.
2. Wash the cells with PBS 48 h post-transfection.
3. Fix the cells with 3.7% formaldehyde for 10 min.
4. Wash off the fixative with PBS.
5. Equilibrate the cells with 1 ml equilibration buffer. In this step, the nuclei can be counterstained with 1 μl of DAPI. Incubate for 5 min at room temperature.
6. Wash out unbound DAPI with equilibration buffer. Do this twice.
7. To mount the coverslip, add a 10 μl droplet of freshly prepared mounting medium to the center of a clean glass slide. Pick up the coverslip from the 24-well plate with a pair of tweezers and blot the edge on a piece of paper towel to remove excess liquid. Turn the coverslip over and carefully place it on the droplet of mounting medium so that the cell side is face down. Avoid the introduction of any air bubbles. Remove any excess mounting medium by gently dabbing the edge of the coverslip with a piece of paper towel.
8. Seal the coverslip by painting a thin layer of clear nail varnish around the edge of the coverslip.
9. Image the coverslip immediately. Acquire 0.2-μm z-slices of the nucleus using the 100× objective in each relevant fluorescent channel.
10. Look for a fluorescent ParB focus in the nucleus that is present in each slice of the z-stack. Images can be processed and represented as maximum intensity projections using freely available software such as ImageJ or Fiji.

3.7 Live Cell Imaging

1. Forty-eight hours prior to imaging, electroporate *ParS* repaired HUVECs to introduce 2 μg of the fluorescently labelled ParB protein expression plasmid (see Subheading 3.4).

2. Seed the electroporated cells in a 0.5-ml droplet of medium on the glass area of a 35-mm MatTek dish.
3. Twenty-four hours post-electroporation, the cells should have adhered to the glass surface. At this point, the MatTek dish can be filled with 2 ml of culture medium.
4. Forty-eight hours post-electroporation, the cells are ready to be imaged. Prepare the microscope by switching on the environmental chamber 30 min prior to imaging.
5. Secure the MatTek dish on the microscope stage and remove the cover. Using the appropriate fluorescent channel, scan for the cells that display fluorescent ParB foci. The imaging process can be set up so that images from a few cells can be acquired in a single acquisition by storing their *x-y* coordinates. For each of these cells, 0.2-μm z-stack positions can also be set.
6. Once the microscope has been set up, carefully open the environmental chamber to add 2 μl TNFα directly to the culture medium to stimulate the assembly of the *SAMD4A*, *TNFAIP2*, and *SLC6A5* multigene complex. Commence the imaging sequence immediately after this. For this particular experiment, the cells were imaged every 30 s for 10 min, in which z-stacks were collected in the bright-field and 488-nm channels (Fig. 4).
7. 4D stacks can be processed with ImageJ or Fiji. It is important that in this process, any drift is corrected for by using landmark features in the field of view of each slice. Measurements on the dynamics of the labelled chromosomal locus can also be made.

4 Notes

1. The site of *ParS* integration is determined by the CRISPR/Cas9 cleavage site. At this stage of the workflow, it is important to consider these sites very carefully so that the insertion does not affect the integrity of the transcript and produce confounding results. This means that gRNA positions should ideally be designed to target intronic regions or nearby regions outside the gene body and regulatory elements. To do this, we advise that well-annotated genomic maps are used to navigate this space, such as those provided by the UCSC genome browser.
2. This web tool does a comprehensive search of the target region for PAM sites required for different Cas orthologs and ranks the returned candidate gRNAs according to various parameters, such as predicted off-targets and cutting efficiency. We limit our searches to the *S. pyogenes* Cas9 variant, which is expressed from the pX330 plasmid that we use. Furthermore, the expression of the gRNA from this plasmid is driven by a Pol

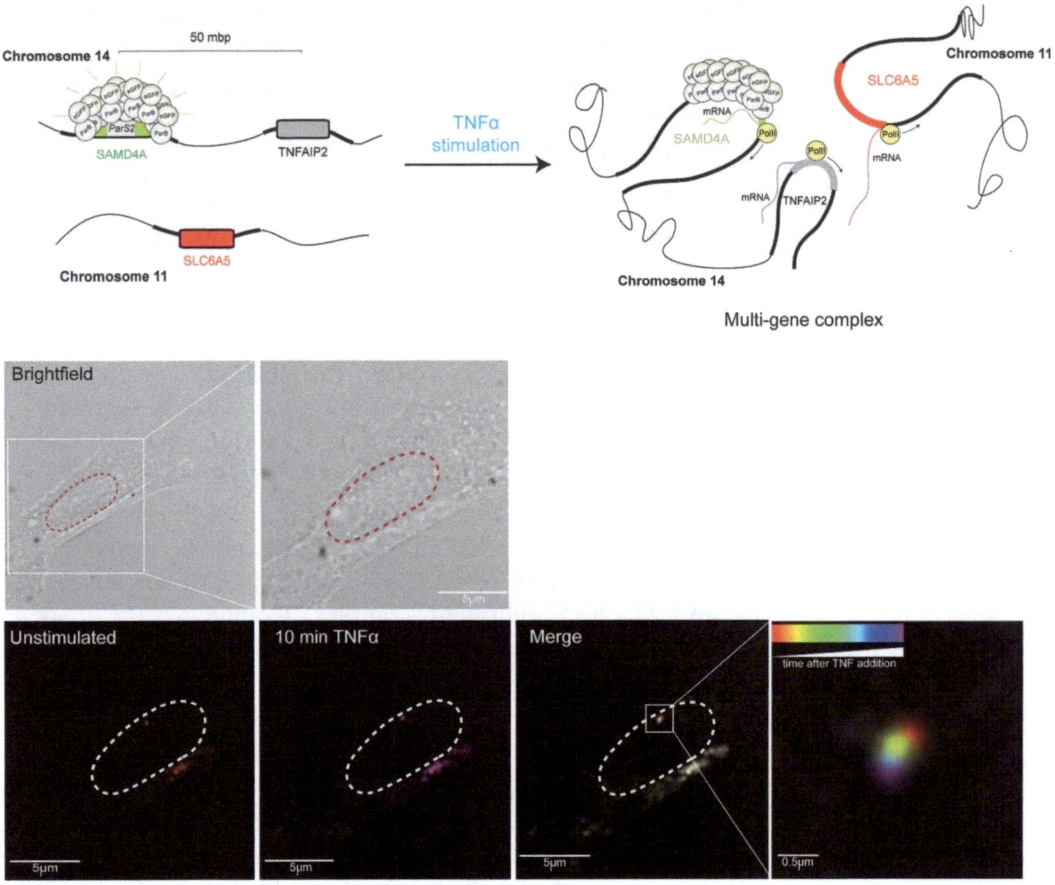

Fig. 4 A multigene complex forms through the establishment of long-range *cis* and *trans* contacts between *SAMD4A*, *TNFAIP2*, and *SLC6A5* in HUVECs when stimulated with TNFα for 10 min. The chromatin dynamics of the *SAMD4A* locus was observed by labelling the site with the ANCHOR system and live cell microscopy. Bars = 5 μm

III U6 promoter and requires a guanine as the first nucleotide for the efficient initiation of transcription. For gRNA candidates that do not begin with a guanine, we simply add this to the start of the sequence to generate a slightly longer 21-nucleotide gRNA without any adverse effects.

3. We generally omit the optional step of Plasmid-Safe exonuclease treatment as we have found that this protocol is highly efficient and produces very few false-positive clones even without it.

4. For lipofection, we generally use 1–2 μg of plasmid with 2.5 μl of Lipofectamine 2000 for 0.8×10^5 retinal pigment epithelial (RPE1) cells per well, seeded 24 h before in a 24-well plate. We follow the exact procedure outlined by the Lipofectamine 2000 protocol.

5. The amplified band should be strong and highly specific. A nonspecific banding pattern will require that PCR conditions be optimized as this will cause false-positive results.
6. For overlap extension PCR assembly, the ends of the parts that need to be fused together should overlap with each other by 20–40 bp. We also recommend that the primers which anneal to the terminal ends of the donor have unique restriction sites embedded in them to allow for the assembled fragment to be cloned into the multiple cloning site of pcDNA3.1(+) for future use. We advise that a well-annotated DNA sequence file is generated for this construct using software that can visually represent these files in a clear way. This will greatly aid in the design process. We use SnapGene for this purpose, which can easily manipulate DNA sequences and simulate cloning and assembly procedures. Other free substitutes could serve this purpose too.
7. The PCR amplification of the homology arms directly from gDNA may be difficult and require substantial effort in optimization. If this process is too cumbersome, the homology arms can be commercially synthesized.
8. For access to plasmids bearing *ParS* sequences, cognate ParB expression vectors, primer, and sequence information, contact NeoVirTech (www.Neovirtech.com; contact@neovirtech.com).

Acknowledgements

We acknowledge contributions from Franck Gallardo (NeoVirTech, Toulouse, France) for the experimental work done optimizing the ANCHOR system. This work was supported by the Council for Industrial and Scientific Research (CSIR, South Africa) and the French National Research Agency (ANR).

References

1. Lieberman-Aiden E, van Berkum NL, Williams L et al (2009) Comprehensive mapping of long-range interactions reveals folding principles of the human genome. Science 326:289–293
2. Dixon JR, Selvaraj S, Yue F et al (2012) Topological domains in mammalian genomes identified by analysis of chromatin interactions. Nature 485:376–380
3. Jin F, Li Y, Dixon JR et al (2013) A high-resolution map of the three-dimensional chromatin interactome in human cells. Nature 503:290–294
4. Fanucchi S, Shibayama Y, Mhlanga MM (2014) Are genes switched on when they kiss? Nucleus 5:103–112
5. Osborne CS, Chakalova L, Brown KE et al (2004) Active genes dynamically colocalize to shared sites of ongoing transcription. Nat Genet 36:1065–1071
6. Schoenfelder S, Sexton T, Chakalova L et al (2010) Preferential associations between co-regulated genes reveal a transcriptional interactome in erythroid cells. Nat Genet 42:53–61

7. Noordermeer D, de Wit E, Klous P et al (2011) Variegated gene expression caused by cell-specific long-range DNA interactions. Nat Cell Biol 13:944–951
8. Li G, Ruan X, Auerbach RK et al (2012) Extensive promoter-centered chromatin interactions provide a topological basis for transcription regulation. Cell 148:84–98
9. Deng W, Lee J, Wang H et al (2012) Controlling long-range genomic interactions at a native locus by targeted tethering of a looping factor. Cell 149:1233–1244
10. Lai F, Orom UA, Cesaroni M et al (2013) Activating RNAs associate with mediator to enhance chromatin architecture and transcription. Nature 494:497–501
11. Peric-Hupkes D, Meuleman W, Pagie L et al (2010) Molecular maps of the reorganization of genome-nuclear lamina interactions during differentiation. Mol Cell 38:603–613
12. Dixon JR, Jung I, Selvaraj S et al (2015) Chromatin architecture reorganization during stem cell differentiation. Nature 518:331–336
13. Ibarra A, Benner C, Tyagi S et al (2016) Nucleoporin-mediated regulation of cell identity genes. Genes Dev 30:2253–2258
14. Isoda T, Moore AJ, He Z et al (2017) Non-coding transcription instructs chromatin folding and compartmentalization to dictate enhancer-promoter communication and T cell fate. Cell 171:103–119.e118
15. Spilianakis CG, Lalioti MD, Town T et al (2005) Interchromosomal associations between alternatively expressed loci. Nature 435:637–645
16. Fullwood MJ, Liu MH, Pan YF et al (2009) An oestrogen-receptor-alpha-bound human chromatin interactome. Nature 462:58–64
17. Papantonis A, Kohro T, Baboo S et al (2012) TNFalpha signals through specialized factories where responsive coding and miRNA genes are transcribed. EMBO J 31:4404–4414
18. Fanucchi S, Shibayama Y, Burd S et al (2013) Chromosomal contact permits transcription between coregulated genes. Cell 155:606–620
19. Raj A, van den Bogaard P, Rifkin SA et al (2008) Imaging individual mRNA molecules using multiple singly labeled probes. Nat Methods 5:877–879
20. Giorgetti L, Galupa R, Nora EP et al (2014) Predictive polymer modeling reveals coupled fluctuations in chromosome conformation and transcription. Cell 157:950–963
21. Tumbar T, Belmont AS (2001) Interphase movements of a DNA chromosome region modulated by VP16 transcriptional activator. Nat Cell Biol 3:134–139
22. Cho WK, Spille JH, Hecht M et al (2018) Mediator and RNA polymerase II clusters associate in transcription-dependent condensates. Science 361:412–415
23. Robinett CC, Straight A, Li G et al (1996) In vivo localization of DNA sequences and visualization of large-scale chromatin organization using lac operator/repressor recognition. J Cell Biol 135:1685–1700
24. Straight AF, Belmont AS, Robinett CC et al (1996) GFP tagging of budding yeast chromosomes reveals that protein-protein interactions can mediate sister chromatid cohesion. Curr Biol 6:1599–1608
25. Kitamura E, Blow JJ, Tanaka TU (2006) Live-cell imaging reveals replication of individual replicons in eukaryotic replication factories. Cell 125:1297–1308
26. Cabal GG, Genovesio A, Rodriguez-Navarro S et al (2006) SAGA interacting factors confine sub-diffusion of transcribed genes to the nuclear envelope. Nature 441:770–773
27. Berger AB, Cabal GG, Fabre E et al (2008) High-resolution statistical mapping reveals gene territories in live yeast. Nat Methods 5:1031–1037
28. Miyanari Y, Ziegler-Birling C, Torres-Padilla ME (2013) Live visualization of chromatin dynamics with fluorescent TALEs. Nat Struct Mol Biol 20:1321–1324
29. Gu B, Swigut T, Spencley A et al (2018) Transcription-coupled changes in nuclear mobility of mammalian cis-regulatory elements. Science 359:1050–1055
30. Dubarry M, Loiodice I, Chen CL et al (2011) Tight protein-DNA interactions favor gene silencing. Genes Dev 25:1365–1370
31. Mariame B, Kappler-Gratias S, Kappler M et al (2018) Real-time visualization and quantification of human cytomegalovirus replication in living cells using the ANCHOR DNA labeling technology. J Virol
32. Germier T, Sylvain A, Silvia K et al (2018) Real-time imaging of specific genomic loci in eukaryotic cells using the ANCHOR DNA labelling system. Methods 142:16–23
33. Saad H, Gallardo F, Dalvai M et al (2014) DNA dynamics during early double-strand break processing revealed by non-intrusive imaging of living cells. PLoS Genet 10: e1004187
34. Ran FA, Hsu PD, Wright J et al (2013) Genome engineering using the CRISPR-Cas9 system. Nat Protoc 8:2281–2308

Chapter 12

Preparing Map of Chromosome Territory Distribution Frequency

Paulina Nastały and Paolo Maiuri

Abstract

The chromosomes in mammalian interphase nuclei are organized into domains called chromosome territories that play a major role in nuclear organization. Here we propose a methodology that combines the use of micro-patterning of adhesive molecules to impose single-cell geometry, with visualization of chromosome territories. This allows obtaining a representative statistical map of the absolute positions of chromosome territories relative to the geometry imposed to the cell population by combining the signal from each cell.

Key words Chromosome territories, Fluorescence in situ hybridization, Immunofluorescence, Confocal microscopy, Micro-patterning, Quantitative imaging, Frequency distribution map

1 Introduction

The relative positions of chromosomes in mammalian interphase nuclei are not random and chromosomes occupy mutually exclusive, globular volumes referred to as chromosome territories [1]. The three-dimensional (3D) organization of eukaryote chromosomes regulates major genome function and nuclear processes [2]. Chromosome positions are heritable during cell cycle and are tissue-specific [3, 4].

It is known that cell geometry dictates the geometry of the cell nucleus [5] and it has already been reported that it influences both gene expression [6] and radial positioning of chromosome territories [7].

Here we describe a methodology that combines imposing single-cell geometry, using micro-patterning of adhesive molecules, with visualization of chromosome territories.

Micro-pattering is a powerful technique used to impose a specific geometry to single cell [8, 9]. This technique allows normalizing the shape of a cell population by forcing all cells plated on

Beatrice Bodega and Chiara Lanzuolo (eds.), *Capturing Chromosome Conformation: Methods and Protocols*, Methods in Molecular Biology, vol. 2157, https://doi.org/10.1007/978-1-0716-0664-3_12,

processed coverslips to acquire the same geometry. Initially, coverslips are covered with a nontoxic cell-repellent surface that is then degraded locally by deep UV light (below 200 nm) passing through a custom-designed photomask [10]. Micro-patterns are after coated with fibronectin, an extracellular matrix protein, in order to form adhesive areas of defined shape, surrounded by a cytophobic surface. Cells are then seeded on micro-patterned coverslips, and by spreading on the adhesive areas, they will acquire the desired geometry. Subsequently, fluorescence in situ hybridization (FISH) is performed using an adapted protocol that enables 3D nuclear structure preservation [11, 12]. Chromosome territories are visualized using commercially available chromosome painting probes. Single cells are then imaged using confocal microscopy. Normalization of the cell population shape obtained with the micro-patterns will finally allow superimposing all single-cell data. Images from each cell are combined to obtain a representative statistical map of the absolute positions of chromosome territories relative to the geometry imposed to the cell population.

2 Materials

2.1 Cell Culture

1. *Cell line*: We routinely use human hTERT-immortalized retinal pigment epithelium cell line, RPE-1 (American Type Culture Collection, ATCC). As ensured by ATCC, this is a human cell line with a modal chromosome number of 46 that occurs in 90% of the cells. Any kind of adherent cell line can be used. You can find the list of recommended cell lines from the 4D Nucleome consortium (*see* **Note 1**).
2. *Culture conditions for hTERT RPE-1 cells*: DMEM/HAM's F12 medium supplemented with 10% fetal bovine serum, 2 mM L-glutamine, 15 mM HEPES, 0.5% sodium pyruvate, and 0.01 mg/ml hygromycin B.

2.2 Micro-patterning of Cells

1. Fibronectin from bovine plasma. Prepare aliquots of 25 μg and store at −20 °C. Working solution: 25 μg/ml in 100 mM $NaHCO_3$ (pH = 8.6); store at 4 °C.
2. Versene solution.
3. 40-μm or 70-μm cell strainer.
4. Centrifuge suitable for cell culture.

2.3 Immunofluorescent Staining Combined with Fluorescence In Situ Hybridization

1. Mouse monoclonal anti-tubulin antibody (clone B-5-1-2, Sigma-Aldrich). Follow the manufacturer's instructions for storage.
2. Secondary anti-mouse Alexa-647-conjugated antibody. Follow the manufacturer's instructions for storage.

3. 1× phosphate buffer saline (1× PBS). Store at 4 °C.
4. *Fixative*: 4% paraformaldehyde in 1× PBS. Aliquot and store at −20 °C.
5. *Blocking solution*: 1% BSA in 1× PBS. Prepare fresh before using.
6. *Permeabilization solution*: 0.1% Triton X in 1× PBS. Prepare fresh before using.
7. *Antibody diluent*: 1% BSA in 1× PBS. Prepare fresh before using.
8. *Permeabilization solution II*: 0.07% Triton X in 0.1 M HCl. Prepare fresh before using.
9. 20× SSC stock solution (3.0 M NaCl, 0.3 M Na citrate) dissolve 175.3 g NaCl and 88.2 g Na citrate in 800 mL distilled water, add distilled water to 1 L, adjust pH to 7.0, and autoclave. Note: store up to 1 year at RT.
10. *Denaturation solution* (pH = 7.4): 50% formamide in 2× SSC. Store at −20 °C.
11. *Hybridization buffer* (pH = 7.0): 50% formamide, 20% dextran in 2× SSC buffer. Aliquot and store at −20 °C.
12. Chromosome painting probes. Store at −20 °C.
13. Rubber cement.
14. *Wash buffer I*: 2× SSC buffer. Prepare fresh before using.
15. *Wash buffer II*: 0.1× SSC buffer. Prepare fresh before using.
16. *DAPI solution*: 5 mg/ml of diamidino-2-phenylindole in DMSO. Working solution: dilute at 1:10,000 in 1× PBS.
17. *Mounting medium*: VECTASHIELD® (Vector Laboratories). Store at 4 °C.
18. Microscopic slides.
19. Transparent nail polish.
20. FISH-dedicated thermoblock suitable for microscopic slides.

2.4 Microscopy

1. Confocal microscope equipped with motorized stage, appropriate lasers, suitable objectives, and software.

2.5 Preparation of Distribution Map

1. ImageJ software [13].

3 Methods

3.1 Micro-patterning of Cells

1. Coat the coverslips (*see* **Note 2**) containing micro-patterns by incubating them for 1 h at 37 °C with fibronectin (*see* **Notes 3** and **4**).

2. For micro-patterning experiments, try not to have too confluent cells. To obtain the best results, split the cells the day before and keep them up to 50% confluence.
3. Remove medium, wash the cells three times with 1× PBS, and add 0.02% EDTA (3 ml for T-25 flask; 5 ml for 100-mm dish).
4. Wait 5–15 min until the cells detach. Control the detachment process under the microscope.
5. Add medium to collect the cells and transfer to a falcon tube. Centrifuge for 3 min at 300 × *g*. Count the cells.
6. If you observe many cell clusters, filter the cells using a cell strainer.
7. Resuspend in medium to obtain a dilution of 10^5 cells in 1 ml.
8. Add a dilution of 10^5 cells in 1 ml.
9. Incubate for 20 min at 37 °C.
10. Check under the microscope cell attachment, and after 20–30 min, wash unattached cells (simultaneously wash out the cell suspension while adding fresh medium).
11. Incubate at 37 °C for 4–6 h.

3.2 Immunofluorescent Staining Combined with Fluorescence In Situ Hybridization

1. Carefully drop a *fixative* onto the cells. Incubate for 10 min. Wash the cells three times for 3 min in 1× PBS.
2. Permeabilize cells for 5 min using the *permeabilization solution I* (*see* **Note 5**). Wash the cells three times for min in 1× PBS.
3. Incubate cells for at least 20 min with *blocking solution*.
4. Prepare the 0.5–1 μg/ml dilution of primary mouse antibody against α-tubulin in *antibody diluent*. Before using the antibody, prepare the titration experiment to choose the optimal dilution. Incubate cells with antibody for 45 min at room temperature. Wash cells three times for 3 min in 1× PBS.
5. Prepare 1:400 dilution of secondary anti-mouse antibody conjugated to Alexa-647 in *antibody diluent*. Incubate cells with antibody for 45 min at room temperature in the dark. Wash the cells three times for 3 min in 1× PBS.
6. Post-fix the cells in *fixative* for 10 min.
7. Incubate the samples in 20% glycerol/1× PBS for at least 60 min (*see* **Note 6**).
8. Submerge each coverslip at a time in liquid nitrogen using small dewar. Withdraw after few seconds and place on a paper towel to defrost. Wait for opaque frozen glycerol to disappear and freeze again. Repeat this procedure three times for each coverslip.
9. Wash the cells for 5 min in 1× PBS.

10. Permeabilize cells for 10 min using the *permeabilization solution II.*
11. Wash the cells for 5 min in 1× PBS.
12. Incubate the cells in *denaturation solution* for 10 min.
13. Prepare the hybridization mixture. For 22 × 22 mm coverslip, use 2.5 μl of each chromosome-painting probe and fill up to 10 μl with hybridization buffer. Gently transfer the mixture onto the microscopic slide. Cover with coverslip and secure with rubber cement.
14. Using the FISH-dedicated thermoblock, denature the slides at 75 °C for 3 min, and allow the probe to hybridize the nuclei for at least 16 h at 37 °C.
15. Perform post-hybridization washes at 45 °C in *wash buffer I* and *wash buffer II* for 10 min each.
16. Wash the slides for 3 min in 1× PBS.
17. Stain the nuclei for 5 min in *DAPI solution.*
18. Mount the coverslips using *mounting medium.* Secure the edges with a transparent nail polish.

3.3 Microscopy

1. Image single cells ensuring that they are well spread within the pattern. It is very convenient to use the confocal microscope with motorized stage in order to select multiple cells, set the z-stack distance and range (*see* **Note** 7), and then let the system acquire multiple positions.
2. Obtain at least ten single cells for each experiment. The optimal number is ≥30 cells/experiment.

3.4 Preparation of Distribution Map

The images are processed using ImageJ software [13].

1. Project all the Z-stacks. Use the "Z Project" command to sum all the slices of the acquired Z-stacks. At this step, you should have one image of each color per cell.
2. Register the images. First decide the registration channel, the channel/color you want to use for the registration. All the remaining channels will be aligned in function of this one. It could be the pattern itself; cytoskeleton components, like actin; or the nucleus, using DAPI. Then choose a reference cell; all the other cells will be registered according to the reference. It is important to choose the reference cell carefully. It has to be as much as possible representative of the population. Run for each cell the TurboReg ImageJ plugin [14]: set the registration channel image of the cell as source and the registration channel image of the reference cell as target. Set "rigid body" algorithm to find the combination of rotation and translation of the source to fit the target. Apply the same roto-translation to the

other channels of the source cell. Repeat the operation for each cell. Because TurboReg is scriptable, all these operations can be automatized in a macro.

3. Combine the registered images. For each channel, combine all the registered images in a single stack. Now you should have a stack of the dimension of your sample for each channel.
4. Segment your channel of interest. Segment the stack corresponding to the channel/color of interest. Use the same algorithm to compute an appropriate threshold value for each image in the stack, e.g., Otsu. Then divide the stack for 255; all pixels over the threshold would then be equal to one. If the goal is to only produce a qualitative distribution and not a quantitative frequency map, skip this step. It is important to notice that at this point, the information associated with the intensity (density) of the signal is lost. Only the frequency of the signal over the threshold on that particular position will be considered.
5. Produce the map. Project the stack of interest making a sum of all the images in the stack. Rescale the image and use the desired lookup table.

4 Notes

1. 4D Nucleome consortium recommended cell lines are listed below:

   https://www.4dnucleome.org/cell-lines.html.
2. To prepare the coverslips with micro-patterns using photolithography, you can follow many protocols [7]. You can also use commercially available ones (e.g., CYTOO or 4Dcell).
3. You can also use any other protein to coat the coverslips e.g. laminin or collagen.
4. In order to visualize micro-patterns, use fluorescently labeled fibronectin or mix fibronectin with fluorescently labeled BSA.
5. Add the permeabilization solution gently in order not to detach the cells.
6. At this step, it is possible to store the cells for several weeks by freezing them at −20 °C in 50% glycerol/1× PBS. Remember to protect them from light. Before proceeding with the protocol, recalibrate the sample in 20% glycerol/1× PBS.
7. We routinely use 60× oil immersion objective and acquire images with z-stacks of 0.3 μm of focal plane.

Acknowledgements

This work was supported by Italian Association for Cancer Research (AIRC) fellowship Project Code: 19416. Paulina Nastaly was supported by Fondazione Umberto Veronesi with a postdoctoral fellowship.

References

1. Cremer T, Cremer M (2010) Chromosome territories. Cold Spring Harb Perspect Biol 2: a003889
2. Schmitt AD, Hu M, Ren B (2016) Genome-wide mapping and analysis of chromosome architecture. Nat Rev Mol Cell Biol 17:743–755
3. Gerlich D, Beaudouin J, Kalbfuss B, Daigle N, Eils R, Ellenberg J (2003) Global chromosome positions are transmitted through mitosis in mammalian cells. Cell 112:751–764
4. Parada L, McQueen PG, Misteli T (2004) Tissue-specific spatial organization of genomes. Genome Biol 5:R44
5. Versaevel M, Grevesse T, Gabriele S (2012) Spatial coordination between cell and nuclear shape within micropatterned endothelial cells. Nat Commun 3:671
6. Jain N, Iyer KV, Kumar A, Shivashankar GV (2013) Cell geometric constraints induce modular gene-expression patterns via redistribution of HDAC3 regulated by actomyosin contractility. Proc Natl Acad Sci U S A 110:11349–11354
7. Wang Y, Nagarajan M, Uhler C, Shivashankar GV (2017) Orientation and repositioning of chromosomes correlate with cell geometry-dependent gene expression. Mol Biol Cell 28:1997–2009
8. Thery M, Racine V, Piel M, Pepin A, Dimitrov A, Chen Y, Sibarita JB, Bornens M (2006) Anisotropy of cell adhesive microenvironment governs cell internal organization and orientation of polarity. Proc Natl Acad Sci U S A 103:19771–19776
9. Thery M (2010) Cell micropatterning as a tool to decipher cell morphogenesis and functions. J Cell Sci 123:4201–4213
10. Azioune A, Storch M, Bornens M, Thery M, Piel M (2009) Simple and rapid process for single cell micro-patterning. Lab On Chip 9:1640–1642
11. Solovei I, Cavallo A, Schermelleh L, Jaunin F, Scasselati C, Cmarko D, Cremer C, Fakan S, Cremer T (2002) Spatial preservation of nuclear chromatin architecture during three-dimensional fluorescence in situ hybridization (3D-FISH). Exp Cell Res 15:10–23
12. Solovei I, Cremer M (2010) 3D-FISH on cultured cells combined with immunostaining. Methods Mol Biol 659:117–126
13. Rueden CT, Schindelin J, Hiner MC et al (2017) ImageJ2: imageJ for the next generation of scientific image data. BMC Bioinformat 18:529
14. Thévenaz P, Ruttimann UE, Unser M (1998) A pyramid approach to subpixel registration based on intensity. IEEE Trans Image Process 7(1):27–41

Chapter 13

Higher-Order Chromatin Organization Using 3D DNA Fluorescent In Situ Hybridization

Quentin Szabo, Giacomo Cavalli, and Frédéric Bantignies

Abstract

A comprehensive analysis of the tridimensional (3D) organization of the genome is crucial to understand gene regulation. Three-dimensional DNA fluorescent in situ hybridization (3D-FISH) is a method of choice to study nuclear organization at the single-cell level. The labeling of DNA loci of interest provides information on their spatial arrangement, such as their location within the nucleus or their relative positioning. The single-cell information of spatial positioning of genomic loci can thus be integrated with functional genomic and epigenomic features, such as gene activity, epigenetic states, or cell population averaged chromatin interaction profiles obtained using chromosome conformation capture methods. Moreover, the development of a diversity of super-resolution (SR) microscopy techniques now allows the study of structural chromatin properties at subdiffraction resolution, making a finer characterization of shapes and volumes possible, as well as allowing the analysis of quantitative intermingling of genomic regions of interest. Here, we present and describe a 3D-FISH protocol adapted for both conventional and SR microscopy such as 3D structured illumination microscopy (3D-SIM), which can be used for the measurement of 3D distances between loci and the analysis of higher-order chromatin structures in cultured *Drosophila* and mammalian cells.

Key words Fluorescent in situ hybridization, Super-resolution microscopy, Chromatin structure, Chromosome folding, Nuclear organization, Single-cell

1 Introduction

Our understanding of the mechanisms underlying genome regulation has been recently expanded by a deeper characterization of the relationships between structural and functional properties of chromatin folding. DNA fluorescent in situ hybridization (FISH) has been a pioneer technique in the study of nuclear organization, revealing, for example, the presence of chromosome territories or the differential location of active and inactive genes [1, 2]. In the last decade, molecular biology experiments have allowed the genome-wide characterization of several chromosomal features, such as chromatin epigenetic status, chromatin interaction profiles,

Beatrice Bodega and Chiara Lanzuolo (eds.), *Capturing Chromosome Conformation: Methods and Protocols*, Methods in Molecular Biology, vol. 2157, https://doi.org/10.1007/978-1-0716-0664-3_13,

or attachment to the nuclear periphery. These methods have provided crucial information, but they cannot address the cell-to-cell variability within populations, and therefore, they do not provide information as to what extent these features reflect the single-cell states. For example, the high-throughput version of the "chromosome conformation capture" method, Hi-C [3], revealed the presence in many species of sub-megabase domains of preferential chromatin interactions, commonly named topologically associating domains (TADs) [4–7]. TADs are often characterized by the enrichment of chromatin interactions within a given genomic region compared to its surrounding chromatin. However, Hi-C experiments represent averaged chromatin interaction profiles typically obtained using millions of cells, hampering the understanding of the structural organization of the chromatin in a single cell. Therefore, FISH appears highly complementary to "chromosome conformation capture" methods in order to decipher how TADs, or any other layer of chromosome organization, fold in individual cells. Recently, different studies have been using FISH to explore the correspondence between single-cell chromatin organizations and genomic-based assays [8–15], showing, for instance, differential degree of condensation for different chromatin states [9], the large-scale organization of chromosomes into active or repressive compartments [13, 15], the absolute contact frequencies between TAD borders [10], or the TAD-based physical organization of chromosomes in *Drosophila* and human cells [8, 12]. In this chapter, we will describe a detailed 3D-FISH protocol that can be applied both in *Drosophila* and in mammalian cells. Two complementary approaches will be presented: 3-color FISH used to systematically measure 3D distances between genomic regions of interest using confocal laser scanning microscopy (CLSM) and 2-color FISH to characterize chromatin structure using 3D-SIM SR microscopy. This FISH protocol can be combined with immunostaining. In addition, we will briefly describe the probe design and present some guidelines for quantitative image analysis.

2 Materials

2.1 Materials for Nick-Translation-Based Probe Production and Labeling

1. Genomic DNA.
2. Primers.
3. GoTaq G2 Hot Start Green Master Mix (Promega, M7422).
4. PCR purification kit.
5. DNA marker and ladder.
6. LE agarose.
7. FISH Tag DNA kits (Invitrogen Life Technologies, F32951 for the multicolor kit).

2.2 Materials for FISH

2.2.1 General Equipment

1. Coverslips 0.170 ± 0.005 mm (also labeled as #1.5H) (Zeiss) (*see* **Note 1**).
2. Inox Tweezers (such as Dumont N°5).
3. Kimwipes Kimtech.
4. 6-well plates for cell culture.
5. Dark humid chamber for slides.
6. Rubber cement.
7. Transparent nail polish.
8. Aluminum block for heating.
9. Water bath with cover (such as VWR digital water bath 5 L).
10. Vacuum.

2.2.2 Solutions

1. 0.01% poly-L-lysine: 1/10 (vol/vol) of 0.1% poly-L-lysine in distilled and deionized water (ddH_2O).
2. PBS.
3. 4% paraformaldehyde (make it fresh): 1/10 (vol/vol) of 10× PBS and 1/4 (vol/vol) of 16% formaldehyde in ddH_2O.
4. 0.5% Triton X-100 (make it fresh): 1/200 (vol/vol) of Triton X-100 in PBS (*see* **Note 2**).
5. 0.1 M HCl: 415 μL of 37% HCl in 50 mL ddH_2O (make it fresh).
6. 2× SSCT: 1/10 (vol/vol) of 20× SSC and 1/1000 (vol/vol) of Tween 20 in ddH_2O.
7. 50% formamide, 2× SSCT (make it fresh): 1/2 (vol/vol) of deionized formamide (Sigma, F-9037) and 1/2 (vol/vol) of 4× SSCT (4× SSCT made from 1/5 [vol/vol] of 20× SSC and 1/500 [vol/vol] of Tween 20 in ddH_2O).
8. 2× SSC: 1/10 (vol/vol) of 20× SSC in ddH_2O (*see* **Note 3**).
9. 0.1× SSC: 1/200 (vol/vol) of 20× SSC in ddH_2O (*see* **Note 3**).
10. 0.2× SSC: 1/100 (vol/vol) of 20× SSC in ddH_2O (*see* **Note 3**).
11. DAPI (4,6-diamidino-2-phenylindole): 1/1000 to 1/10,000 of DAPI stock solution (stock solution at 1 mg/mL in 180 mM Tris-HCl, pH 7.5, stored at −20 °C) diluted in PBS (*see* **Note 4**).

2.2.3 Denaturation, Hybridization, and Mounting

1. FISH hybridization buffer (FHB): 1/2 (vol/vol) of deionized formamide, 1/10 (vol/vol) of 20× SSC, 1/5 (vol/vol) of 50% dextran sulfate solution (made from dextran sulfate Mw 500,000), sheared Salmon Sperm DNA at final concentration

0.5 mg/mL, volume adjusted in ddH_2O. Store FHB in aliquots at −20 °C (*see* **Note 5**).

2. RNase A stock in 10 mg/mL solutions at −20 °C.
3. Probes.
4. Vectashield.

2.3 Materials for Immunostaining (If Combined with FISH, See Note 17)

1. PBT: 1/1000 (vol/vol) Tween 20 diluted in PBS.
2. Blocking solution: 2% BSA diluted in PBT.

2.4 Software

1. Primer design: Primer 3 (for the nick-translation-based probes, *see* Subheading 3.1.1).
2. Image analysis: A wide range of software can be used for FISH analysis (*see* Subheading 3.4).

3 Methods

3.1 Probe Design and Production

Choose the genomic region of interest. After selecting the region, different approaches can be used to obtain DNA fluorescent probes. Here, we will use two different methods. The first one consists in amplifying the genomic region of interest using PCR, followed by direct labeling of DNA fragments by nick translation. The second one uses the Oligopaint technology, developed in the laboratory of C-T Wu (Harvard Medical School Boston) and described in [16, 17]. An example of probe design based on Hi-C maps and using these two methods is shown in Fig. 1. Note that the presented protocol also works with probes derived from Bacterial Artificial Chromosomes (BAC).

3.1.1 Probes Labeled by Nick Translation

Extract the genomic sequence of interest covering 10–12 kb (for *Drosophila* probes) or 20–24 kb (for mammalian probes) and subdivide the sequence into contiguous 2-kb segments, i.e., representing a total of 5–6 fragments for *Drosophila* and 10–12 fragments for mammals (*see* **Note 6**). It is crucial to avoid including repeat sequences such as transposable elements. For each segment, design PCR primers using Primer3 software (http://primer3.ut.ee/) in order to obtain amplicons of 1.2–1.7 kb, again avoiding repeat sequences by using the appropriate reference genome. After purification, the products can be re-amplified using a second round of PCR in order to increase the concentration and purified again. PCR fragments are then directly labeled using the FISH Tag DNA kit, starting one given probe with 1 μg of total DNA from every PCR fragments pooled in equimolar amount. Follow the procedure describe in the FISH Tag DNA kit protocol (*see* **Note 7**).

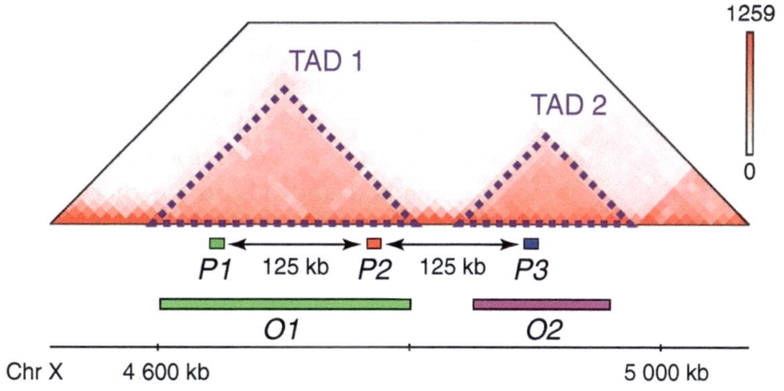

Fig. 1 Example of FISH experimental design to study TAD folding. Hi-C map (visualized with Juicebox [22]) from *Drosophila* S2R+ cells representing a specific chromosomal region along with the location of FISH probes. TADs are highlighted with dashed lines. *P1*, *P2*, and *P3* probes were produced using the nick translation method starting from 6 PCR fragments each and were labeled with Alexa-488, Alexa-555, and Alexa-647 fluorophores, respectively. P1–P2 and P2–P3 are at equivalent genomic distances, i.e., 125 kb. Three control probes were also produced by alternatively labeling the PCR fragments of the *P2* region with Alexa-488, Alexa-555, and Alexa-647 fluorophores (2 PCR fragments used for each color, *see* Fig. 3). *O1* and *O2* Oligopaint probes were labeled with Alexa-488 and ATTO-565, respectively. On the one hand, this design with the nick-translation-based probes allows studying 3D distances between loci inside TADs versus between loci located in two adjacent TADs (*see* Fig. 3). On the other hand, this design with the Oligopaint probes allows studying some physical properties of TADs in single cells, such as their size or the degree of overlap between two adjacent TADs (*see* Fig. 4)

3.1.2 Oligopaint Probes

To produce Oligopaint probes, we recommend to follow the procedures described by Beliveau et al. [16, 17]. Additional information can be found on the Oligopaint website: https://oligopaints.hms.harvard.edu/. Briefly, this technique consists in producing a series of DNA oligos composed of a homology genomic sequence targeting the region of interest and a non-genomic "MainStreet" tail labeled with a fluorophore. The MainStreet tails can be targeted with secondary oligos carrying additional fluorophores to increase the fluorescence signal.

3.1.3 Control Probes

The very nature of light, or sometimes, the design of the microscope, introduces biases and systematic errors during image acquisition. Depending on the type of the analysis, it can thus be necessary to assess errors such as chromatic aberrations by producing a set of probes labeled simultaneously with different fluorophores and/or produce color-swapped probes (two sets of probes where fluorophores are swapped) (*see* Subheading 3.4.5). This is more accurate than simply imaging fluorescent labeled beads in mounting medium because the optical path between the control and the actual experiment environments is the same. In the case of nick-translation-based probes, the series of the 1.2–1.7-kb fragments of one probe can be alternatively labeled with the different fluorophores, i.e., in a 6-fragment scenario and a 3-color FISH experiment, one color for fragments 1 and 4, another color for fragments 2 and 5, and another color again for fragments 3 and 6. For Oligopaints, secondary oligos labeled with a different fluorophore than the primary ones can be used.

3.2 FISH Procedure

3.2.1 FISH

This protocol is designed for FISH on cells attached on coverslips (*see* **Note 8**). Individual coverslips are placed within a well of a 6-well plate. For each incubation step, 3 mL (or 4 mL for washes) of the appropriate solution is deposited within the well and aspirated after the incubation time using vacuum before the next step. Unless specific indications, incubations are performed at room temperature. After hybridization (**step 17**), coverslips must be protected from light.

1. Prepare coverslips: deposit each coverslip in a well of a 6-well plate, rinse with 100% ethanol, allow coverslips to dry for 5–15 min, incubate for 5 min in 0.01% poly-L-lysine, and allow coverslips to dry again (*see* **Note 9**).
2. Deposit coverslips within a humid chamber, i.e., with water deposited in the bottom of the chamber without touching the coverslips (*see* **Note 10**).
3. Prepare a cell suspension of ~1.5–2 × 10^6 cells/mL in growth media and apply 100 μL of the cell suspension in the center of each coverslip. Allow cells to adhere to coverslips for approximately 1 h (*see* **Note 10**).
4. Take coverslips with tweezers, plunge them briefly in a recipient containing PBS, and deposit them (cells facing up) within wells of a 6-well plate (*see* **Note 10**).
5. Fix for 10 min in 4% paraformaldehyde.
6. Wash three times for at least 2 min each in PBS.
7. Incubate for 10 min in 0.5% Triton X-100.
8. Wash three times for at least 2 min each in PBS.
9. Incubate for 10 min in 0.1 M HCl.
10. Wash twice for 2 min each in 2× SSCT.
11. Incubate for at least 30 min in 50% formamide, 2× SSCT (*see* **Note 11**).
12. Prepare probe mixture: for nick translation probes, mix 15–30 ng of each probe with 0.8 μL of RNase A in FHB for a total volume of 15–25 μL (*see* **Notes 12** and **13**); for Oligopaint probes, mix probes (each at ~1–3 μM final concentration)

with the same amount of their secondary oligo and 0.8 μL of RNase A in FHB for a total volume of 15–25 μL. For both nick translation and Oligopaint probes, keep the highest possible ratio of FHB/total volume (*see* **Note 14**).

13. Use tweezers to take out coverslips from wells, lay the coverslips with cells facing up, and drain off the excess of liquid with Kimwipes without touching the cells.
14. Add probe mixtures directly on coverslips.
15. Invert coverslips onto slides, i.e., cells and probe mixture between the coverslip and the slide, and seal coverslips on slides with rubber cement. Allow rubber cement to air-dry about 10–15 min.
16. Denature DNA by putting slides, coverslips facing up, for 3 min at 78 °C or 80 °C on a heating block immersed in a water bath.
17. Remove slides from the heating block, and deposit them in a dark humid chamber for O/N hybridization (14–20 h), at 37–52 °C (*see* **Note 15**).
18. Use tweezers to peel off rubber cement and deposit coverslips (cells facing up) into wells containing 2× SSC (*see* **Note 16**).
19. Wash three times for 5 min each at 37 °C in 2× SSC by putting the 6-well plate floating within a water bath (*see* **Note 16**).
20. Wash three times for 5 min each at 45 °C in 0.1× SSC by putting the 6-well plate floating within a water bath (*see* **Note 16**).
21. Wash twice for 2 min each in PBS (*see* **Note 17**).
22. Incubate for 10 min with DAPI (*see* **Note 4** for concentrations), with gentle agitation.
23. Wash three times for 2 min each in PBS (*see* **Note 18**).
24. Use tweezers to take out coverslips from wells, lay the coverslips with cells facing up, and drain off the excess of liquid with Kimwipes without touching the cells.
25. Deposit one drop of Vectashield (~12 μL) directly on coverslips.
26. Invert coverslips onto slides, i.e., cells and Vectashield between the coverslip and the slide, and seal coverslips on slides with nail polish. Allow nail polish to air-dry for ~1 h (*see* **Note 19**).
27. Slides can be stored up to 2 weeks in the dark at 4 °C before imaging.

3.2.2 FISH Combined with Immunostaining

To combine FISH with immunostaining, proceed as follows after **step 21** of Subheading 3.2.1 of the FISH procedure.

1. Wash three times for at least 1 min each in PBT.

2. Incubate in blocking solution for 30–60 min.
3. Use tweezers to take out coverslips from wells, lay the coverslips with cells facing up, and drain off the excess of liquid with Kimwipes without touching the cells.
4. Add primary antibody diluted in blocking solution (total volume of ~20 μL) directly on coverslips.
5. Invert coverslips onto slides, i.e., cells and antibody solution between the coverslip and the slide, and incubate for 120 min at room temperature or overnight at 4 °C in a dark humid chamber.
6. Wash for 1 min, 3 min, and twice 5 min in PBT.
7. Use tweezers to take out coverslips from wells, lay the coverslips with cells facing up, and drain off the excess of liquid with Kimwipes without touching the cells.
8. Add secondary antibody diluted in blocking solution (total volume of ~20 μL) directly on coverslips.
9. Invert coverslips onto slides, i.e., cells and antibody solution between the coverslip and the slide, and incubate for 60 min at room temperature in a dark humid chamber.
10. Wash for 1 min, 3 min, and twice 5 min in PBT.
11. Proceed as described in Subheading 3.2.1 from **step 21** (included).

3.3 Image Acquisition

Conventional wide-field microscopy, CLSM, and SR methods such as 3D-SIM can be used to image 3D-FISH experiments. A good image quality with a good resolution and signal-to-noise ratio (SNR) is essential for analysis. Use objectives with the highest numerical aperture (NA); they will optimize both signal and resolution. To maximize the signal, use bright and photostable fluorophores. Increase the SNR by increasing exposure time and laser power rather than using detector gain. While doing so, keep under control the eventual photobleaching. The images should use as much as possible the dynamic range of the camera, avoiding saturated pixels. To take advantage of the resolution offered by the microscope, the sampling distances in lateral (*xy*) and axial (*z*) directions should ideally be less than half the theoretical resolution accordingly to the Nyquist criterion. Most modern microscopes do provide tools to calculate those sampling densities.

1. *CLSM*: Here, CLSM images were acquired with a Leica SP8 microscope equipped with a ×63/1.4NA Plan Apochromat oil immersion objective and photomultiplier tube and hybrid detectors, for a pixel size of 59 nm and z-step size of 300 nm.
2. *3D-SIM*: To successfully perform 3D-SIM imaging, we strongly recommend to refer to the protocol published by

Demmerle et al. [18], which provides comprehensive and detailed guidelines to obtain super-resolution images of high quality. Here, 3D-SIM images were acquired with an OMX V4 microscope (GE Healthcare) equipped with a 100×/1.4NA PSF Plan Super Apochromat oil immersion objective (Olympus), electron-multiplying charge-coupled device (EMCCD) (Photometrics) cameras, for a pixel size of 80 nm and z-step size of 125 nm. Z-stacks were acquired with five phases and three angles per image plane. Raw images were reconstructed using SoftWoRx v6.5 or 7.0 (GE Healthcare), using channel-specific optical transfer functions (pixel size of reconstructed images: 40 nm). The quality of super-resolution reconstructed images was assessed using the SIMcheck plugin [19] on Fiji software [20]. Alignments of colors used for FISH probes was done with Chromagnon [21], using as reference image Oligopaint FISH probe simultaneously labeled with Alexa-488 and ATTO-565 fluorophores (*see* Subheading 3.1.3).

3.4 Image Analysis

The following part is a basic guideline for FISH analysis. Quantitative image analysis can be addressed using a variety of software packages. For example, Image J is a widely used, free and open-source software enriched by many plug-ins. Other commercial software packages, such as Imaris or Volocity, allow image analysis with a polished and user-friendly graphic user interface. Programming using, for example, Matlab, Python, or ImageJ's own scripting interfaces, offers other advantages such as a high versatility and adaptability to specific aims. All the following steps of this part can be addressed using a wide range of software or programs and do not require particular knowledge in programing for image analysis.

3.4.1 Image Preprocessing

Before starting the analysis, data can first be processed in order to reduce the noise of microscopic images (Fig. 2a, b). Image analysis software often proposes different types of filters, such as Gaussian, median, or average filters. For example, a 3 × 3 average filter will replace each pixel value with the neighboring mean in a 3 × 3 pixel region. This will have a blurring effect on the image and thus reduce the noise, which can facilitate further segmentation processes. It will, however, come at the cost of losing resolution. Other more sophisticated techniques, such as image deconvolution, will increase SNR and resolution at the same time, but a certain experience is necessary to avoid the introduction of artifacts.

3.4.2 Segmentation

To separate and identify the labeled objects, an intensity threshold value is defined in order to obtain a binary image in which pixels will be divided into two classes, one under the threshold corresponding to the background and one above the threshold corresponding to the labeled objects. To define the threshold, it can be helpful to visualize the grayscale intensity value distribution

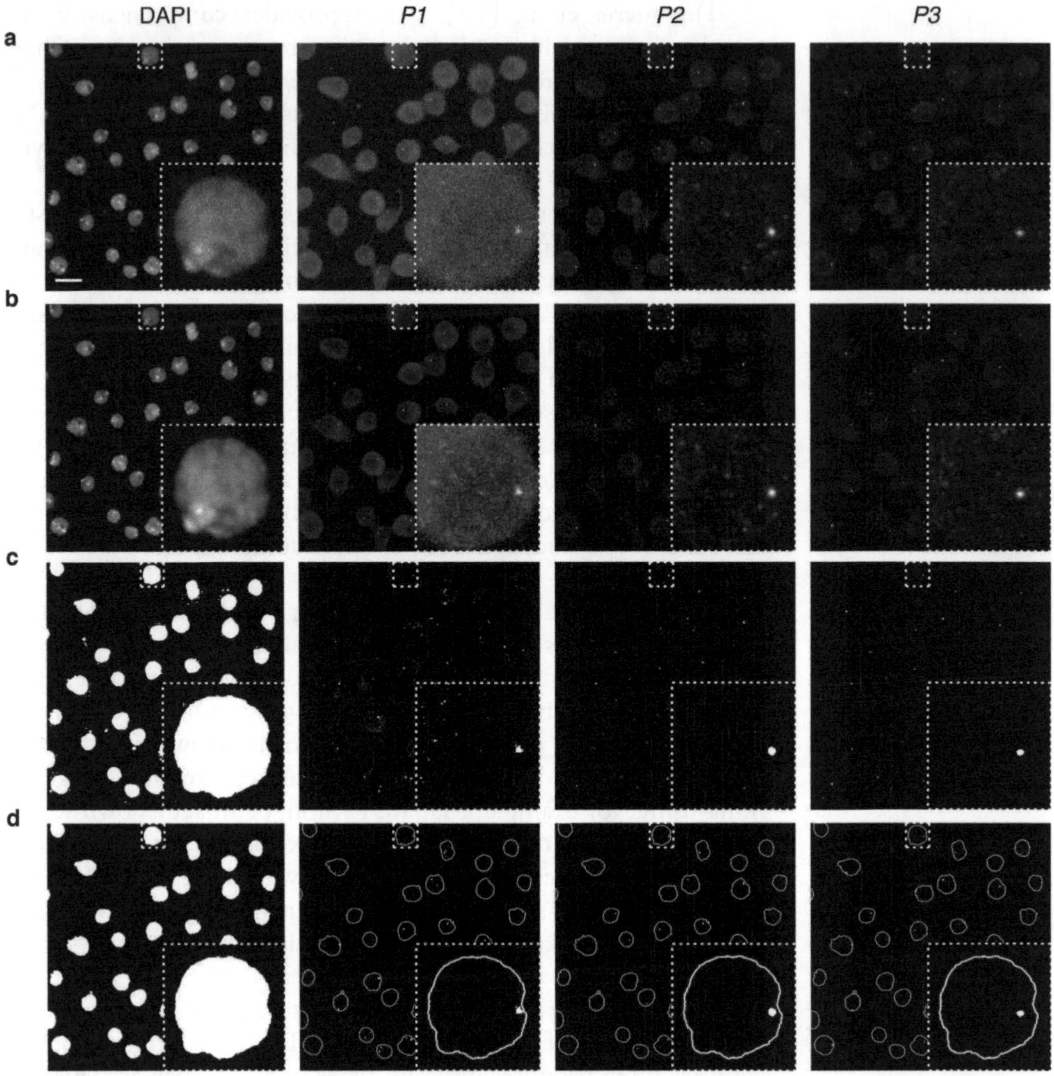

Fig. 2 Example of image processing and segmentation filtering. Each column represents a maximum projection of an individual channel from a 3-color FISH experiment (DAPI and *P1*, *P2*, *P3* probes shown in Fig. 1), along with a magnification of the region within the square. (**a**) Raw images. (**b**) Image preprocessing using a smoothing filter to reduce the noise. (**c**) Image segmentation. (**d**) Filtering of segmented objects using size criteria to discard too small and too large objects as well as application of a DAPI mask to the FISH channels in order to discard any segmented FISH signals located outside of the DAPI segmented objects. Scale bar: 10 μm

of the image using histograms. Image analysis software often provides automatic thresholding algorithms (such as Otsu's method) that can be tested in order to obtain satisfying segmentation. However, it may be difficult to obtain acceptable thresholding using an automatic method, and a threshold value may be manually defined. It is important to have a visual feedback to verify the quality of the segmentation. Examples of segmentations are shown in Fig. 2c. Volume range can be defined to discard too small or too large objects, and it can also be very useful to segment DAPI staining in order to define nuclei in which FISH signals will be analyzed (DAPI mask), i.e., FISH signals outside DAPI segmented objects will be automatically discarded for the analysis (Fig. 2d). This can be particularly helpful if some nonspecific residual fluorescent signal remains outside the nuclei. Moreover, it can be useful to identify individual nuclei for further single-cell analysis. However, if nuclei are too crowded, it may be difficult to automatically segment individual nuclei.

3.4.3 FISH Object Identification

FISH objects can be segmented using intensity-based routines as presented above. Keep in mind that structures of loci of tens to hundreds of kb cannot generally be resolved using conventional wide-field or CLSM microscopy (~200-nm lateral and ~500-nm axial resolution). In this case, segmented objects are larger than the real genomic loci. Loci can thus be defined by the centroids of segmented FISH objects (Fig. 3b). Local intensity maxima approaches can also be used, which identify high-intensity pixels that stand out from the surrounding pixels by more than a defined intensity value. However, if the labeled genomic regions are large enough, FISH segmentation can also provide information such as the volume of the regions and their intermingling (Fig. 4), although the segmented volume may not accurately correspond to the absolute volume of the real genomic loci (the error increasing with decreasing volumes). As volumes of segmented objects strongly depend on the threshold value, special care must be taken to interpret this type of analysis. Depending on the quality of the FISH probes, the SNR, and the efficiency of the intensity-based FISH identification, some nuclei may display false-positive signals, and on the other hand, not all nuclei may display FISH objects. If individual nuclei have been segmented, an efficient way to minimize errors is to define a minimum and maximum number of FISH objects per nucleus such that nuclei that do not respect these criteria will be discarded for the analysis. Further data processing can also be used to ensure proper analysis (*see* the following step).

3.4.4 3D Distance Measurements

After FISH object identification, their *x*, *y*, and *z* coordinates are extracted. If objects were identified as segmented objects instead of single voxels, these coordinates can be defined by their centroids, which allows subpixel resolution. The Euclidean distance is then calculated between coordinates of the FISH signals of different colors, allowing, for example, identification of nearest neighbors using the shortest 3D distance for a given FISH object (Fig. 3c). Alternatively, computing the Euclidean distance transform of binary images can provide nearest distances between segmented objects. Keeping only the mutual nearest neighbors between the

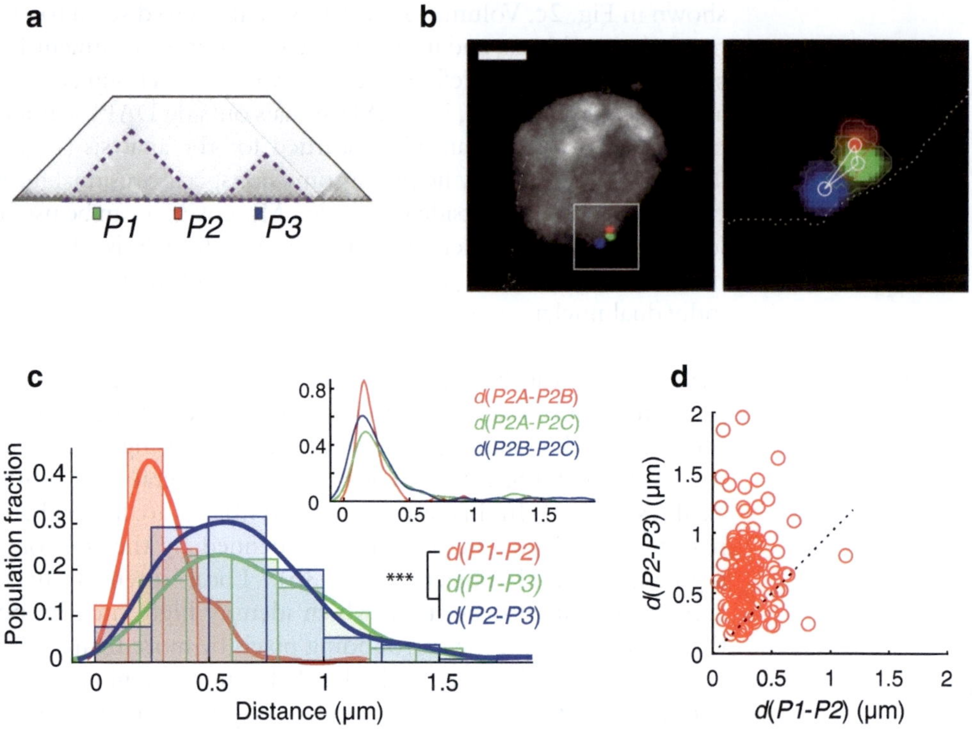

Fig. 3 Example of 3D-FISH analysis using CLSM. (**a**) *P1* (in green), *P2* (in red), and *P3* (in blue) probes were designed to measure 3D distances within a TAD [*d*(*P1–P2*)] and between two adjacent TADs [*d*(*P1–P3*) and *d* (*P2–P3*)]. (**b**) Left, example of CLSM imaging of DAPI (in gray) and FISH probes in S2R+ cells; right, zoom in view of the square showing processed image. Segmented object contours are shown (dashed line for DAPI, segmentation using Otsu's method automatic thresholding; colored lines for FISH, segmentation using a defined threshold value from intensity distribution and visual inspection), and circles indicate centroids of segmented FISH objects (mutual nearest neighbors linked to each other with white lines). Maximum intensity projections are shown. Scale bar: 2 μm. (**c**) Histograms and distribution fits of mutual nearest neighbor 3D distances (from 130 nuclei). On top right, mutual nearest neighbor 3D distance distribution fits between the three control probes (*P2a*, *P2b*, and *P2c* for Alexa-488, Alexa-555, and Alexa-647, respectively, from 49 nuclei). The differences in distance distribution observed in the actual experiments (*P1*, *P2*, and *P3* probes) are not observed with the control probes, indicating that they are not due to technical bias. $***P < 0.0001$ with Kruskal-Wallis and Dunn's multiple comparisons tests. (**d**) Single-cell paired distance analysis, with the *P1–P2* distance compared to the *P2–P3* distance from each triplets of mutual nearest FISH neighbors. The *P1–P2* distance is shorter than the *P2–P3* distance in 85% of the analyzed nuclei. This analysis indicates that the probes located in the same TAD are closer compare to the probes located into adjacent TADs in a large majority of single cells, although they are at the same genomic distance

different channels can also remove outliers. For example, the nearest distance of a FISH object within a nucleus missing the FISH object from the other color may actually represent a distance between FISH objects from different nuclei or unpaired loci. The distance distribution can also inform on the presence of outliers, and a distance cutoff can thus be used to remove them. Moreover, in 3-color FISH experiments, having triplets of mutually nearest

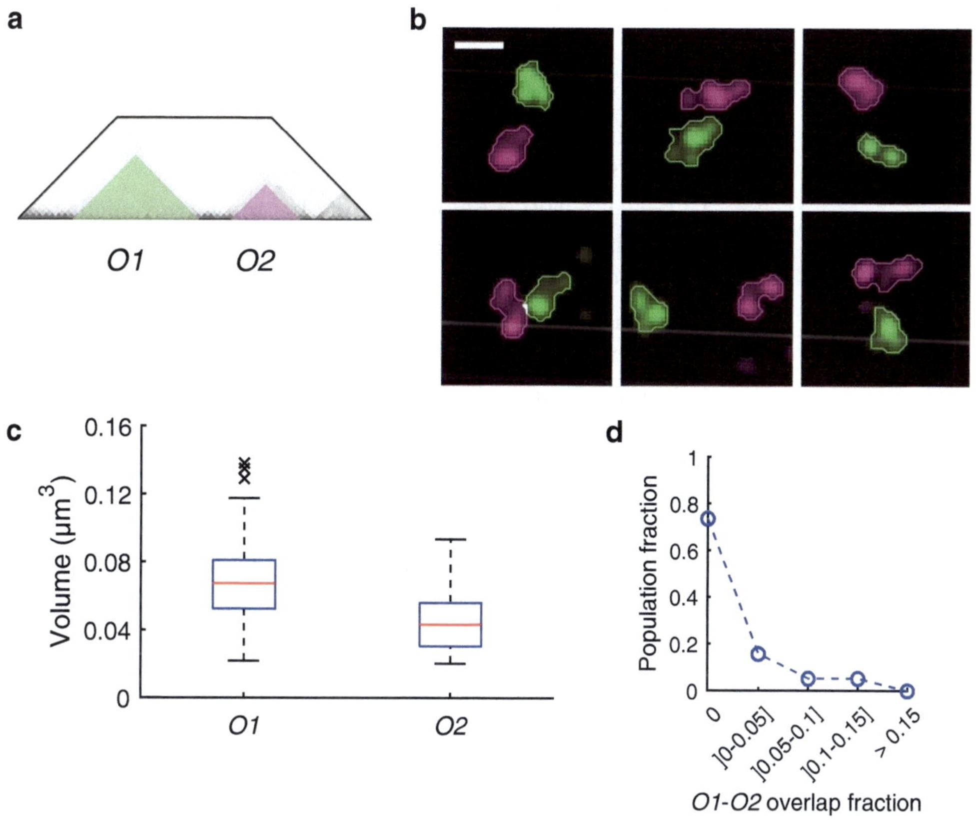

Fig. 4 Example of 3D-FISH analysis using 3D-SIM. (**a**) *O1* (in green) and *O2* (in magenta) probes were used to label two adjacent TADs. (**b**) Examples of 3D-SIM imaging of the probes with their segmentation (Otsu's method automatic thresholding) contour in S2R+ cells. Maximum intensity projections are shown, scale bar: 500 nm. (**c**) Super-resolution 3D-SIM allows volume quantification of the segmented probes (from 38 nuclei). (**d**) Volume overlap fraction (using Jaccard index) between *O1* and *O2* probes. This analysis illustrates the physical partitioning of the genome into TADs in *Drosophila* cells

neighbors allows paired analysis of distances, i.e., distance between one pair of colors compared to the other pairs from the same nucleus (Fig. 3d).

3.4.5 Control Measurements

Errors such as chromatic aberrations or other biases induced by the different wavelengths of the probes, which can lead to systematic shifts between the different probes, have to be assessed. A way to do so is to perform the same experiment and analysis from control experiments with a set of probes labeled simultaneously with the different fluorophores. This control can be used to correct the positioning of target loci and/or to estimate the error of the measurement and calculate the distance distribution corresponding to a truly co-localized genomic region (Fig. 3c).

4 Notes

1. The thickness of the support of the cells is crucial for optimal imaging, especially with SR microscopy. We recommend to use coverslips of 170 ± 5 μm.
2. Triton X-100 is highly viscous. It can be convenient to do an initial 1/10 (vol/vol) solution in ddH$_2$O before further dilutions.
3. Prepare the SSC buffers according to the chosen washing steps of the FISH procedure (*see* Subheading 3.2.1 and **Note 16**).
4. For conventional microscopy, we recommend a final DAPI concentration of 0.1 μg/mL, and for 3D-SIM, we recommend to increase up to 1 μg/mL. When using DAPI, we recommend a relatively fresh batch. DAPI staining is very efficient and may last for many years; however, the quality of a DAPI staining for 3D-SIM is remarkably reduced using aged DAPI.
5. First, prepare a 50% dextran sulfate solution dissolving 5 g of dextran sulfate with 7 mL of prewarmed ddH$_2$O at 60 °C (total volume of 10 mL) in a 50-mL falcon tube. Solution can be stored at 4 °C.
6. The higher is the number of fragments, the better will be the fluorescent signal. Five to six PCR fragments in *Drosophila* and 10–12 PCR fragments in mouse generally ensure good signals, although this number may eventually be reduced.
7. The labeling of probes is described in the FISH Tag DNA kit; however, we adapted the nick translation reaction to optimize using a pool of 1.2–1.7-kb DNA fragments, representing 1 μg of total DNA. For this, we use 1.4–1.7 μL of DNase I working solution for a 50 μL nick translation reaction during 90–120 min. We also recommend to first run a time course test of nick translation, for which the reaction is performed in the same manner, but 5 μL is taken out from the reaction mix every 30 min (from 0 to 120 min). As soon as 5 μL aliquots are taken out, vortex well to inhibit DNase I (you can also add EDTA), and freeze at −20 °C before running every aliquot in gel electrophoresis. Gels should display smears decreasing gradually in size with time. Smears ranging from approximately 100 to 300 bp represent the optimal condition for the nick translation reaction. You can also further increase or decrease the amount of DNase I in order to reach this optimal condition for a 90- to 120-min reaction. At the end of the labeling procedure, the efficiency of dye incorporation will depend on the fluorophore, and good probes should have approximately 5–10 pmol/μL of A488 dye, 1–3 pmol/μL of A555 dye, and 4–8 pmol/μL of A647 dye.

8. Because of the smaller thickness, we recommend to use coverslips rather than slides for optimal imaging (especially for SR microscopy). However, this protocol can also be used with cells attached to slides. In this case, incubation steps are performed with slides deposited in Coplin jars instead of coverslips deposited in 6-well plates.
9. Alternatively, grow cells on autoclaved and coated coverslips, with each coverslip within a well of a 6-well plate.
10. If cells were directly grown in 6-well plates, aspirate the media with a vacuum and rinse with PBS instead of **steps 2–4**.
11. Alternatively, incubate for at least 20 min in 50% formamide, 2× SSCT, and then incubate for 20 min in 50% formamide, 2× SSCT, at 60 °C by putting the 6-well plate floating in a water bath. We obtain better results with mammalian cells using these steps instead.
12. If you use nick-translation-based probes, denature probes in probe mixture for 10 min at 80 °C and put them on ice until usage.
13. If you use mammalian cells and nick-translation-based probes, include within the probe mixture Cot-1 DNA. For example, for mouse cells, use mouse Cot-1 DNA (Invitrogen Life Technologies, 18440-016) at a final concentration of 0.1–0.2 μg/μ L to suppress potential hybridization with repetitive sequences.
14. The higher the FHB/total volume ratio is, the better is the hybridization. A typical experiment uses approximately 2 μL of probes plus 0.8 μL of RNase, which gives a ~8–9/10 ratio for 20 μL of total volume. Do not go lower than a 3/4 ratio of FHB/total volume.
15. The hybridization temperature may vary depending on the probes, with stringency increasing with temperature. We obtain efficient hybridization at 37 °C using both nick-translation based probes and Oligopaints with 42-mer genomic sequences in *Drosophila* cells. We often do our hybridization at 42 °C using Oligopaints with 35-41-mer genomic sequences in mammalian cells. For Oligopaints, we recommend to refer to [17] to choose the optimal hybridization temperature.
16. This note is an alternative of **18–20** washing steps. Use tweezers to peel off rubber cement and deposit coverslips (cells facing up) into wells containing 2× SSCT. Wash for 15 min at 60 °C in 2× SSCT, then wash for 10 min at room temperature in 2× SSCT, and then wash for 10 min at room temperature in 0.2× SSC. We obtain better results with mammalian cells using this washing procedure instead.
17. FISH can be followed by immunostaining against proteins of interest (*see* Subheading 3.2.2). However, the FISH procedure

can alter protein integrity and/or epitopes, and immunostaining may not be suitable with some proteins or antibodies.

18. If samples are going to be imaged using 3D-SIM, extensive washing of DAPI is crucial. Wash four times for at least 5 min each instead.
19. Manage the amount of Vectashield so that there is not too much. Clean the excess of mounting medium before sealing.

Acknowledgements

We thank Julio Mateos-Langerak and Thierry Cheutin for critical reading of the manuscript. We thank the Montpellier Resources Imagerie facility (BioCampus Montpellier, CNRS, INSERM, Université de Montpellier). Q.S. is supported by the French Ministry of Higher Education and Research and La Ligue Nationale Contre le Cancer. F.B. and G.C. are supported by the CNRS. Research at the G.C. laboratory is supported by grants from the CNRS, the European Research Council (2017-AdG No 788972, 3DEpi) the European Union's Horizon 2020 research and innovation program under grant agreement no 676556 (MuG), the Agence Nationale de la Recherche (ANR-15-CE12-0006 EpiDevoMath), the Fondation pour la Recherche Médicale (DEI20151234396), the INSERM, the French National Cancer Institute (INCa), the MSDAVENIR Reseach fund (MSDAVENIR DS-2018-0041 GENE-IGH) and the Laboratory of Excellence EpiGenMed.

References

1. Bickmore WA (2013) The spatial organization of the human genome. Annu Rev Genomics Hum Genet 14:67–84. https://doi.org/10.1146/annurev-genom-091212-153515
2. Cremer T, Cremer M (2010) Chromosome territories. Cold Spring Harb Perspect Biol 2 (3):a003889. https://doi.org/10.1101/cshperspect.a003889
3. Lieberman-Aiden E, van Berkum NL, Williams L, Imakaev M, Ragoczy T, Telling A, Amit I, Lajoie BR, Sabo PJ, Dorschner MO, Sandstrom R, Bernstein B, Bender MA, Groudine M, Gnirke A, Stamatoyannopoulos J, Mirny LA, Lander ES, Dekker J (2009) Comprehensive mapping of long-range interactions reveals folding principles of the human genome. Science 326 (5950):289–293. https://doi.org/10.1126/science.1181369
4. Dixon JR, Selvaraj S, Yue F, Kim A, Li Y, Shen Y, Hu M, Liu JS, Ren B (2012) Topological domains in mammalian genomes identified by analysis of chromatin interactions. Nature 485(7398):376–380. https://doi.org/10.1038/nature11082
5. Hou C, Li L, Qin ZS, Corces VG (2012) Gene density, transcription, and insulators contribute to the partition of the Drosophila genome into physical domains. Mol Cell 48(3):471–484. https://doi.org/10.1016/j.molcel.2012.08.031
6. Nora EP, Lajoie BR, Schulz EG, Giorgetti L, Okamoto I, Servant N, Piolot T, van Berkum NL, Meisig J, Sedat J, Gribnau J, Barillot E, Bluthgen N, Dekker J, Heard E (2012) Spatial partitioning of the regulatory landscape of the X-inactivation centre. Nature 485 (7398):381–385. https://doi.org/10.1038/nature11049
7. Sexton T, Yaffe E, Kenigsberg E, Bantignies F, Leblanc B, Hoichman M, Parrinello H, Tanay A, Cavalli G (2012) Three-dimensional

folding and functional organization principles of the Drosophila genome. Cell 148:458–472. https://doi.org/10.1016/j.cell.2012.01.010

8. Bintu B, Mateo LJ, Su JH, Sinnott-Armstrong NA, Parker M, Kinrot S, Yamaya K, Boettiger AN, Zhuang X (2018) Super-resolution chromatin tracing reveals domains and cooperative interactions in single cells. Science 362(6413). https://doi.org/10.1126/science.aau1783
9. Boettiger AN, Bintu B, Moffitt JR, Wang S, Beliveau BJ, Fudenberg G, Imakaev M, Mirny LA, Wu CT, Zhuang X (2016) Super-resolution imaging reveals distinct chromatin folding for different epigenetic states. Nature 529(7586):418–422. https://doi.org/10.1038/nature16496
10. Cattoni DI, Cardozo Gizzi AM, Georgieva M, Di Stefano M, Valeri A, Chamousset D, Houbron C, Dejardin S, Fiche JB, Gonzalez I, Chang JM, Sexton T, Marti-Renom MA, Bantignies F, Cavalli G, Nollmann M (2017) Single-cell absolute contact probability detection reveals chromosomes are organized by multiple low-frequency yet specific interactions. Nat Commun 8(1):1753. https://doi.org/10.1038/s41467-017-01962-x
11. Fudenberg G, Imakaev M (2017) FISH-ing for captured contacts: towards reconciling FISH and 3C. Nat Methods 14(7):673–678. https://doi.org/10.1038/nmeth.4329
12. Szabo Q, Jost D, Chang JM, Cattoni DI, Papadopoulos GL, Bonev B, Sexton T, Gurgo J, Jacquier C, Nollmann M, Bantignies F, Cavalli G (2018) TADs are 3D structural units of higher-order chromosome organization in Drosophila. Sci Adv 4(2):eaar8082. https://doi.org/10.1126/sciadv.aar8082
13. Wang S, Su JH, Beliveau BJ, Bintu B, Moffitt JR, Wu CT, Zhuang X (2016) Spatial organization of chromatin domains and compartments in single chromosomes. Science 353 (6299):598–602. https://doi.org/10.1126/science.aaf8084
14. Williamson I, Berlivet S, Eskeland R, Boyle S, Illingworth RS, Paquette D, Dostie J, Bickmore WA (2014) Spatial genome organization: contrasting views from chromosome conformation capture and fluorescence in situ hybridization. Genes Dev 28(24):2778–2791. https://doi.org/10.1101/gad.251694.114
15. Nir G, Farabella I, Perez Estrada C, Ebeling CG, Beliveau BJ, Sasaki HM, Lee SD, Nguyen SC, McCole RB, Chattoraj S, Erceg J, AlHaj Abed J, Martins NMC, Nguyen HQ, Hannan MA, Russell S, Durand NC, Rao SSP, Kishi JY, Soler-Vila P, Di Pierro M, Onuchic JN, Callahan SP, Schreiner JM, Stuckey JA, Yin P, Aiden EL, Marti-Renom MA, Wu CT (2018) Walking along chromosomes with super-resolution imaging, contact maps, and integrative modeling. PLoS Genet 14(12):e1007872. https://doi.org/10.1371/journal.pgen.1007872
16. Beliveau BJ, Boettiger AN, Avendano MS, Jungmann R, McCole RB, Joyce EF, Kim-Kiselak C, Bantignies F, Fonseka CY, Erceg J, Hannan MA, Hoang HG, Colognori D, Lee JT, Shih WM, Yin P, Zhuang X, Wu CT (2015) Single-molecule super-resolution imaging of chromosomes and in situ haplotype visualization using Oligopaint FISH probes. Nat Commun 6:7147. https://doi.org/10.1038/ncomms8147
17. Beliveau BJ, Kishi JY, Nir G, Sasaki HM, Saka SK, Nguyen SC, Wu CT, Yin P (2018) OligoMiner provides a rapid, flexible environment for the design of genome-scale oligonucleotide in situ hybridization probes. Proc Natl Acad Sci U S A 115(10):E2183–E2192. https://doi.org/10.1073/pnas.1714530115
18. Demmerle J, Innocent C, North AJ, Ball G, Muller M, Miron E, Matsuda A, Dobbie IM, Markaki Y, Schermelleh L (2017) Strategic and practical guidelines for successful structured illumination microscopy. Nat Protoc 12 (5):988–1010. https://doi.org/10.1038/nprot.2017.019
19. Ball G, Demmerle J, Kaufmann R, Davis I, Dobbie IM, Schermelleh L (2015) SIMcheck: a toolbox for successful super-resolution structured illumination microscopy. Sci Rep 5:15915. https://doi.org/10.1038/srep15915
20. Schindelin J, Arganda-Carreras I, Frise E, Kaynig V, Longair M, Pietzsch T, Preibisch S, Rueden C, Saalfeld S, Schmid B, Tinevez JY, White DJ, Hartenstein V, Eliceiri K, Tomancak P, Cardona A (2012) Fiji: an open-source platform for biological-image analysis. Nat Methods 9(7):676–682. https://doi.org/10.1038/nmeth.2019
21. Matsuda A, Schermelleh L, Hirano Y, Haraguchi T, Hiraoka Y (2018) Accurate and fiducial-marker-free correction for three-dimensional chromatic shift in biological fluorescence microscopy. Sci Rep 8(1):7583. https://doi.org/10.1038/s41598-018-25922-7
22. Durand NC, Robinson JT, Shamim MS, Machol I, Mesirov JP, Lander ES, Aiden EL (2016) Juicebox provides a visualization system for Hi-C contact maps with unlimited zoom. Cell Syst 3(1):99–101. https://doi.org/10.1016/j.cels.2015.07.012

Chapter 14

3D Immuno-DNA Fluorescence In Situ Hybridization (FISH) for Detection of HIV-1 and Cellular Genes in Primary CD4$^+$ T Cells

Bojana Lucic, Julia Wegner, Mia Stanic, Katharina Laurence Jost, and Marina Lusic

Abstract

Fluorescence in situ hybridization (FISH) is a powerful, broadly used microscopy-based technique that leverages fluorescently labeled nucleic acid probes to detect parts of the genome inside metaphase or interphase cell nuclei. In recent years, different methodologies developed to visualize genome topology and spatial relationships between genes have gained much attention as instruments to decode the relationship between chromatin structure and function. In addition to chromosome conformation capture–based techniques, highly multiplexed forms of FISH combined with high-throughput and super-resolution microscopy are used to map and spatially define contact frequencies between different genomic regions. All these approaches have strongly contributed to our knowledge of how the human genome is packed in the cell nucleus.

In this chapter, we describe detailed step-by-step protocols for 3D immuno-DNA FISH detection of genes and Human immunodeficiency virus 1 (HIV-1) provirus in primary CD4$^+$ T cells from healthy donors, or cells infected in vitro with the virus. Our multicolor 3D-FISH technique allows, by using up to three fluorophores, visualization of spatial positioning of loci inside a 3D cell nucleus.

Key words HIV-1, Nucleus, CD4$^+$ T cells, FISH

1 Introduction

Within nuclei, eukaryotic genome is hierarchically organized, from nucleosomes to topologically associated domains (TADs) and chromosomal territories, and this genomic organization is crucial for the coordinated function of the complex chromatin polymer [1, 2]. Fluorescent in situ hybridization (FISH) approaches allow the three-dimensional (3D) architecture of genomes to be visualized at single-cell level.

While FISH applications are broad and range from karyotyping for the diagnosis of chromosomal abnormalities to detection of

Beatrice Bodega and Chiara Lanzuolo (eds.), *Capturing Chromosome Conformation: Methods and Protocols*, Methods in Molecular Biology, vol. 2157, https://doi.org/10.1007/978-1-0716-0664-3_14,

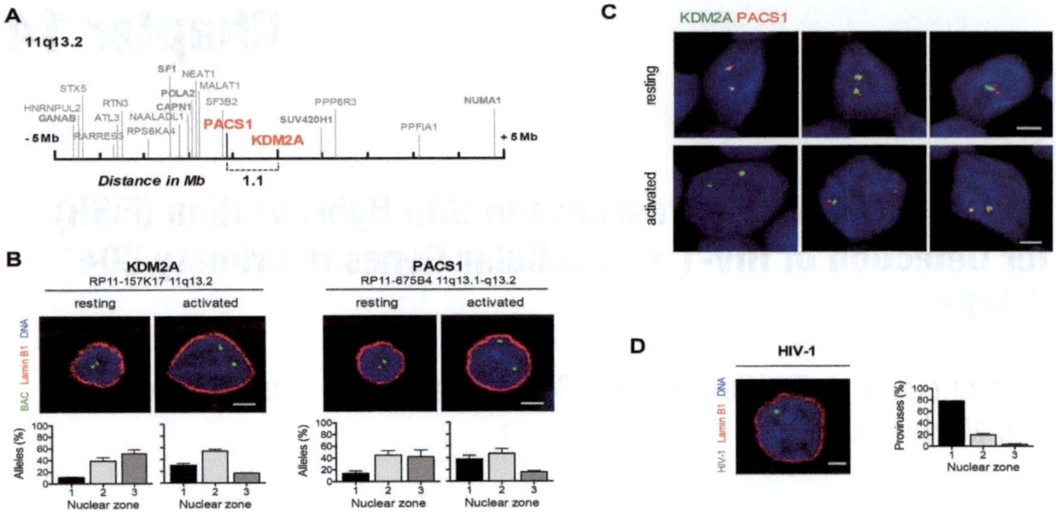

Fig. 1 (**a**) Schematic representation of chromosomal region 11q13.2 within 10 Mb: recurrent integration genes (RIGs) in bold red and single HIV-1 integration sites in plain grey. (**b**) Three-dimensional immuno-DNA FISH of *KDM2A* and *PACS1* in resting and activated (anti-CD3/anti-CD28 beads, IL-2 for 48 h) $CD4^+$ T cells (green, BAC/gene probe; red, lamin B1; blue, DNA staining with Hoechst 33342; scale bar represents 2 μm). Bar plots represent distribution of alleles in 3 nuclear zones of equal surface as in [4]. (**c**) Three-dimensional dual-color DNA FISH of *KDM2A* and *PACS1* in resting and activated (anti-CD3/anti-CD28 beads, IL-2 for 48 h) $CD4^+$ T cells (green, *KDM2A* gene; red, *PACS1* gene; blue, DNA staining with Hoechst 33342). (**d**) 3D immuno-DNA FISH representative image of HIV-1 in activated $CD4^+$ T cell (green, HIV-1 probe; red, lamin B1; blue, DNA staining with Hoechst 33342; scale bar represents 2 μm). Bar plots represent distribution of alleles in 3 nuclear zones of equal surface as in [4]

single-nucleotide polymorphisms, here we focus on visualizing specific genes and HIV-1 provirus in primary $CD4^+$ T cells. We adapted the protocol originally developed by Solovei and Cramer [3] to inspect the spatial positioning of the HIV-1 provirus on one hand, as well as the distribution in 3D nuclear space of human genes into which the virus integrates. By combining immunofluorescence (IF) for nuclear lamina and 3D DNA FISH, either for provirus or its recurrent integration genes (commercially available bacterial artificial chromosome [BAC]-based probes), we inspect the nuclear positioning of the HIV-1 and genes into which it frequently integrates (Fig. 1) [4, 5].

This protocol provides a detailed description of the individual procedures and methods, including primary cell fixation, different cell permeabilization procedures, DNA probe generation and hybridization, and the resolution of the FISH signals in primary T cells. The main advantage of this protocol is the combined detection of preserved lamin protein complexes (or other nuclear structures, such as promyelocytic leukemia (PML) nuclear bodies) and

DNA of interest. The general principles of the FISH technique described herein should be, in principle, applicable to other mammalian cells.

2 Materials

2.1 Reagents and Equipment

- Culture plates (24 wells).
- Coverslips ($d = 12$ mm).
- Microscope slides.
- Ethanol (EtOH).
- Polyethyleneimine (PEI).
- Paraformaldehyde (PFA) electron microscopy grade.
- Dulbecco phosphate-buffered saline (PBS) w/o Mg^{2+} w/o Ca^{2+}.
- Bovine serum albumin (BSA).
- Fetal bovine serum (FBS).
- Anti-lamin B1 ab16048 (rabbit).
- Alexa Fluor 647 anti-rabbit IgG (goat).
- Ethylene glycol bis(succinimidylsuccinate) (EGS).
- Glycerol.
- Triton X-100.
- Tween-20.
- Tris-HCl pH 7.5.
- Sodium chloride (NaCl).
- Sodium citrate tribasic dihydrate.
- Sodium acetate (NaAc).
- Hydrochloric acid (HCl).
- Formamide, deionized.
- Dextran sulfate 50% solution.
- DMSO.
- Saponin.
- PureLink RNase A 20 mg/ml.
- illustra MicroSpin G-25 columns.
- TSA Plus Fluorescein kit.
- UltraPure Agarose.
- 50× TAE buffer (Tris-acetate-EDTA).

- Biotin-16-dUTP.
- Cot-1 DNA (human).
- Herring sperm DNA.
- DNA ladder 100 bp.
- DIG-nick translation mix.
- Anti-digoxigenin-fluorescein.
- Nick translation mix.
- Nick translation kit.
- SpectrumGreen dUTP.
- SpectrumOrange dUTP.
- Midori Green Advanced DNA stain.
- Streptavidin-HRP.
- Parafilm M.
- Hoechst 33342.
- Mowiol 4-88.
- Water bath at 80 °C.
- Incubator at 37 °C.
- Desktop centrifuge.
- Thermoblock with cooling system at 15 °C.
- Hotplate Cimarec.

2.2 Buffers and Solutions

1. 20× SSC: NaCl 175.3 g (in stock solution 3 M), Na_3 citrate × $2H_2O$ 88.2 g (in stock solution 300 mM) in 800 ml ddH_2O. Adjust pH to 7.0 with 14 N (or similar) HCl; adjust the volume to 1 L.
2. Hybridization buffer: 50 ml deionized formamide (4 °C), 10 ml 20× SSC, 20 ml sterile water; adjust pH to exactly 7.0, 20 ml of 50% dextran sulfate in 100 ml final volume. Stir, autoclave, aliquot, and store at 4 °C.
3. Hybridization solution: 50 ml deionized formamide (4 °C), 10 ml 20× SSC, 20 ml sterile water. Adjust pH to exactly 7.0. Store at room temperature, protected from light.
4. TNB blocking buffer: 0.1 M Tris-HCl, pH 7.5, 0.15 M NaCl, 0.5% blocking reagent (supplied in the kit).

   Heat TNB gradually to 60 °C, until completely dissolved. Aliquot and store at −20 °C.
5. TNT wash buffer: 0.1 M Tris-HCl, pH 7.5, 0.15 M NaCl, 0.05% Tween-20.

3 Methods

3.1 DNA FISH with Detergent Permeabilization

3.1.1 Cell Preparation (Work Under the Laminar Flow Hood). Please see Note 1.

1. Dip coverslips in 70% ethanol solution; let them dry in an upright position.
2. Apply 150 μl of PEI solution (0.5 mg/ml) directly on the coverslip (use a 10-cm dish with Parafilm for this procedure); incubate for 45–60 min at RT.
3. Take off PEI, add ddH_2O to wash 2×, and let coverslips air-dry for 2–3 h; put them into a 24-well plate.
4. Concentrate cells to 2×10^6/ml in 5% FBS/PBS, add 150 μl (3×10^5) to each coverslip, and leave them to adhere for 1 h at 37 °C.
5. Remove remaining suspension, add 0.3× PBS, and incubate for 20 s (hypotonic shock).
6. Remove PBS, add 4% PFA/PBS, incubate for 8 min, add two drops of 0.5% Triton X-100/0.1% Tween-20/PBS, and incubate for further 2 min.
7. Wash 4 × 5′ with 1× PBS.
8. Permeabilize for 10 min in PBS+ 0.5% Triton X-100.
9. Wash 3 × 5′ with PBST (0.1% Tween-20).
10. Store in 500 μl of 20% glycerol/PBS at 4 °C up to 1 month.

3.1.2 Immunofluorescence (IF) and FISH Cell Permeabilization. Please see Note 1.

1. Wash 3 × 5′ with PBS.
2. Block in 4% BSA/PBS for 30′ at RT.
3. Add primary antibody in 1% BSA/PBS for 1–2 h at RT or overnight at 4 °C.
4. Wash 3 × 5′ with PBST (0.1% Tween).
5. Add secondary antibody in 1% BSA/PBS for 1 h at RT.
6. Wash 3 × 5′ with PBST (0.1% Tween).
7. Add 1:1000 EGS/PBS (stock: 0.5 M in DMSO); incubate for 10 min.
8. Wash 3 × 5′ with PBST (0.1% Tween).
9. Add 0.5% Triton/0.5% saponin/PBS and incubate for 10 min.
10. Wash 3 × 5′ with PBST (0.1% Tween).
11. Add 0.1 M HCl, and incubate for 10 min.
12. Wash 3 × 5′ with PBST (0.1% Tween).
13. Add 0.5% Triton/0.5% saponin/PBS and incubate for 10 min.
14. Wash 3 × 5′ with PBST (0.1% Tween).
15. Equilibrate for 5 min in 2× SSC.

16. Only for HIV-1 FISH: perform RNase A digestion with 100 μg/ml RNase A in 2× SSC for 30 min at 37 °C.
17. Wash 2 × 5′ with 2× SSC.
18. Leave slides in hybridization solution (at least 3 h at room temperature, better overnight in the fridge).

3.1.3 DNA Probe Labeling

Setup of Indirect BAC Probe Labeling with Digoxigenin (DIG-Nick Translation Mix)

1. Add 3 μg of BAC DNA to ddH_2O up to the final volume of 16 μl. Please see Note 2.
2. Add 4 μl DIG-nick translation mix and centrifuge briefly (contains labeled and unlabeled dNTPs, DNA polymerase I, and DNase I).
3. Incubate at 15 °C using thermoblock with cooling system or PCR machine with the lid heating off.
4. Put the samples on ice.
5. Check the digestion on a 1% agarose gel using 1.5 μl of the reaction. A smear with a majority around 300–400 bp should be seen. If the DNA fragments are still above the optimal size, incubate for an additional 15–30 min and check the probe again. Please see Note 3.
6. Clean the probe using an illustra MicroSpin G-25 column according to the protocol.
7. Continue with the probe precipitation (Subheading 3.1.4).

Setup of Direct BAC Probe Labeling (Abbott)

1. Prepare 0.2 mM SpectrumGreen, SpectrumOrange, or SpectrumRed dUTP by adding 10 μl of 1 mM dUTP to 40 μl nuclease-free water.
2. Prepare 0.1 mM dTTP by adding 10 μl of 0.3 mM dTTP to 20 μl nuclease-free water.
3. Prepare 0.1 mM dNTP mix by adding 10 μl of 0.3 mM dATP, 0.3 mM dCTP, and 0.3 mM dGTP.
4. Add all the components to the tube in the order listed below. Briefly centrifuge and vortex the tube before adding the enzyme:

22.5 to *x* μl	Nuclease-free water
x μl	for 1 μg extracted DNA
2.5 μl	0.2 mM SpectrumOrange/SpectrumGreen/ SpectrumRed dUTP
5 μl	0.1 mM dTTP
10 μl	0.1 mM dNTP mix

(continued)

5 μl	5× nick translation buffer
5 μl	Nick translation enzymes
Final volume: 50 μl	

5. Incubate between 8 and 16 h at 15 °C using thermoblock with cooling system or PCR machine with the lid heating off.
6. (to be determined empirically for each DNA probe)
7. Clean the probe using an illustra MicroSpin G-25 column according to the protocol.
8. Check the probe size by agarose gel electrophoresis as described in Subheading Setup of Indirect BAC Probe Labeling with Digoxigenin (DIG-Nick Translation Mix) and continue with precipitation when optimal size is achieved (Subheading 3.1.4). Please see Note 3.

Setup of Indirect HIV-1 Probe Labeling with Biotin (Nick Translation Mix)

1. Prepare 50 μl of a 5× concentrated labeling mix, by combining in a sterile reaction tube:

5 μl	2.5 mM dATP
5 μl	2.5 mM dCTP
5 μl	2.5 mM dGTP
3.4 μl	2.5 mM dTTP
4 μl	1 mM biotin-dUTP
27.6 μl	H_2O

2. Add all the components to the tube in the order listed below.
   (a) 3 μg of HIV-1 DNA (GenBank accession number K03455.1)
   (b) 4 μl of nucleotide mix prepared with biotin (*see* above)
   (c) 4 μl nick translation mix (if the mix is old, take at least 5 μl)
   (d) +H_2O to a final amount of 20 μl.
3. Incubate at 15 °C using thermoblock with cooling system or PCR machine with the lid heating off.
4. Control on a 1% agarose gel (load 1.5 μl of the reaction) and use the 100-bp marker (keep samples on ice). Size should be under 500 bp ranging ideally with the main smear between 200 and 400 bp (smaller than that can result in unspecific binding). Please see Note 3.
5. Clean the probe using an illustra MicroSpin G-25 column according to the manufacturer's instructions.

3.1.4 Probe Precipitation

1. To clean the labeled probe from unbound labeled dUTPs that might not have been removed sufficiently by the column purification, add the following components:

Cot-1 DNA (human if human cells are used)	10 µl
Fish sperm DNA	10 µl
3 M NaAc, pH 5.2	40 µl
Pure EtOH 100%, ice cold	400 µl

2. Precipitate the probe at −20 °C overnight.
3. Centrifuge at 4 °C and 14,000 rpm (18,000 g) for 30′.
4. Remove supernatant (pellet should be clearly visible).
5. Add 800 µl of ice-cold 70% EtOH.
6. Centrifuge at 4 °C and 14,000 rpm (18,000 g) for 10′.
7. Remove supernatant and let pellet air-dry.
8. Add 10 µl of superclean formamide (small bottle at 4 °C, aliquots set aside) to dissolve the pellet; shake for 15′ at 37 °C.
9. Add 10 µl 4×SSC/20% Dextran sulfate (stored at −20 °C).
10. Denature the probe at 95 °C for 5 min, keep it on ice for 1–2 min, and then store it at −20 °C.

3.1.5 Probe Hybridization

1. Before applying the probe to slides (1–2 µl for HIV-1 probes, 2–4 µl for BACs), add up to 6 µl with hybridization buffer, denature the probe at 95 °C for 5 min, and then keep it on ice for 1–2 min.
2. Spot the probe on the coverslip and then put on top of it the cover slides that have been taken out from the hybridization solution (FA/2×SSC) just a moment before.
3. Use the metal chamber or seal with rubber cement.
4. Denature the probe with the cells in water bath at 80 °C for 8 min and then transfer the metal chamber to the water bath or incubator at 37 °C for 48–72 h. Please see Note 4 and 5.

3.1.6 Probe Detection

For Digoxigenin-Labeled Probes

1. Wash 4× in 2×SSC at 37 °C for 5′.
2. Wash 1× in 2×SSC at RT for 10′.
3. Wash 2× in 0.5× SSC at 56 °C for 5′.
4. Block in 4% BSA in 2×SSC at RT for 30′.

5. Detect with anti-DIG-FITC (1:500) in 1% BSA in 2×SSC for 1 h at 37 °C.
6. Wash 4× with 2xSSCT for 5′.
7. Counterstain with Hoechst.
8. Mount slides with Mowiol. Please see Note 6.

For Direct Labeled Probes

1. Wash 4× in 2×SSC at 37 °C for 5′.
2. Wash 1× in 2×SSC at RT for 10′.
3. Wash 2× in 0.5× SSC at 56 °C for 5′.
4. Counterstain with Hoechst.
5. Wash 2× 5′ in PBS.
6. Mount slides with Mowiol. Please see Note 6.

For HIV-1 Biotin-Labeled Probes

1. Wash 4× in 2×SSC at 37 °C for 5′.
2. Wash 1× in 2×SSC at RT for 10′.
3. Wash 2× in 0.5× SSC at 56 °C for 5′.
4. Block with TSA blocking buffer (TNB) for 45′ at RT.
5. Dilute SA-HRP 1:1500 in TNB and incubate for 30–45′ at 37 °C.
6. Wash 4× with TNT wash buffer for 5′.
7. Dilute TSA fluorescent amplification reagent 1:500 in TSA Plus and incubate for 5′ at RT.
8. Wash 5× with TNT for 5′.
9. Counterstain with Hoechst.
10. Mount slides with Mowiol. Please see Note 6.

3.2 DNA FISH with Freeze/Thaw Permeabilization

This is an alternative protocol applied for hard to permeabilize cells, but also for T cells.

Cell preparation as in Subheading 3.1.2. until EGS antibody cross-linking step (**step** 7). Please see Note 1.

8. Wash 3 × 5′ with PBST (0.1% Tween).
9. Add 0.5% Triton/PBS, incubate for 10 min.
10. Wash 3 × 5′ with PBST (0.1% Tween).
11. Incubate cells in 20% glycerol/PBS for 45 min (cells can also be stored overnight at this point).
12. Prepare two containers for 20% glycerol in PBS and liquid nitrogen (N_2).
13. Perform five cycles of freezing and thawing the slides, by immersing them into liquid N_2 for a few seconds (freeze).
14. Place the slides on a tissue in the vertical position for a few seconds (thaw). Keep the slides protected from light.

15. After each freeze/thaw cycle, dip the coverslip in 20% glycerol in PBS before freezing them again.
16. Wash 3 × 5′ with PBS.
17. Add 0.1 M HCl, and incubate for 10 min.
18. Wash 3 × 5′ with PBST (0.1% Tween).
19. Add 0.5% Triton/PBS and incubate for 10 min.
20. Wash 3 × 5′ with PBST (0.1% Tween).
21. Equilibrate for 5 min in 2× SSC.
22. Only for HIV-1 FISH: perform RNase A digestion with 100 μg/ml RNase A in 2× SSC for 30 min at 37 °C.
23. Wash 2 × 5′ with 2× SSC.
24. Leave slides in hybridization solution before proceeding with the hybridization (at least 3 h at room temperature, better overnight in the fridge).

For all following steps, *see* protocol 3.1.4, 3.1.5, and 3.1.6.

4 Notes

1. The most critical incubation times to keep are during the cell fixation and cell permeabilization steps. Therefore, to obtain reproducible results, it is also crucial to prepare all the chemicals and materials needed immediately prior to their use.
2. For the optimal probe generation, high-quality BAC DNA is necessary for the reproducible labeling reactions.
3. The size of the generated labeled probe is critical for optimal hybridization and low background signal.
4. Lifting the slides after the hybridization process should be performed with extra care, as cell/nucleus rupture occurs frequently due to the mechanical forces.
5. Optimal humidity of the chamber is necessary for the preservation of cellular morphology, protein structures, and DNA throughout the procedure.
6. Proceed to imaging in the first week after the procedure has been performed.

References

1. Bickmore WA (2013) The spatial organization of the human genome. Annu Rev Genomics Hum Genet 14:67–84. https://doi.org/10.1146/annurev-genom-091212-153515
2. Szabo Q, Bantignies F, Cavalli G (2019) Principles of genome folding into topologically associating domains. Sci Adv 5:eaaw1668. https://doi.org/10.1126/sciadv.aaw1668

3. Solovei I, Cremer M (2010) 3D-FISH on cultured cells combined with immunostaining. Methods Mol Biol 659:117–126. https://doi.org/10.1007/978-1-60761-789-1_8
4. Marini B, Kertesz-Farkas A, Ali H, Lucic B, Lisek K, Manganaro L et al (2015) Nuclear architecture dictates HIV-1 integration site selection. Nature 521:227–231. https://doi.org/10.1038/nature14226
5. Bojana Lucic, Heng-Chang Chen, Maja Kuzman, Eduard Zorita, Julia Wegner, Vera Minneker, Wei Wang, Raffaele Fronza, Stefanie Laufs, Manfred Schmidt, Ralph Stadhouders, Vassilis Roukos, Kristian Vlahovicek, Guillaume J. Filion, Marina Lusic, (2019) Spatially clustered loci with multiple enhancers are frequent targets of HIV-1 integration. Nature Communications 10(1)

Chapter 15

Visualization of Nuclear and Cytoplasmic Long Noncoding RNAs at Single-Cell Level by RNA-FISH

Tiziana Santini, Julie Martone, and Monica Ballarino

Abstract

The RNA fluorescence in situ hybridization (RNA-FISH) methodology offers an attractive strategy to deepen our knowledge on the long noncoding RNA biology. In this chapter, we provide a comprehensive overview of the current RNA-FISH protocols available for imaging nuclear and cytoplasmic lncRNAs within cells or tissues. We describe a multicolor approach optimized for the simultaneous visualization of these transcripts with their specific molecular interactors, such as proteins or DNA sequences. Common challenges faced by this methodology such as cell-type specific permeabilization, target accessibility, image acquisition, and post-acquisition analyses are also discussed.

Key words RNA-FISH, Immunofluorescence, In situ hybridization, ncRNA, lncRNA, Single-cell detection

1 Introduction

The finding of functional long noncoding RNAs (lncRNAs) has questioned the dogma that, in higher organisms, proteins are the only determinants for phenotypes [1]. Indeed, functional studies performed through the generation of knockout mice have provided convincing evidence for the requirement of specific long noncoding RNAs for organ development and function [2–8]. Long ncRNAs represent a class of transcripts longer than 200 nucleotides, in most cases with no protein-coding capacity [9, 10]. They participate in gene expression regulation by interacting with different ribonucleoprotein complexes and exert their function on a diverse range of cellular processes, such as cell growth, differentiation, and apoptosis [11, 12]. In line with their crucial role in physiology, the dysregulation of lncRNA expression was found to be associated with various disorders, including neurodegeneration, cancer, and muscle diseases [13, 14].

Beatrice Bodega and Chiara Lanzuolo (eds.), *Capturing Chromosome Conformation: Methods and Protocols*, Methods in Molecular Biology, vol. 2157, https://doi.org/10.1007/978-1-0716-0664-3_15,

The specific subcellular distribution, nuclear or cytoplasmic, is a critical aspect for lncRNA functions. The majority of them is found to be enriched in the nucleus and in particular associated with chromatin [15–17]. LncRNAs by interacting with DNA, RNA, and proteins can impact the epigenetic state of the chromatin, transcription, and RNA processing and also contribute to the 3D organization of chromatin domains [18]. Several archetypes of lncRNAs have been also described in the cytoplasm, which mainly controls mRNA turnover, translation, and protein stability [19].

The possibility to obtain spatial-temporal profiles of lncRNA expression in diverse physiological contexts (i.e., cells, tissues, and organs) represents a critical step to gain insight into their mechanisms of action. However, the conventional methodologies established for the study of mRNAs lack enough accuracy for lncRNA studies. In fact, most of these species exhibit low steady-state levels and spatially restricted patterns of expression that cannot be precisely defined by cell population-based RNA analyses. Instead, a deeper analysis of lncRNA subcellular localization can positively impact, for instance, the ability to knock down their expression, which is a preliminary step toward the phenotypic/functional analysis. In this perspective, the use of RNA in situ hybridization technologies (by Chromogenic or Fluorescent detection systems) provides researchers an unprecedented opportunity to improve molecular characterizations at a single-cell resolution. The use of imaging-based technologies is particularly advantageous when the chromatin-binding features of nuclear-retained lncRNAs are studied. In this respect, further optimization of the RNA Fluorescent in situ hybridization (RNA-FISH) protocols using combinatorial labeling may allow the simultaneous detection of the lncRNAs with specific chromatin loci (DNA/RNA-FISH) or protein partners (IF/RNA-FISH).

In this chapter, we provide step-by-step instructions to visualize lncRNAs in adherent or suspension cells (Subheading 3.1) with subcompartment resolution. We describe examples of RNA in situ hybridization technologies applied to the visualization of two functional muscle-specific lncRNAs, *Charme* enriched in the nucleus (Subheading 3.2) and linc-MD1 expressed in the cytoplasm (Subheading 3.3) of differentiated myotubes [8, 20]. With respect to earlier studies [21, 22], the imaging procedures have been improved for the simultaneous visualization of these noncoding species with specific interactors, such as proteins (Subheading 3.4) or DNA elements (Subheading 3.5). Suggestions on how to overcome common obstacles in relation to cell-type-specific fixation and permeabilization to improve target accessibility and during image acquisition and post-acquisition analyses (Subheading 3.6) are also provided.

2 Materials

2.1 RNA-FISH Strategies and Probe Design

The optimal signal-to-noise ratio (S/N) depends on the number of the probes used for the detection or the amplification power of the detection system. Here, the multi-probe and single-probe strategies were used, respectively, for the detection of nuclear and cytoplasmic lncRNAs (Fig. 1). For the multi-probe strategy, we modified previously described protocols [17, 22, 23] using a smaller set (n = 4–8) of custom 20-nt-long DNA oligonucleotides. These oligos can be either marked with Cy3 fluorophore (direct method) or biotin haptens (indirect method). In the indirect method, the biotin label is detected with Cy3 or fluorescein-conjugated anti-biotin antibodies. This strategy allows improving the fluorescent signals, in respect to background, when only a limited number of probes are available. Probes for the direct method were designed using the online Stellaris' design software (Probe Designer) setting 2–3 nt as minimal probe spacing. To minimize aspecific cross-hybridization, probe sequences are checked by local alignment (NCBI-BLASTN) in the genomic +transcript database optimized for "somewhat similar sequences"; oligos showing complementary cross-annealing (at least two same off-targets; query cover >50%) must be discarded. The use of a

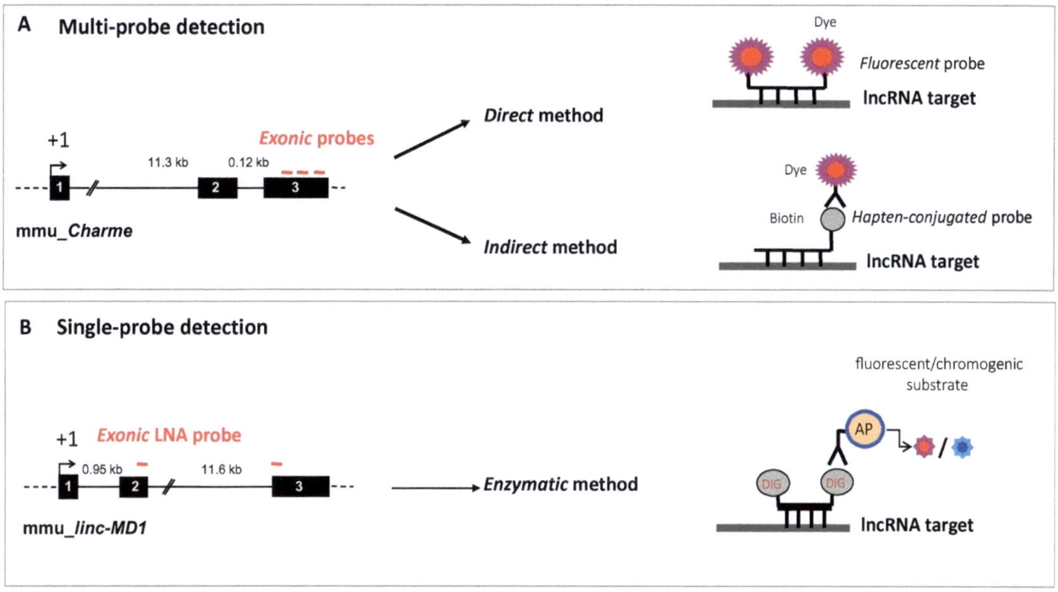

Fig. 1 *In situ* hybridization strategies for the visualization of nuclear and cytoplasmic long noncoding RNAs. (**a**) Multi-probe strategy was performed [8] using 20-nucleotide long single-strand DNA probes complementary to the *Charme* exon 3 sequences. Probes are labeled at both 5′- and 3′-ends with fluorophores (*direct* method) or with biotin at the 5′- or 3′-end (*indirect* method). (**b**) Single-probe detection was performed [20] using digoxigenin (DIG)-conjugated LNA probes complementary to the *linc*-MD1 exon 2/3 splice junction. To note, fluorescent or chromogenic substrates were used, respectively, for the hybridization in cell cultures or tissue cryosections

spinning disk confocal microscope allows a fast image acquisition of these signals and provides a benefit of reduced photo-oxidation. A laser scanning microscope may work as well.

The LNA probe 3′–5′ DIG-conjugated used in the single-probe detection strategy was designed by Exiqon's Probe Designer against the splice junction of mature lncRNA. Sequences, which were found to be similar to miRNAs by miRBase search tool, were discarded. The LNA™ technology combined with enzymatic amplification (alkaline phosphatase + substrate) was previously used for the chromogenic detection of miRNAs on tissue sample [24, 25], but it can be applied to improve the detection of long ncRNAs also in cell culture [20, 26]. In general, enzymatic amplification improves the signal intensity, although sometimes also the noise level increases. For this reason, the use of Fast Red combined with confocal analysis provides a higher subcellular resolution in respect to the chromogenic NBT/BCIP substrate used for bright-field microscopy.

2.2 Reagents

1. DPBS: nuclease-free PBS with $MgCl_2$ and $CaCl_2$.
2. PBSA: nuclease-free PBS without $MgCl_2$ and $CaCl_2$.
3. UltraPure DNase/RNase-free water.
4. 8% PFA aqueous solution (Electron Microscopy Sciences, 157–8.
5. Deionized formamide (Sigma-Aldrich, 47671). Store in small aliquot at −20 °C.
6. Dextran sulfate sodium salt, average Mw >500,000 (Sigma-Aldrich, D8906).
7. 200 mM vanadyl ribonucleoside complexes (VRC) (Sigma-Aldrich, R3380).
8. Endogenous biotin-blocking kit (component A+ component B) (Thermo Fisher Scientific, E21390).
9. ProLong Diamond Antifade (Thermo Fisher Scientific, P36965).
10. Bovine serum albumin lyophilized powder (BSA).
11. Mouse monoclonal anti-biotin-Cy3 antibody (Jackson Immunoresearch, 200-162-211).
12. Mouse monoclonal anti-digoxigenin-Alexa Fluor® 488 antibody (Jackson Immunoresearch, 200-542-156).
13. Mouse monoclonal anti-SC35 (phospho) antibody (Abcam, ab11826).
14. Mouse monoclonal anti-MHC antibody from hybridoma supernatant (MF20).
15. Polyclonal sheep anti-digoxigenin-AP antibody (Roche, 11093274910).

16. Collagen type 1 from rat tail.
17. Poly-L-lysine hydrobromide.
18. RNase A, DNase, and protease-free.
19. Salmon Sperm DNA Solution (Invitrogen, 15632011).
20. Mouse Cot-1 DNA (ThermoFisher Scientific, 18440016).
21. Glycogen, molecular biology grade.
22. 3 M sodium acetate (pH 5.5).
23. 96% ethanol.
24. Triton X-100.
25. ≥99.5% glycerol solution.
26. 100 mM dNTP (A/T/C/G) solutions.
27. 1 mM Green 496-dUTP (Enzo Life Sciences 42831).
28. 1 mM digoxigenin-11-dUTP, alkali-stable (Roche, 11093088910).
29. 10 U/ μl DNA polymerase I.
30. DNase I recombinant RNase-free.
31. ≥99.0% 2-mercaptoethanol.
32. 37% hydrochloric acid reagent.
33. 1-Methylimidazole for synthesis (Sigma-Aldrich, 805852).
34. EDC (N-(3-dimethylaminopropyl)-N′-ethylcarbodiimide hydrochloride) (Sigma-Aldrich, 03450).
35. ≥99.0% triethanolamine (Sigma-Aldrich, 90279).
36. ≥99% acetic anhydride ReagentPlus® (Sigma-Aldrich, 320102).
37. NBT/BCIP powder reagents: NBT (Sigma-Aldrich, N6876), BCIP (Sigma-Aldrich, B6777).
38. (-)-tetramisole hydrochloride (Levamisole) (Sigma-Aldrich, L9756). Resuspend in water and store in small aliquot at −20 °C.
39. Blocking reagent, powder (Merck, 11096176001).
40. HNPP/Fast Red TR detection set (Merck, 11758888001).
41. Normal goat serum. Heat-inactive at 56 °C for 30 min. Sterilize by filtering. Store in small aliquot at −20 °C.
42. Agarose powder, molecular biology grade.
43. DNA gel loading dye.
44. DNA staining dye.
45. Low-range DNA ladder.

2.3 Solutions

1. 4% PFA solutions (v/v): in a chemical hood, dilute 8% PFA aqueous solution in DPBS (PFA/DPBS) or nuclease-free H_2O. Store at 4 °C up to 1 month.
2. Ethanol solution (v/v): 50%, 70%, 100% ethanol in nuclease-free water.
3. Saline-sodium citrate buffer 20× (SSC): dissolve 175.3 g of NaCl and 88.2 g of sodium citrate in 800 ml of distilled H_2O. Adjust the pH to 7.0 with a few drops of 1 M HCl. Adjust the volume to 1 L with nuclease-free H_2O. Sterilize by filtering. It can be stored at 4 °C for more than 1 year.
4. TBS: 50 mM Tris-HCl buffer (pH 7.8–8), 150 mM NaCl, nuclease-free water. Sterilize by filtering. It can be stored at 4 °C. Check pH periodically.
5. Glycine buffer: 3% (w/v) of glycine powder in DPBS. Sterilize by filtering. Store at 4 °C.
6. Pre-hybridization buffer I: 10% deionized formamide, 2× SSC in nuclease-free water. It can be stored at 4 °C for 1 month.
7. Hybridization buffer I: 10% deionized formamide, 2× SSC, 10% (w/v) Dextran sulfate, 2 mM VRC in nuclease-free water. Make fresh each time.
8. DAPI: dissolve 4′,6-diamidine-2′-phenylindole dihydrochloride powder in PBSD or 2× SSC to obtain 1 μg/ml final solution. Store at 4 °C in the dark.
9. TN buffer: 100 mM Tris-HCl buffer (pH 7.5), 150 mM NaCl. Sterilize by filtering. Store at 4 °C.
10. 4% BSA/TN buffer: dissolve BSA powder in TN buffer to obtain 4% w/v final solution.

    Stir at room temperature and sterilize by filtering. Store at 4 °C.
11. IF blocking buffer: 2% (w/v) of BSA, 2 mM VRC in TN buffer. Make fresh each time.
12. 10× nick translation buffer: 500 mM Tris-HCl buffer (pH 7.8–8), 50 mM $MgCl_2$, 0.5 mg/ml BSA. Sterilize by filtering. Keep at 4 °C for 1 month.
13. Pre-hybridization buffer II: 50% deionized formamide, 2× SSC in nuclease-free water. Make fresh each time.
14. Hybridization buffer II: 50% deionized formamide, 2× SSC, 10% (w/v) Dextran sulfate. Make fresh each time.
15. Pepsin solution: dissolve pepsin lyophilized powder (Merck, P6887) in nuclease-free water to obtain 10% w/v final solution. Store in small aliquot at −20 °C.
16. Methylimidazole solution: 0.13 M 1-methylimidazole, 300 mM NaCl in TBS.

Adjust to pH 8.0 with HCl. Make fresh each time.

17. EDC solution: dissolve 31 mg of EDC powder in methylimidazole solution. Make fresh immediately before use.
18. Pre-hybridization buffer III: 25% deionized formamide, 2× SSC, 0.2% Triton X-100 in nuclease-free water. Make fresh each time.
19. Hybridization buffer III: 50% deionized formamide, 2× SSC, 10% (w/v) Dextran sulfate, 10 μg/ml *E. coli* tRNA, 500 μg/ml Salmon Sperm DNA in nuclease-free water. Store in small aliquot at −20 °C.
20. FISH blocking reagent solution: dissolve 0.5% (w/v) blocking reagent in TNT buffer. Store in small aliquot at −20 °C.
21. TNT buffer: 100 mM Tris-HCl buffer (pH 7.5), 150 mM NaCl, 0.1% Triton X-100 in nuclease-free water. Sterilize by filtering. Store at room temperature.
22. TMN buffer: 100 mM Tris base buffer (pH 9.5), 50 mM $MgCl_2$, 500 mM NaCl in nuclease-free water. Sterilize by filtering. Store at 4 °C.
23. Pre-hybridization buffer IV: 25% deionized formamide, 4× SSC, 0.1% Triton X-100 in nuclease-free water. Make fresh each time.
24. Hybridization buffer IV: 4× SSC, 10% (w/v) Dextran sulfate, 250 μg/ml *E. coli* tRNA in nuclease-free water. Make fresh each time.
25. Detection buffer: 100 mM Tris-HCl buffer (pH 7.5), 100 mM NaCl, 20 mM $MgCl_2$ in dH_2O. Sterilize by filtering. It can be stored at 4 °C.
26. HNPP/Fast Red TR solution: 20 μl of HNPP, 20 μl of Fast Red TR solution in 2 ml of detection buffer. Make fresh each time immediately before use. Protect from light.
27. NBT/BCIP solution: dissolve 0.375 mg/ml of NBT and 0.188 mg/ml of BCIP in TMN buffer. Make fresh each time immediately before use.
28. Agarose gel electrophoresis solution: mix 1 g of agarose powder in 100 ml of 1× electrophoresis buffer. Dissolve by boiling on hot plate stirrer or microwave oven.
29. 10× electrophoresis buffer: 1 M Tris, 0.9 M boric acid, 0.01 M EDTA in dH_2O.

2.4 Equipment and Supplies (Fig. 2)

1. 35-mm cell culture dishes.
2. 24-well cell culture plates.
3. Standard glass slides (25 × 75 mm).

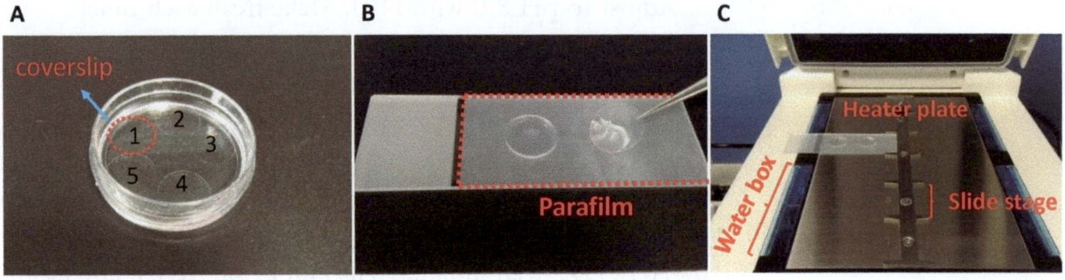

Fig. 2 Equipment and coverslip handling in microscopy. (**a**) Typical disposition of five 12-mm-diameter coverslips inside a 35-mm Petri dish. (**b**) Microscope glass slides covered with Parafilm and used for overnight hybridization. Round coverslips are gently slipped on the probe-MIX drop, with cells facing down. (**c**) A humidified heater plate is shown together with the numbered slide stages and the water-containing boxes for humidity control

4. 12-mm-diameter round glass coverslips 0,13–0.17 mm thickness.
5. Parafilm.
6. Tweezers.
7. Horizontal gel electrophoresis machine.
8. UV transilluminator.
9. Hot plate with magnetic stirrer.
10. Gel electrophoresis power supply.
11. No-wrinkle rubber cement (e.g., Elmer's).
12. Automatic slide hybridizer machine (e.g., Top Brite, Resnova s.r.l.).
13. Bright-field microscope.
14. Laser scanning confocal or LED-based spinning disk confocal microscope.
15. High-numerical-aperture (NA) oil-immersion apochromatic objectives.
16. High-resolution and sensitivity CCD camera.
17. Software for image collection and Z-stack analyses (e.g., Metamorph, Imaris, Fiji, Huygens).

3 Methods

3.1 Culturing and Fixation Protocols for Adherent and Suspension Cells

3.1.1 Adherent Cells

1. Place sterile glass coverslips inside a 35-mm Petri dish (Fig. 2a) (*see* **Note 1**).
2. Grow adherent cells on the coated coverslips with 2 ml of media (*see* **Note 2**).
3. Aspirate culture media and wash the cells three times for 5 min each with 2 ml of sterile-filtered DPBS (with calcium and magnesium) at room temperature (*see* **Note 3**).
4. Discard DPBS and fix cell samples in 1.5 ml of 4% PFA/DPBS. Incubate at 4 °C for 20 min.
5. Aspirate the fixative solution and wash the cells three times for 5 min each with 1 ml DPBS at room temperature.
6. Replace DPBS with 1.5 ml of glycine buffer. Incubate at room temperature for 10 min (*see* **Note 4**).
7. Wash with DPBS and dehydrate in 2 ml of ice-cold ethanol series (50%, 70%, 100%). Incubate at room temperature for 5 min.
8. Discard the last ethanol solution (*see* **Note 5**) and proceed with the following steps.

3.1.2 Suspension Cells

1. Transfer the cell suspension (4×10^6 cultured cells) to a 15-ml centrifuge tube and spin down at 115 × *g* for 5 min at room temperature (*see* **Note 6**).
2. Aspirate culture media and gently resuspend the cells in 1 ml of cold DPBS (*see* **Note 7**).
3. Add 1 ml of 4% PFA/DPBS to the sample for a final concentration of 2%. Mix the cell solution gently by tapping the bottom of the centrifuge tube. Incubate at 4 °C for 10 min.
4. Spin down the fixed cells as in **step 1** to pellet. Discard the fixative solution and resuspend the pellet gently in 2 ml of RNase-free H_2O.
5. Spin down the cells as in **step 1**. Discard the supernatant and resuspend the pellet in 1 ml of 70% ethanol.
6. Spin down the cells as in **step 1**. Discard the supernatant and resuspend the pellet in 400 μl of absolute ethanol.
7. Deposit a drop of 10 μl of the cell suspension on round-glass coverslips with rotary movements of the pipette tip (*see* **Note 8**).
8. Let the cells dry for at least 5 min, and then put the coverslips in a Petri dish using tweezers.
9. Add 2 ml of absolute ethanol and store at −20 °C for up to 1 month.

3.2 Visualization of Nuclear Long ncRNAs by RNA-FISH

3.2.1 RNA-FISH with Fluorescent Probes: The "Direct Hybridization Method" (Fig. 3a)

1. Transfer carefully the coverslips cell face up on a 24-multiwell plate by using tweezers (*see* **Note 9**).
2. Rehydrate cells by adding 1 ml of descendent ice-cold ethanol series (100%, 70%, 50%) being careful not to disturb the cells. Let the cells stand at room temperature for 5 min (*see* **Note 10**).
3. Discard ethanol and add 500 μl of DPBS. Let the cells stand at room temperature for 10 min (*see* **Note 11**).
4. Aspirate DPBS and permeabilize the cells on ice in 0.5% Triton X-100/2 mM VRC/DPBS for 5 min (*see* **Note 12**).
5. Wash the cells three times for at least 5 min each with 1 ml DPBS at room temperature.
6. Keep the coverslips at 4 °C. Proceed with the RNase A pretreatment for the negative control by incubating the sample in 400 ng/ml RNase A/PBSA at 37 °C for 1 h (*see* **Note 13**).
7. Wash the cells three times for at least 5 min each with 1 ml DPBS at room temperature.
8. Add 500 μl of 2× SSC to all the coverslips and let them stand 5 min at room temperature.
9. Discard SSC and add 500 μl of pre-hybridization buffer I. Incubate at 37 °C for 15 min.
10. Meanwhile, prepare the hybridization buffer I plus probes (probe-MIX) solution (*see* **Note 14**).

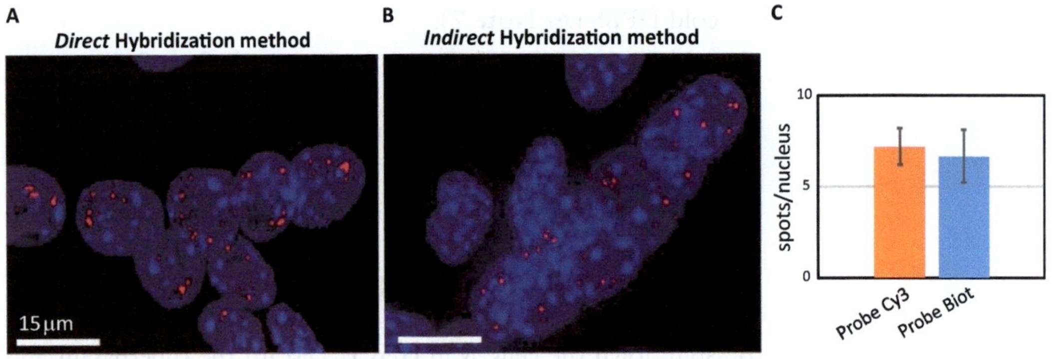

Fig. 3 Visualization of nuclear lncRNAs by RNA-FISH: detection of *Charme* in murine C_2C_{12} myotubes. (**a**) Representative confocal image from RNA-FISH experiments performed with the *direct* method using 5× fluorescent DNA oligonucleotides (Cy3, 5′-3′-ends labeled). The RNA molecules are displayed as well-defined punctate signals (red spots) inside the nuclei (DAPI, blu staining). (**b**) Representative confocal image from RNA-FISH experiments performed with the *indirect* method using 8× biotin 3′-end labeled DNA oligonucleotides in the first step of detection and then visualized with Cy3-conjugated anti-biotin antibody. (**c**) Histogram shows the mean number of the RNA-FISH spots per nucleus. Imaging of *Charme* with *direct* (blu) or *indirect* (orange) methods. The similar range of mean values denotes the same sensibility and reproducibility of the two strategies. *See* text and reference [8] for details on *Charme* lncRNA

11. Deposit 80 μl of probe-MIX on a 25 × 75 mm microscope slide coated with Parafilm. Place the coverslips onto drop (Fig. 2b) with cells facing down (*see* **Note 15**).
12. Put the slides in the slide hybridizer machine (Fig. 2c) and incubate overnight at 37 °C (*see* **Note 16**).
13. The next day, transfer the coverslips into a 24-multiwell plate. Gently wash the coverslips with 1 ml of 2× SSC for 5 min at 37 °C and then with 1 ml of 1× SSC for 5 min at room temperature (*see* **Note 17**).
14. Discard the last washing solution. Then add 300 μl of DAPI diluted in 2× SSC at the final concentration of 1 μg/ml. Incubate at room temperature for 3 min.
15. Aspirate the DAPI solution and gently wash the coverslips at room temperature with 1 ml of 2× SSC for 5 min.
16. Mount slides. Acquire images as in Subheading 3.6.

3.2.2 RNA-FISH with Biotinylated Probes: The "Indirect Hybridization Method" (Fig. 3b)

1. Proceed as in Subheading 3.2.1, until **step** 7.
2. Place coverslips onto a microscope slide coated with Parafilm cells facing up (*see* **Note 18**).
3. Add 45 μl of streptavidin solution (component A) plus 2 mM VRC. Incubate at 37 °C in a slide hybridizer machine for 20 min.
4. Transfer the coverslips in a 24-multiwell plate and wash three times for 5 min each with 1 ml DPBS at room temperature.
5. Place again the coverslips onto a microscope slide and add 45 μl of biotin solution (component B) plus 2 mM VRC. Incubate the coverslips at 37 °C in a slide hybridizer machine for 20 min.
6. Wash the coverslips as in **step 4**.
7. Proceed as in Subheading 3.2.1, **steps 8–10**.
8. The next day, gently transfer the coverslips into a 24-multiwell plate and wash twice for 5 min each with 1 ml 2× SSC at room temperature.
9. Post-fix the cells with 4% PFA/DPBS for 5 min at room temperature (*see* **Note 19**).
10. Discard PFA and wash three times for 3 min each with 1 ml DPBS at room temperature.
11. Add 500 μl of TN buffer and incubate at room temperature for 10 min.
12. Set up the coverslips as in **step 2**. Add 45 μl of 1:150 diluted Cy3-conjugated anti-biotin antibody in 4% BSA/TN buffer. Incubate at room temperature in a humid box for 1 h (*see* **Note 20**).

13. Wash three times for 5 min each with 1 ml of TN buffer at room temperature.
14. Discard TN buffer and add 500 μl of 2× SSC. Incubate at room temperature for 5 min.
15. Proceed with DAPI staining as in Subheading 3.2.1, **steps 14** and **15**.
16. Mount slides and acquire images as in Subheading 3.6.

3.3 Visualization of Cytoplasmic Long ncRNAs by RNA-ISH (Fig. 4)

3.3.1 Detection on Tissue Cryosections or Cell Cultures

1. Collect frozen (7-μm-thick) sections on microscope slides coated with poly-L-lysine and stored at −80 °C until use.
2. Fix tissues in 4% PFA/DPBS at 4 °C for 20 min (*see* **Note 21**).
3. Wash two times for 5 min each with DPBS at room temperature (*see* **Note 22**).
4. Remove DPBS and wash two times for 5 min each with TBS buffer at room temperature (*see* **Note 23**).
5. Incubate two times for 10 min each with freshly prepared methylimidazole solution at room temperature.
6. Remove methylimidazole solution and replace it with EDC solution. Incubate at room temperature for 2 h (*see* **Note 24**).
7. Incubate two times for 5 min with 0.2% glycine/TBS at room temperature.
8. Wash twice for 5 min each with TBS buffer at room temperature.
9. Proceed to acetylation by adding freshly prepared 0.1 M triethanolamine/0.25% acetic anhydride at room temperature for 30 min (*see* **Note 25**).
10. Rinse the slides with TBS at room temperature for 5 min.
11. Add pre-hybridization buffer III/2mM VRC and incubate at room temperature for 1 h (*see* **Note 26**). Meanwhile, prepare the hybridization buffer III/2mM VRC plus probes (probe-MIX).
12. Denature at 80 °C on heater block and keep on ice for 5 min. Leave at 37 °C until use (*see* **Note 27**).
13. Remove pre-hybridization buffer III and add probe-MIX. Incubate in slide hybridizer overnight at 37 °C (*see* **Note 28**).
14. The next day, wash the slides with 5× SSC at room temperature for 5 min.
15. Wash with 1× SSC at 45 °C for 15 min. Wash at room temperature for 10 min with: 1× SSC/2% BSA, 0.2× SSC/2% BSA and 2× SSC (*see* **Note 29**).
16. Rinse with 1 ml of TN buffer at room temperature for 5 min.

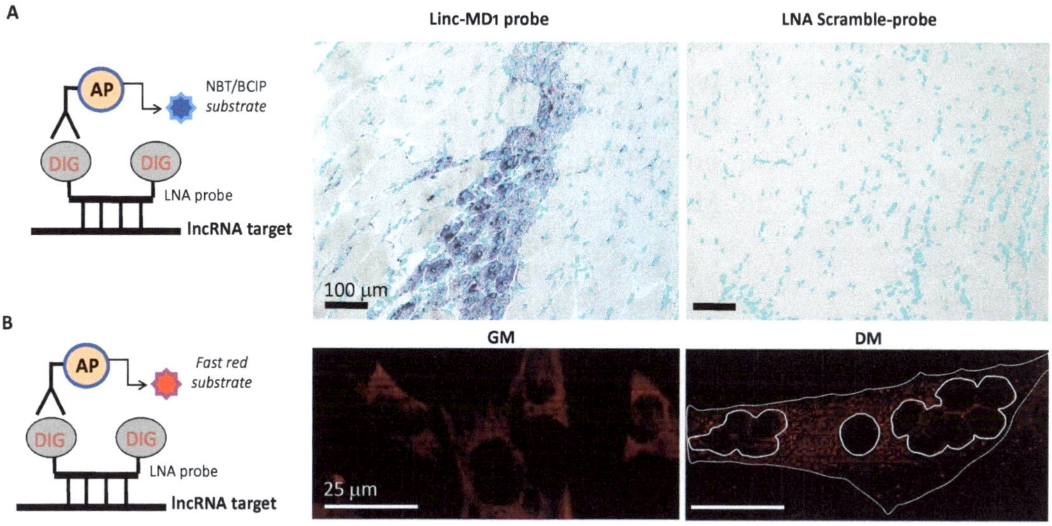

Fig. 4 Visualization of cytoplasmic lncRNAs by RNA-in situ hybridization (ISH): detection of *Linc-MD1* in murine C_2C_{12} myotubes. (**a**) Chromogenic detection performed on murine *mdx* tibialis cryosections using double DIG-labeled LNA probes and NBT/BCIP as chromogenic alkaline phosphatase (AP) substrate. This chemical compound gives a very stable dark blue dye, visible with a bright-field microscope when a specific probe for *linc-MD1* was used (left panel). LNA-scramble probe provides a negative control to measure the level of background staining within the tissue (right panel). (**b**) Representative example of LNA-based RNA-FISH in mouse myoblasts (left, GM) where linc-MD1 is not expressed (diffuse signal) and myotubes (right, DM) where discrete spots are visible in the cytoplasmic compartment. Fluorescent detection was obtained using the AP-Fast Red substrate. Dashed line: myotube profile. Solid line: nuclei. *See* text and reference [20] for details on *Linc-MD1* lncRNA. Images of panel **a** were reprinted from Cell, 2011, 147(2), Cesana M, Cacchiarelli D, Legnini I, Santini T, Sthandier O, Chinappi M, Tramontano A, Bozzoni I., "A Long Noncoding RNA Controls Muscle Differentiation by Functioning as a Competing Endogenous RNA", Pages 358–69, Copyright (2011), with permission from Elsevier

17. Add the FISH blocking reagent and incubate at room temperature for 30 min (*see* **Note 30**).
18. Discard the solution. Add 1:100 diluted AP-conjugated anti-digoxigenin antibody in FISH blocking reagent solution. Incubate inside a humid box at room temperature for 2 h (*see* **Note 31**).
19. Wash two times for 10 min each with TNT buffer at room temperature.
20. Wash three times for 10 min each with TMN buffer at room temperature (*see* **Note 32**).
21. Discard TMN buffer and add NBT/BCIP solution +2 mM levamisole. Incubate in the dark at room temperature for 2 h (*see* **Note 33**).
22. Discard the solution and wash twice for 10 min with 1 ml of nuclease-free H_2O at room temperature.

23. Post-fix sections in 4% PFA/nuclease-free H_2O for 10 min at room temperature.
24. Wash three times for 5 min each with dH_2O at room temperature.
25. Proceed as in Subheading 3.2.1, **steps 14–15**.
26. Mount slides and acquire images as in Subheading 3.6.

3.4 Combining the "Direct Hybridization Method" with IF (Fig. 5)

3.4.1 Perform Immunofluorescence (IF) Before Hybridization

1. Use cell culturing and fixing conditions as in Subheading 3.1.1 (*see* **Note 34**).
2. Rehydrate cells as in Subheading 3.2.1, **steps 2** and **3**.
3. Permeabilize and wash the cells as in Subheading 3.2.1, **steps 4** and **5** (*see* **Notes 10** and **35**).
4. Incubate with 1 ml of IF blocking buffer for 15 min at room temperature (*see* **Note 36**). Do not wash.

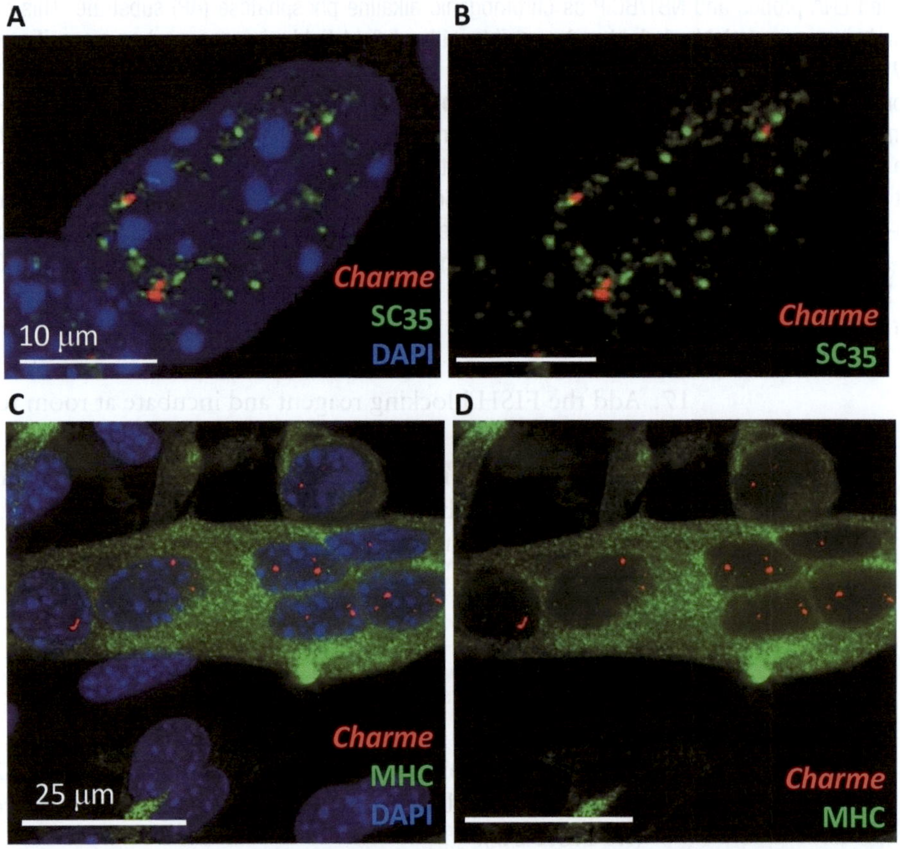

Fig. 5 Combining the *direct* RNA-FISH method with immunofluorescence (IF). Representative examples of confocal images obtained by combining RNA-FISH (*Charme*, red spots) and IF (green signals) methodologies in murine C_2C_{12} myotubes. IF staining is performed with nuclear (SC35, **a** and **b**) or cytoplasmic (Myosin heavy chain, **c** and **d**) proteins. *See* text and reference [8] for details on *Charme* lncRNA

5. Using tweezers, transfer the coverslips onto a microscope slide as in Subheading 3.2.2, **step 2**.
6. Add 45 μl of the specific primary antibody opportunely diluted in IF blocking buffer. Incubate overnight at 4 °C in a humid box (*see* **Note 37**).
7. Transfer the coverslips into a 24-multiwell plate. Wash three times for 5 min each at room temperature with 1 ml TN buffer.
8. Place again the coverslips onto a microscope slide. Add 45 μl of the fluorescent secondary antibody. Incubate in a humid box at room temperature for 45 min (*see* **Note 38**).
9. In a 24-multiwell plate, wash the coverslips at room temperature for 5 min each with 1 ml TN buffer. Repeat washing five times.
10. Post-fix the cells in 4% PFA/DPBS for 5 min at room temperature (*see* Note **39**).
11. In a 24-multiwell plate, wash the coverslips at room temperature for 5 min each with 1 ml PBSA. Repeat washing five times.
12. Proceed with Subheading 3.2.1, starting from **step 6**.

3.4.2 Perform Immunofluorescence (IF) After Hybridization

1. Proceed with the "direct hybridization method" as in Subheading 3.2.1, until **step 13**.
2. Post-fix the cells in 4% PFA/DPBS at room temperature for 5 min.
3. Discard PFA solution and wash for 3 min each with 1 ml of DPBS at room temperature. Repeat washing three times.
4. Proceed as in Subheading 3.4.1, **steps 4–8**.
5. Transfer the coverslips in a 24-multiwell plate. Wash the coverslips for 5 min with 1 ml PBSA at room temperature. Repeat washing five times.
6. Stain the nuclei for 3 min at room temperature with 300 μl of DAPI.
7. Discard DAPI. Wash the coverslips with 1 ml of PBSA at room temperature for 5 min.
8. Mount slides and acquire images as in Subheading 3.6.

3.5 Combining the "Direct Hybridization Method" with DNA-FISH

3.5.1 Preparation of the DNA-FISH Probes

1. Dilute 2.5 μg of the DNA template in 28 μl of RNase-free H_2O. Keep on ice (*see* **Note 40**).
2. Add 5 μl of 10× nick translation buffer.
3. Add 5 μl of dNTP mix (0.5 mM each).
4. Add 2.5 μl of dTTP/labelled-dUTP mix (1 mM each) (*see* **Note 41**).
5. Add 5 μl of 100 mM 2-mercaptoethanol (*see* **Note 42**).

6. Keep 10 min on ice.
7. Meanwhile, dilute 1:30 the DNase I (10 U/μl) in ice-cold DNase I buffer. Keep 5 min on ice.
8. Add 1 μl of diluted DNase I to the nick translation mix. Keep on ice (*see* **Note 43**).
9. Add 1 μl of 10 U/μl DNA polymerase I.
10. Mix gently with pipette and incubate at 16 °C for 2 h (*see* **Note 44**).
11. Stop the reaction by adding 3 μl of 0.2 mM EDTA.
12. Check the size of the labelled DNA fragments on 1% agarose gel (*see* **Note 45**).
13. Add the following reagents to 100 ng of DNA probes: 3.5 μg of Cot-1 DNA, 6 μg of 10 mg/ml Salmon Sperm DNA, 1 μl of glycogen 20 mg/ml, 10 μl of sodium acetate (3 M), 300 μl of absolute ethanol.
14. Mix and incubate for 1 h at −80 °C.
15. Spin down at (15,871 × *g*) at 4 °C for 30 min to pellet.
16. Discard supernatant. Wash twice for 5 min each at 4 °C with 300 μl of 70% ethanol.
17. Dry the pellet and resuspend in 5 μl of hybridization buffer II (*see* **Note 46**).
18. Store at −20 °C until use.

3.5.2 Preparation of Cells for DNA-FISH

1. Use cell culturing and fixing conditions as in Subheading 3.1.1.
2. Rehydrate cells as in Subheading 3.2.1, **steps 2** and **3**.
3. Aspirate DPBS and permeabilize the cells in 0.3% Triton X-100/2 mM VRC/DPBS for 10 min on ice.
4. Wash three times for 5 min each with 1 ml DPBS at room temperature.
5. Aspirate DPBS and add 20% glycerol/2 mM VRC/DPBS for 20 min at room temperature.
6. Discard the solution and add 50% glycerol/2 mM VRC/DPBS. Store at −20 °C at least overnight (*see* **Note 47**).

3.5.3 Performing DNA-FISH Before RNA-FISH (Fig. 6)

1. Equilibrate cells at room temperature for 20 min.
2. Discard the 50% glycerol solution. Add 20% glycerol/2 mM VRC/DPBS and incubate for 20 min at room temperature.
3. Freeze/thaw cells four times in dry ice (*see* **Note 48**).
4. Wash the cells at room temperature in a 24-multiwell plate five times for 5 min each in 1 ml DPBS.
5. Discard DPBS. Incubate with 0.01 M HCl/0.1% pepsin/H_2O at room temperature for 2 min (*see* **Note 49**).

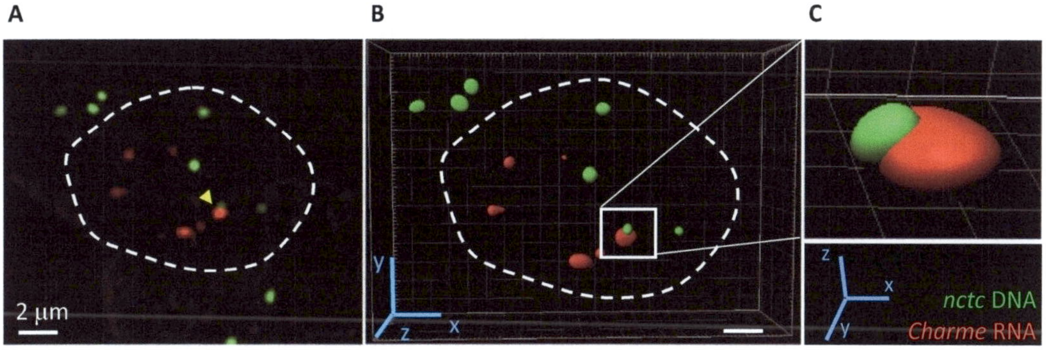

Fig. 6 Sequential DNA-RNA-FISH analyses performed using direct fluorescent probes. (**a**) Confocal image of *nctc* DNA-FISH (green spots) and *Charme* RNA-FISH (red spots) signals in the nucleus of murine myotubes at day = 2.5 of myogenic differentiation. (**b**) Isosurface rendering of (**a**) using the Imaris software (Bitplane). (**c**) Magnified insert of the white box in (**b**) after axis rotation showing the spatial proximity of the DNA-RNA-FISH spots according to the spatial resolution of conventional light microscopy. *See* text and reference [8] for details on *Charme* lncRNA

6. Discard the supernatant. Incubate at room temperature with 0.01 M HCl/0.2% pepsin/H_2O for 4 min.
7. Aspirate and wash quickly three times with 1 ml of H_2O at room temperature.
8. Post-fix the cells in 4% PFA/DPBS at room temperature for 3 min.
9. Wash cells three times for 3 min each in 1 ml of DPBS at room temperature.
10. Aspirate DPBS and add 2× SSC. Let stand for 10 min at room temperature.
11. Discard the 2× SSC and add 500 μl of pre-hybridization buffer II. Incubate at room temperature at least 1 h.
12. Without washing, dehydrate cells as Subheading 3.1.1, **step 7** (*see* **Note 50**).
13. Air-dry for 20 min at room temperature.
14. Meanwhile, equilibrate probe solution for 10 min at room temperature.
15. Deposit 5 μl of probe solution on a 25 × 75 mm microscope slide and place the coverslip onto drop with cells facing down.
16. Seal the coverslips with rubber cement and leave them completely dry at room temperature (*see* **Note 51**).
17. Place microscope slide on heater block set to 78–80 °C covered with an aluminum foil. Denature for 4–5 min (*see* **Note 52**).
18. Put the slides immediately in the slide hybridizer machine and incubate at 37 °C overnight.

19. The next day, transfer the coverslips into a 24-multiwell plate. Wash the coverslips in the dark two times at 37 °C for 5 min with 1 ml 2× SSC, 1× SSC, 4× SSC, and 2× SSC.
20. Discard the last wash and add hybridization buffer I. Incubate at room temperature for 10 min.
21. Proceed with RNA-FISH as in Subheading 3.2.1, **steps 9–16** (*see* **Note 53**).

3.6 Imaging and Post-acquisition Analyses

3.6.1 Mounting the Microscope Slides

1. Retrieve the coverslips from each well with tweezer and drain the excess of buffer.
2. Add 10 μl of mountant on clean microscope slides and place coverslip cells facing down onto the drop.
3. Flip the slides onto adsorbent paper to remove the excess of medium.
4. Leave at room temperature for 24 h or keep at −20 °C in the dark for prolonged storage (*see* **Note 54**).

3.6.2 Acquisition and Image Processing

1. Place the microscope slide on a confocal inverted microscope, coverslips facing down.
2. For high magnification analysis, apply a drop of immersion oil on top of the 60X or 100X objective (designed for use with immersion oil), and focalize cells in bright-field or DAPI channels.
3. Set the Z-spacing at least to 200 nm in acquisition software and collect the emission fluorescence from a fluorophore at time at each step (sequential acquisition). In our conditions, exposure time setted in a confocal spinning disk microscope for DAPI/FITC/TRITC/Cy5 dye are 100/500/300/500 ms, respectively. It is important to check the acquisition parameters in order to capture all the physical informations of the image. Modify the acquisition area, image size and zoom factor according to the optical properties of the microscope in order to improve image resolution and to reduce the sampling size (voxel size). Save the acquisition as 16-bit Tiff files (*see* **Note 55**).
4. To eliminate background, set the intensity threshold using MetaMorph (Molecular Devices) or ImageJ software. To do that, you can subtrack to the total intensity range of the image, the lower threshold values obtained from the sample hybridized without probes or in other negative RNA-FISH control (Fig. 4 and 7).
5. Record exposure time and threshold setting for each fluorophore, in order to apply the same parameters to all images.
6. To obtain a reliable 3D quantification data analysis, improve the contrast and the resolution of the signals by Deconvolution

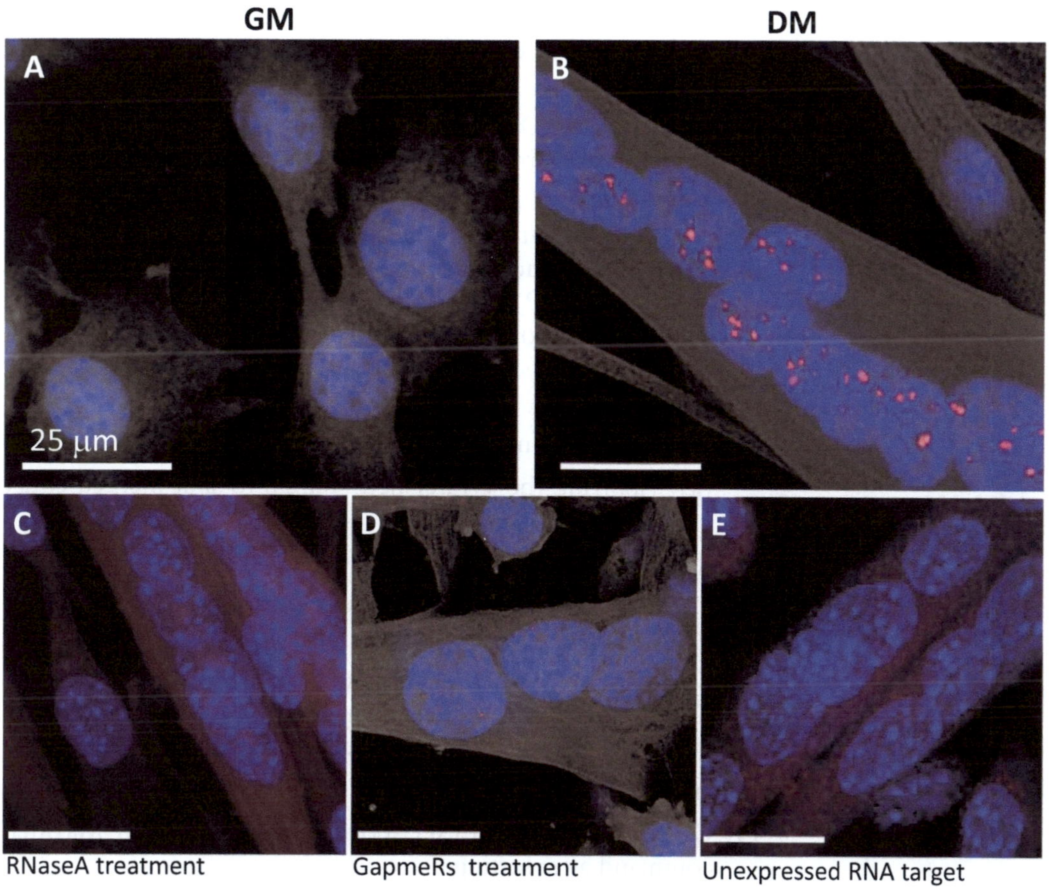

Fig. 7 RNA-FISH controls to test the specificity of *Charme* nuclear signals. RNA-FISH performed in (**a**) murine C_2C_{12} myoblasts (GM condition) where *Charme* is not expressed and in (**b**) differentiated C_2C_{12} myotubes (DM condition) where *Charme* is expressed. A significant reduction of the lncRNA spots was detected in myotubes treated with (**c**) RNase performed prior to hybridization, (**d**) antisense LNA GapmeRs against *Charme* transcript, or (**e**) by using probes targeting an unexpressed target (GFP mRNA). *See* text and reference [8] for details on *Charme* lncRNA

softwares (e.g., Huygens, SVI). Deconvolution can restore images that are spoilt by diffraction and aberration of the microscope lens. Main steps are as follows:

(a) To obtain experimental PSF from latex bead images (point spread function) or theoretical PSF (e.g., Nyquist calculator) using the parameters of the microscope.

(b) To set Deconvolution algorithm (e.g., classic maximum likelihood estimation in Huygens software).

(c) To set restoration parameters (maximum number of interaction, signal-to-noise ratio, background threshold).

7. To enhance the FISH signal over background, it is possible to apply a 3D Laplacian of Gaussian filter (also available in the Fiji plugin).
8. Proceed with 3D image analysis using a software (e.g., Imaris, Bitplane) that provides tools for surface rendering and colocalization studies.
9. For semiautomatic quantification of cell nuclei or RNA-FISH spots, the Z-stack can be merged in a maximum intensity projection (MIP) and processed by FIJI software to perform manual or automatic threshold before binary conversion. (Image→ Adjust→ Threshold; Process→ Binary→ Make Binary).
10. The number of nuclei/field can be quantified by:
    (a) Applying the "Watershed" plugin to split merged edged of close nuclei.
    (b) Running the "Analyze Particles" command upon.
        - Setting the counting parameters (size, 50-Infinity; Circularity, 0–1; Show, Outlines).
        - Selecting "Display results," "Clear results," "Summarize," and "Add to manager".

    Count, area, and average size are provided as text windows and the outlined particles are visualized as binary images.
11. Quantify the RNA-FISH spots by running "Find Maxima" command after setting:
    (a) "Noise tolerance" (10–100)
    (b) "Output type" (single points).

4 Notes

1. Potential troubles with focusing and sharpness can be avoided by seeding the cells on thin supports. The thickness of the coverslips is particularly important when images are captured with high-resolution optical microscopes. Indeed, variations of only a few micrometers reduce image quality and result in optical aberrations. For these reasons, we recommend the use of coverslips at 0.13–0.17-mm thickness (#1.0–1.5) for high-numerical-aperture oil objectives. It is also a good practice to store the coverslips, until use, at room temperature in absolute ethanol to remove any trace of fat and protect them from dust. Before use, coverslip sterilization must be performed after ethanol evaporation in a laminar flow hood by UV irradiation. To obtain the maximum number of technical replicates for

single plating, up to five coverslips can be located inside single 35-mm-diameter Petri dishes (Fig. 2a).

2. Adherent cells do not grow over glass very well. For this reason, the culture dish surface needs to be precoated before seeding to create an extracellular environment, which allows the cells to anchor the plate. To this purpose, 100 μg/ml poly-L-Lysine or 0.4 mg/ml collagen is used for muscle cells, while 0.01% poly-L-ornithine/20 μg/ml murine laminin in DPBS is preferred for neuronal cell lines.

   Incubation time with the appropriate coating solution is essential to obtain a stable layer. Normally, the coverslips are incubated at 4 °C overnight. The day after, the coating solution is discarded and the plate is left to dry in a laminar flow hood. Washing is not required at this step. It is possible to store the dried coverslips aseptically up to 1 month at 4 °C. For laminin-based coating, coverslips are incubated at 37 °C overnight. They can be also kept in this solution up to 1 week at 4 °C, sealed with Parafilm to prevent evaporation and contamination. Be very careful to prevent the laminin coating solution from drying out as it might result in laminin degradation.

   Chambered cover glasses can be also used. However, we often observed nonhomogeneous *seeding* densities and a tendency of the cells to crowd the edges of the chamber, both with manual and commercial pretreated supports.

   Many cells, especially primary and ESC-derived motor neuron cell cultures and cocultures, had difficulty staying attached for a long time to coated-glass supports. In these particular situations, it is preferable to grow cells directly on precoated plastic Petri dishes. After staining and mounting with thin coverslips, observation at the inverted microscope can be performed by overturning the Petri dish to put the objective in direct contact with glass.

3. Washing with DPBS stabilizes the cell attachment on support. In fact, magnesium (Mg2+) and calcium (Ca2+) are divalent cations essential to aid integrin-mediated cell binding.

4. Glycine quenches autofluorescence, by binding free reactive aldehydes in PFA.

5. Cells can be stored dry at room temperature after ethanol washes. Cells are stable at −20 °C in 2 ml of absolute ethanol for up to 6 months.

6. This is a centrifugation speed used for blood cell cultures. Optimization is required when other types of cells are used.

7. Never use vortex to resuspend cell pellets, but gently tap the bottom of the tube.

8. For suspension cells, poly-L-lysine-coated coverslips are used. Fast evaporation of ethanol in the drop and rotatory

movements allow to uniformly spread cells without the use of expensive mechanical procedures (such as cytocentrifuge) that may damage and alter the overall cell morphology. Moreover, in this way, we can proceed with the same RNA-FISH protocol used for adherent cells.

9. Work aseptically to prevent bacterial contaminations. It is also a good practice to keep an RNase-free environment by cleaning all the surfaces, plastic containers, and pipettes with RNase decontaminants. To this purpose, 30-min incubation with 3% hydrogen peroxide (H_2O_2) or commercial cleaning solutions can be performed. Moreover, all the solutions should be freshly prepared by using ultrafiltered and nuclease-free water.

10. The number of coverslips depends on some considerations. Take into account, for instance, the number of coverslips that are needed for negative controls, especially during the first setup of the protocol.

    The qualitative consistency of the post-hybridization signal (Fig. 7a) can be ascertained by using:

    (a) Samples obtained from cell types, tissues, or differentiation time points where the lncRNA is physiologically not expressed and hybridized with probe sets specific for the lncRNA target. This quality control is required to check for aspecific background signals (Fig. 7b).

    (b) Samples pretreated with RNase A and hybridized with probe sets specific for the lncRNA target. This quality control is required to check if the signal is RNA-specific (Fig. 7c).

    (c) Samples hybridized with probe sets specific for the lncRNA target obtained from cells where the lncRNA of interest is knocked down. This quality control is required to check if the signal is target-specific (Fig. 7d).

    (d) Samples hybridized with probe sets specific for a transcript expressed in unrelated species, such as the green fluorescent protein (GFP) mRNA from *Aequorea victoria* jellyfish. This quality control is required for monitoring off-target hybridization events (Fig. 7e).

11. The volume of buffers and solutions is adjusted to single 12-mm-diameter cover glasses. Be sure to soak the cells completely by adapting proportionally these volumes when other supports are used. Do not let cells dry out at any step of the protocol as it may lead to aspecific background signals.

12. Cell permeabilization is a critical point and requires optimization to make the lncRNA target as much as possible available for hybridization, without altering cell morphology and the topology of specific subcellular compartments. Incubation

with alcohols (i.e., ethanol) should be sufficient to permeabilize the cells. However, the use of detergents might be necessary to treat the inner membranes of differentiating myotubes or tissues especially when hybridization probes are longer than 20 base pairs. Buffers containing 0.5% or 0.2% Triton are used for nuclear or cytoplasmic lncRNAs, respectively. A mild incubation with pepsin or protease III can be also used as alternative to increase permeabilization. In fact, pretreatment with proteases contributes to unmask hidden sequence by disrupting the cross-linking of the RNA target with proteins. However, this strategy must to be finely calibrated (*see* also Subheading 3.5.3 and **Note 49**) as it may cause cell detachment during the staining. In all cases, the addition of VRC (vanadyl ribonucleoside complex) is required to inhibit the activity of possible contaminating ribonucleases.

13. RNase A digestion can be carried out with 2 ml of warm solution in 35-mm-diameter Petri dishes sealed with Parafilm and placed in a standard thermoblock at 37 °C.
14. We normally resuspend the probes in nuclease-free H_2O to a 100 μM concentration of stock solution and store at −20 °C. The working concentration must to be empirically determined, dependently from cell type, subcellular localization, and pretreatment processing. We generally test three different probe concentrations (100, 50, and 25 nM). The amount of probes that guarantees greater signal reproducibility and signal-to-noise ratio is then chosen for the subsequent steps.
15. Avoid air bubbles by placing one side of the coverslip on the probe-MIX drop and then gently slide it. Hybridization does not occur and cells dry if air bubbles are presents. Be careful also with the order of the coverslips: if the experiment involves more than one sample (for example, different cell lines, a time course from a single cell line, or simply different probe sets), sign the identity of the coverslips at the side of each microscope slide. Make sure that the coverslips are far enough from each other: it is advisable not to put more than three coverslips (12 mm diameter) on a single 25 × 75 mm microscope slide.
16. For a successful and reproducible hybridization assay, it is essential to prevent evaporation and abrupt temperature changes. Evaporation might:
    (a) Increase probe concentration, leading to aspecific hybridization events.
    (b) Overdry the cells, which leads to high autofluorescence.
    (c) Damage the subcellular structures.

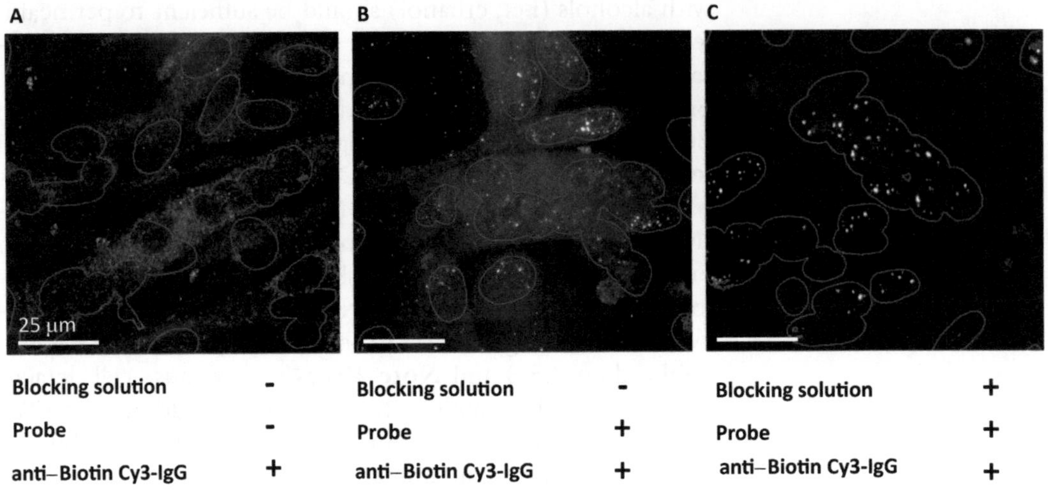

Fig. 8 Quenching of the endogenous biotin in the *indirect* hybridization method. Endogenous biotin signals from C_2C_{12} myotubes prior (**a**, **b**) and after (**c**) quenching with the blocking reagent. Signals derive exclusively from biotin-conjugated oligonucleotides. Blue solid line: nuclei

These troubles can be avoided by using the automatic slide hybridizer machine that allows a rigorous control of temperature and humidity of the hybridization chamber.

17. To avoid cell damage during the transfer of the coverslips, add 50 μl of 2× SSC under the glass, grasp gently with tweezers, and then place the coverslips into the multiwell plate with cells facing up.
18. Biotin is a physiological component of the cell, which may produce artefacts and aspecific signals by interfering with the exogenous biotin-conjugated oligonucleotides [27]. To avoid background, the endogenous biotin must be masked by using sequential steps of incubation with streptavidin (or avidin) and biotin to saturate all the free biotin-binding sites of streptavidin. Although these solutions can be manually prepared in a less expensive way, the percentage of the reagents must to be finely defined to avoid overblocking or high background. For reliable background suppression, the use of a commercial blocking kit (Thermo Fisher Scientific) is preferable (Fig. 8).
19. A second step of PFA fixation after in situ hybridization improves the oligonucleotide retention on the RNA target during immunofluorescence detection. This was also noticed when a similar strategy was applied in the combined immunofluorescence-DNA-FISH approach [28].
20. Blocking buffer contains biotin contaminations (normal serum, casein, and nonfat dry milk). BSA fraction V is a good substitute as it is biotin-free. Generally, immunodetection of

biotin occurs in 1 h, but it can be extended up to 3–4 h. The humid box can be obtained from a slide box equipped with bench paper wet in TN buffer on the bottom.

21. Mark areas to be stained with a wax pen before adding solutions. Adjust the buffer volumes to cover the whole section. For lncRNA detection on cell cultures, proceed as in Subheading 3.2.1 up to **step 4**.
22. Tilt the slides on adsorbent paper to discard the solutions. As formaldehyde-fixed frozen sections tend to detach from the slides, the use of staining containers (i.e., Coplin jars) is not recommended.
23. Replace PBS with TBS to reduce inorganic phosphates in the sample, which could inhibit the alkaline phosphatases.
24. 1-Ethyl-3-3-dimethylaminopropyl-carbodiimide (EDC) is a cross-linking agent that, in the presence of imidazole, immobilizes the 5′-ends of DNA and RNA to primary amines [24].
25. Triethanolamine and acetic anhydride lead to acetylation of the free protein amine groups. This pretreatment reduces the aspecific binding of the probes to positively charged residues in the sample.
26. For cell culture samples, add pre-hybridization buffer IV/2mM VRC and incubate at 37 °C for 30 min.
27. To check if the signal is RNA-specific, a commercial scramble LNA™ probe must be used as negative control. We generally use 80 ng of LNA™ probe in 100 μl of hybridization buffer III. For cell cultures, dilute the probe-MIX in hybridization buffer IV/2mM VRC.
28. Add 20 μl of probe to each tissue slice and spread with a piece of Parafilm. For cell cultures, incubate on microscope slides at 37 °C for 1 h as in Subheading 3.2.1, **step 11**.
29. Temperature and salt concentration of hybridization and washing buffers must be determined empirically. For cell culture samples, transfer the coverslips into a 24-multiwell plate and wash at 37 °C for 5 min with 1 ml of preheated 4× SSC, 2× SSC, 1× SSC.
30. This step can be omitted for cells in culture.
31. For cell culture samples, add 45 μl of 1:500 diluted AP-conjugated anti-digoxigenin antibody in 1% normal goat serum/2 mM levamisole/TN buffer. Incubate in humid box at 4 °C overnight.
32. For cell culture samples, aspirate TMN buffer and add 1 ml of detection buffer. Keep at room temperature for 15 min.

33. Levamisole inhibits endogenous alkaline phosphatases (AP) and does not disturb the AP isoenzyme used in the enzymatic detection. For cell culture samples, discard the detection buffer. Then add 500 μl of HNPP/Fast Red TR solution. Incubate at room temperature in the dark for 15 min.

34. Immunofluorescence can be performed *before* or *after* RNA-FISH. Choosing one strategy over the other depends on the protein and on the antibody used. In fact, the formamide-denaturing activity may affect the integrity of the epitopes and the affinity of primary antibodies. For these reasons, it is good practice to optimize the IF staining by testing samples *before* and *after* formamide incubation.

35. To preserve RNA integrity, all the solutions should contain RNase A inhibitors (such as 2 mM vanadyl ribonucleoside complex or RNasin).

36. To prevent the aspecific binding of antibodies, different blocking reagents should be used. A final concentration of 2% or 4% BSA is used for cells and tissues, respectively.

37. The incubation time of the primary antibody requires optimization. We observed that low temperature (4 °C) and long incubation times (18–20 h) are required for the efficient visualization of nuclear proteins after in situ hybridization.

38. In the RNA-FISH/IF colocalization analyses, excitation and emission spectra of the fluorophores conjugated to probes or secondary antibodies must not overlap. Microscope must be equipped with a set of filters that reduce potential cross-talk and bleedthrough.

39. A second step is necessary to fix the IF immune complex to the proteins within the cell. This procedure allows stabilizing the immunocomplex during the subsequent RNA-FISH steps.

40. DNA-FISH can be performed *before* or *after* RNA-FISH. The first approach, performed in RNase-free conditions, is better at preserving the topology of the signals for reliable 3D multichannel colocalization analyses. A DNA denaturation step is always required for efficient hybridization. BAC clones and PCR products (at least 10 Kb long) allow a more stable binding and offer a more reproducible copy number evaluation.

41. Probe labeling is performed with fluorophore- or hapten-conjugated (dUTP) nucleotides (in the latter case an immunofluorescence staining may be performed as in **Note 53**). To achieve efficient incorporation of labeled dUTP, the DNA template must be purified using a commercial extraction kit to remove salts, bacterial/genomic DNA, and protein contaminations that can inhibit the DNA polymerase I enzymatic

activity. Phenol/chloroform-based extraction is not recommended.

42. Diluted 2-mercaptoethanol is stable not longer than 1 month at 4 °C.
43. The DNase I enzymatic activity can change from lot to lot. To improve reproducibility, the correct dilution is found by testing serial dilution of the DNase I stock on DNA templates.
44. Transfer at 4 °C on a thermal cycler or ice water bath.
45. After nick translation, the DNA appears like a smear in which the size range depends on DNase I concentration/incubation. A size range of 100–300 Kb is preferable to favor the permeation of probes across cell membranes.
46. Do not air-dry the pellet but resuspend it by gently tapping the bottom of the tube. Do not vortex. If the pellet is hard to resuspend, incubate at 37 °C for 1 h.
47. For DNA- and RNA in situ hybridization, cells can be stored in 50% glycerol/2 mM VRC/DPBS up to 1 month. Glycerol is a cryoprotectant which prevents cellular damages caused by freezing [29].
48. Repeated freezing in glycerol helps make DNA more accessible to probes [30]. Transfer the coverslips from the glycerol solution to a dry ice ethanol bath with cells facing up. Wait until glycerol freezes (few seconds) and then thaw on bench paper. Put the coverslips in the glycerol solution again and repeat the procedure four times.
49. Digestion with proteases can be used to permeabilize cell or tissue sections under optimized conditions. Parameters in Subheading 3.5.3 are calibrated for protein-rich cytoplasms (i.e., differentiated muscle cells) and skeletal/cardiac tissue sections. Examine the samples by microscopy to check for undesirable morphology changes [30].
50. Discard the pre-hybridization buffer and proceed immediately to wash with ice-cold ethanol.
51. A thin layer of vacuum grease can be used as an alternative to rubber cement. Add the probe solution and then place the coverslip with cells facing down. Finally, apply light pressure to seal the edges.
52. Genomic DNA and probes are denatured simultaneously before the overnight incubation at 37 °C. This temperature has been adapted for muscle cell cultures. Optimization is required when other types of cells are used.
53. If digoxigenin-labeled probes are used, the coverslips are incubated *after the* post-hybridization *washes* in 45 μl of 1:250 diluted Alexa Fluor 488-conjugated anti-digoxigenin antibody

in 4% BSA/4× SSC for 1 h at room temperature. Then wash three times for 10 min each with 2× SSC. Proceed as Subheading 3.2.1, **steps 14–16**.

54. This procedure is appropriate when aqueous-based mounting media are used and drying is prevented by sealing the coverslips with nail polish. When solvent-based media are used, samples must be dehydrated and cleared with organic solvent (i.e., xylene) before mounting. Avoid bubbles, as they may lead to fluorescence artifacts.

    Check the formulation of mounting media (i.e., refractive index [RI], compatibility with fluorophores, photoprotective properties, and buffer composition) before each experiment. For example, if you have Cy stained fixed cells to be viewed on inverted high-numerical-aperture objectives, you might choose a glycerol-based hardening mount medium with high RI (>1.4) containing antifading additives. Buffers should be phenylenediamine-free, an fading blocker agent which destroys immunofluorescence of Cy2 dyes.

55. It is necessary to know the features of the fluorescent dyes in the sample when multicolor imaging is performed. If the emission spectra of fluorophores do not overlap, simultaneous imaging can be applied by exciting and acquiring all fluorophores at the same time. Perform the acquisition in according with Nyquist criterion, in order to avoid undersampling events that leading to non-optimal resolution. Is possible to use Nyquist Calculator (https://svi.nl/NyquistCalculator) to calculate the ideal sampling rate.

Acknowledgments

This work was partially supported by grants from Sapienza University (prot. RM11715C7C8176C1 and RM11916B7A39DCE5) and FFABR 2017 to M.B. *Panel a* of Fig. 4 was reprinted from Cell, 2011, 147(2), Cesana M, Cacchiarelli D, Legnini I, Santini T, Sthandier O, Chinappi M, Tramontano A, Bozzoni I., "A Long Noncoding RNA Controls Muscle Differentiation by Functioning as a Competing Endogenous RNA", Pages 358-69, Copyright (2011), with permission from Elsevier. License number: 4398770836312. The authors declare no competing financial interests.

References

1. Carninci P, Kasukawa T, Katayama S, Gough J, Frith MC, Maeda N, Oyama R, Ravasi T, Lenhard B, Wells C et al (2005) The transcriptional landscape of the mammalian genome. Science 309:1559–1563
2. Ripoche MA, Kress C, Poirier F, Dandolo L (1997) Deletion of the H19 transcription unit reveals the existence of a putative imprinting control element. Genes Dev 11:1596–1604

3. Moseley ML, Zu T, Ikeda Y, Gao W, Mosemiller AK, Daughters RS, Chen G, Weatherspoon MR, Clark HB, Ebner TJ et al (2006) Bidirectional expression of CUG and CAG expansion transcripts and intranuclear polyglutamine inclusions in spinocerebellar ataxia type 8. Nat Genet 38:758–769
4. Anguera MC, Ma W, Clift D, Namekawa S, Kelleher RJ, Lee JT (2011) Tsx produces a long noncoding RNA and has general functions in the germline, stem cells, and brain. PLoS Genet 7:e1002248
5. Nakagawa S, Naganuma T, Shioi G, Hirose T (2011) Paraspeckles are subpopulation-specific nuclear bodies that are not essential in mice. J Cell Biol 193:31–39
6. Zhang B, Arun G, Mao YS, Lazar Z, Hung G, Bhattacharjee G, Xiao X, Booth CJ, Wu J, Zhang C et al (2012) The lncRNA Malat1 is dispensable for mouse development but its transcription plays a cis-regulatory role in the adult. Cell Rep 2:111–123
7. Sauvageau M, Goff LA, Lodato S, Bonev B, Groff AF, Gerhardinger C, Sanchez-Gomez DB, Hacisuleyman E, Li E, Spence M et al (2013) Multiple knockout mouse models reveal lincRNAs are required for life and brain development. elife 2:e01749
8. Ballarino M, Cipriano A, Tita R, Santini T, Desideri F, Morlando M, Colantoni A, Carrieri C, Nicoletti C, Musarò A et al (2018) Deficiency in the nuclear long noncoding RNA *Charme* causes myogenic defects and heart remodeling in mice. EMBO J 37(18):pii: e99697
9. Rinn JL, Chang HY (2012) Genome regulation by long noncoding RNAs. Annu Rev Biochem 81:145–166
10. Ulitsky I, Bartel DP (2013) LincRNAs: genomics, evolution, and mechanisms. Cell 154:26–46
11. Fatica A, Bozzoni I (2014) Long non-coding RNAs: new players in cell differentiation and development. Nat Rev Genet 15:7–21
12. Li J, Tian H, Yang J, Gong Z (2016) Long noncoding RNAs regulate cell growth, proliferation, and apoptosis. DNA Cell Biol 35:459–470
13. Batista PJ, Chang HY (2013) Long noncoding RNAs: cellular address codes in development and disease. Cell 152:1298–1307
14. Ballarino M, Morlando M, Fatica A, Bozzoni I (2016) Non-coding RNAs in muscle differentiation and musculoskeletal disease. J Clin Invest 126:2021–2030
15. Derrien T, Johnson R, Bussotti G, Tanzer A, Djebali S, Tilgner H, Guernec G, Martin D, Merkel A, Knowles DG et al (2012) The GENCODE v7 catalog of human long noncoding RNAs: analysis of their gene structure, evolution, and expression. Genome Res 22:1775–1789
16. Cabili MN, Dunagin MC, McClanahan PD, Biaesch A, Padovan-Merhar O, Regev A, Rinn JL, Raj A (2015) Localization and abundance analysis of human lncRNAs at single-cell and single-molecule resolution. Genome Biol 16:20
17. Khalil AM, Guttman M, Huarte M, Garber M, Raj A, Rivea Morales D, Thomas K, Presser A, Bernstein BE, van Oudenaarden A et al (2009) Many human large intergenic noncoding RNAs associate with chromatin-modifying complexes and affect gene expression. Proc Natl Acad Sci U S A 106:11667–11672
18. Sun Q, Hao Q, Prasanth KV (2018) Nuclear long noncoding RNAs: key regulators of gene expression. Trends Genet 34:142–157
19. Noh JH, Kim KM, McClusky WG, Abdelmohsen K, Gorospe M (2018) Cytoplasmic functions of long noncoding RNAs. Wiley Interdiscip Rev RNA 9:e1471
20. Cesana M, Cacchiarelli D, Legnini I, Santini T, Sthandier O, Chinappi M, Tramontano A, Bozzoni I (2011) A long noncoding RNA controls muscle differentiation by functioning as a competing endogenous RNA. Cell 147:358–369
21. Femino AM, Fay FS, Fogarty K, Singer RH (1998) Visualization of single RNA transcripts in situ. Science 280:585–590
22. Raj A, van den Bogaard P, Rifkin SA, van Oudenaarden A, Tyagi S (2008) Imaging individual mRNA molecules using multiple singly labeled probes. Nat Methods 5:877–879
23. Itzkovitz S, Lyubimova A, Blat IC, Maynard M, van Es J, Lees J, Jacks T, Clevers H, van Oudenaarden A (2011) Single-molecule transcript counting of stem-cell markers in the mouse intestine. Nat Cell Biol 14:106–114
24. Pena JT, Sohn-Lee C, Rouhanifard SH, Ludwig J, Hafner M, Mihailovic A, Lim C, Holoch D, Berninger P, Zavolan M, Tuschl T (2009) miRNA in situ hybridization in mammalian tissues fixed with formaldehyde and EDC. Nat Methods 6:139–141
25. Cacchiarelli D, Martone J, Girardi E, Cesana M, Incitti T, Morlando M, Nicoletti C, Santini T, Sthandier O, Barberi L, Auricchio A, Musarò A, Bozzoni I (2010) MicroRNAs involved in molecular circuitries relevant for the Duchenne muscular dystrophy pathogenesis are controlled by the dystrophin/nNOS pathway. Cell Metab 12:341–351

26. Soares RJ, Maglieri G, Gutschner T, Diederichs S, Lund AH, Nielsen BS, Holmstrøm K (2018) Evaluation of fluorescence in situ hybridization techniques to study long non-coding RNA expression in cultured cells. Nucleic Acids Res 46(1):e4
27. Wood GS, Warnke R (1981) Suppression of endogenous avidin-binding activity in tissues and its relevance to biotin-avidin detection systems. J Histochem Cytochem 29:1196–1204
28. Chaumeil J, Micsinai M, Skok JA (2013) Combined Immunofluorescence and DNA FISH on 3D-preserved interphase nuclei to study changes in 3D nuclear organization. J Vis Exp (72):e50087
29. Solovei I (2002) FISH on three-dimensionally preserved nuclei. In: Beatty B, Mai S, Squire J (eds) FISH: a practical approach. Oxford University Press, Oxford
30. Cremer M, Grasser F, Lanctôt C, Müller S, Neusser M, Zinner R, Solovei I, Cremer T (2008) Multicolor 3D fluorescence in situ hybridization for imaging interphase chromosomes. Methods Mol Biol 463:205–239

Chapter 16

3D COMBO chrRNA–DNA–ImmunoFISH

Federica Marasca, Alice Cortesi, and Beatrice Bodega

Abstract

Epigenetic mechanisms govern the quality, the stability, and the responsiveness of transcriptional programs to the environment. This regulation is ensured via the concerted action of different players (transcription factors, "reader" and "writer" enzymes, histone marks, structural proteins, noncoding regulatory RNAs) that flow in the 3D organization of the genome. Indeed, nuclear architecture participates in the punctual and cell-type-specific regulation of transcription. Hence, the fine dissection of these mechanisms will allow a deeper understanding of the gene expression machinery. In this chapter, we propose a challenging imaging-based method to study the reciprocal interactions between chromatin-associated RNAs, genomic loci, and chromatin compartment with a procedure of 3D COMBO chrRNA–DNA–ImmunoFISH, specifically developed to preserve the nuclear integrity and topology of human primary T cells. We believe that our protocol will contribute to the improvement of epigenetic studies on the 3D nuclear structure of T cell subsets, possibly shedding light on the still hidden epigenetic players responsible for the great plasticity and functional diversification exerted by T cells.

Key words 3D COMBO RNA-DNA-Immuno FISH, Chromatin associated RNAs, Epigenetics, 3D genome organization, Human primary T cells, Transposable elements

1 Introduction

An extraordinary combination of mechanisms acts on chromatin to regulate the huge amount of informations stored in the genome, orchestrating cell-specific transcriptional patterns, cell plasticity and adaptability to the environment in time and in space. Classically, it is possible to discriminate between three hierarchically strictly interconnected levels at which epigenetic control of the genome takes place: (1) targeting of nuclear factors that regulate the transcriptional state [1]; (2) presence of epigenetic marks as DNA methylation, covalent post-translational modifications (PTMs) of histones, histone variants, and chromatin-associated proteins that "read" or "write" previous marks [2, 3]; and (3) the three-dimensional (3D) organization of chromatin in the nucleus that reflects nuclear metabolisms [4–6]. As an additional and

Beatrice Bodega and Chiara Lanzuolo (eds.), *Capturing Chromosome Conformation: Methods and Protocols*, Methods in Molecular Biology, vol. 2157, https://doi.org/10.1007/978-1-0716-0664-3_16,

interconnected level, the regulatory function at chromatin of noncoding RNAs (ncRNAs), as long intergenic ncRNAs (lincRNA) is becoming prominent [7–11].

Many laboratories shed light on the plethora of ncRNAs epigenetic functions; intriguingly, their impressive diversification is partially explained by the discovery that 83% of lincRNAs contains transposable elements (TEs), interspersed repeat sequences that expanded during evolution and constitute almost the half of the human genome [12]. It is indeed demonstrated that TE RNAs can be stably chromatin-associated, regulating chromatin accessibility and the transcription of a subset of specific genes in mESCs forming complex with chromatin remodelers and suggesting novel ncRNA-like functions [13–17].

Genome-wide Next-Generation Sequencing and imaging-based approaches at increasing resolution have been concomitantly improved and diversified to obtain the deepest dissection of the mechanisms that govern the genome structure and transcription in various contexts. For this purpose, many different imaging methodologies have been developed to visualize directly, and at single-cell level, the reciprocal dynamics between DNA genomic loci, RNAs, and proteins [18–20]; here, we briefly retrace the history and development of these technologies.

Fluorescent in situ hybridization (FISH) is a technique used for the quantification and spatial detection of nucleic acids; established in the 1970s [21], it is based on labelled, short, nucleic acid probes complementary to their target sequence in fixed cells or tissues [18–20]. DNA FISH enables various genetic, cytological, and diagnostic screenings [22, 23] and today is the elitist technique to study and validate spatial interactions and dynamics within many different genomic loci [24–28]. RNA FISH allows the detection of primary transcripts and ncRNAs, offering important advantages to study transcript dynamics and localization within the cellular population [26, 29–31]. Improvements of the technique led to the detection and quantification of single mRNA molecules [32] and to the cotemporal visualization of many multiple targets [33] with automated and computational approaches [34]. Finally, the integration of imaging data with omics datasets [35] offers a comprehensive view of the transcriptional diversification among cells. These features have been combined with recent technological advances in microscopy, conveying the application of superresolution to FISH [36–39], or rather reaching the examination of nucleic acids in live imaging using CRISPR–Cas approaches [16, 40]. Immunofluorescence (IF) is an ancient technique [41] that allows detection of antigen (protein) through specific antibodies that are directly labeled or further detected through secondary-labelled antibodies [42]. IF aims at providing information about the "protein-based structure" within the cells, reaching unprecedented resolution with superresolution microscopy and live cell imaging [43].

In this chapter, we provide a protocol for combining DNA FISH, RNA FISH, and IF that we have developed and applied for our studies in primary human T cells, aiming to highlight mutual 3D dynamics between genomic loci, chromatin-associated ncRNA (as the RNA of TEs), and chromatin compartment (e.g., histone marks). Such approach has been developed on preserved nuclei, which required special demands (as the avoidance of alcoholic scales) to maintain nuclear morphology and nucleic acid integrity and at the same time allow probe and antibody accessibility [44]. T lymphocytes, given their low cytoplasm abundance and the highly compacted nuclei, had required particular adjustment in terms of fixation and pretreatment [45]. The protocol provided below describes all materials, reagents, and steps required for a 3D COMBO of noncoding chromatin-associated RNA–DNA and ImmunoFISH (3D COMBO chrRNA–DNA–ImmunoFISH) along with controls and notes that could be useful upon particular experimental needs.

2 Materials

2.1 Cell Type

Ex vivo sorted, human primary T cells.

2.2 Sorting Strategy

Mononuclear cells will be isolated from peripheral blood of healthy donors by Ficoll-Hypaque density-gradient centrifugation. Among them, $CD4^+$Naive subset will be isolated by fluorescence-activated cell sorting (FACS) according to the surface receptor ($CD4^+$ $CD45RA^+$ $CD45RO^-$).

2.3 Equipment

1. Cell culture hood (i.e., biosafety cabinet).
2. Incubator set at 37 °C or 60 °C with an oscillating shaker.
3. Water bath set at 37 °C or 39 °C.
4. Hot block set at 50 °C or 75 °C.
5. Refrigerated centrifuge.
6. Thermoblock.
7. Orbital shaker.
8. Qubit fluorometer.
9. Micropipettes (1–10, 2–20, 20–200, 200–1000 μL).
10. Pipette aid.
11. Fine tip forceps.
12. Glass coverslips (10 mm).
13. Microscope slides.

14. Dry ice.
15. Rubber cement.
16. Metallic box.

2.4 Disposables

1. 6- and 24-multiwell plates.
2. Sterile plastic pipettes (5 mL, 10 mL, 25 mL).
3. Filter pipette tips (0.5–10, 2–20, 20–200, 200–1000 μL).
4. Conical tubes (15 and 50 mL).
5. 1.5-mL tubes.

2.5 Chemicals and Reagents

1. DEPC H_2O.
2. Poly-L-lysine 0.1% (w/v).
3. Paraformaldehyde (PFA).
4. 1× PBS (phosphate-buffered saline).
5. 100% Tween 20.
6. 100% Triton X-100.
7. 100% NP-40.
8. 100% glycerol.
9. Ribonucleoside vanadyl complexes (RVC).
10. 0.1 N HCl in DEPC H_2O.
11. 100% formamide at pH 7.
12. UltraPure bovine serum albumin (BSA).
13. Goat serum.
14. Biotin RNA labeling mix 10×.
15. dNTPs (separated mix; 100 mM each).
16. Digoxigenin-11-dUTP.
17. 100 mM β-mercaptoethanol.
18. 100 mM $MgCl_2$.
19. 1 M Tris–HCl pH 7.8.
20. 1 M Tris–HCl pH 8.
21. 5 M NaCl.
22. Human *Cot1* DNA.
23. Salmon sperm DNA.
24. Yeast tRNA.
25. 3 M Na acetate pH 5.2.
26. 0.5 M EDTA pH 8.0.
27. 100%, 70% ethanol (in DEPC water).

2.6 Enzymes, Kits, and Antibodies

1. MAXIscript T7 transcription kit.
2. Qubit RNA HS Assay Kit.
3. Phusion High-Fidelity DNA Polymerase.
4. PCR purification kit.
5. Qubit dsDNA HS Assay Kit.
6. DNA polymerase I.
7. Amplification grade DNAse I.
8. RNase Cocktail.
9. TSA Plus Fluorescent kit Cy3.5.
10. Streptavidin HRP.
11. Anti-digoxigenin 488, made from goat.
12. Primary antibodies of interest.
13. Alexa Fluor 647 fluorescent secondary antibody (goat anti mouse/rabbit).
14. Antifade prolong mounting.

2.7 Solutions (See Note 1)

1. PFA 3%, 4% (1× PBS/0.1% Tween 20, pH 7.0, filtered).
2. 20× SSC: 175.3 g NaCl, 88.2 g trisodium citrate in 1 L of DEPC H_2O. Autoclave and filter (*see* **Note 2**).
3. TBN: 0.1 M Tris–HCl pH 8, 0.150 M NaCl, DEPC H_2O.
4. TBN/BSA: 0.1 M Tris–HCl pH 8, 0.150 M NaCl, DEPC H_2O, 4% BSA.
5. TNT: 0.1 M Tris–HCl pH 8, 0.150 M NaCl, 0.1% NP-40, DEPC H_2O.
6. TNT/BSA: 0.1 M Tris–HCl pH 8, 0.150 M NaCl, 0.1% NP-40, DEPC H_2O, 4% BSA.

3 Methods

3.1 3D chrRNA FISH (3D RNA FISH for Chromatin-Associated RNAs)

3.1.1 Probe Preparation and Labelling Procedures

We use in vitro transcribed, biotin-labelled, antisense riboprobes. PCR amplicons of 1000 pb, covering the entire sequence of interest, are subjected to in vitro transcription and then to RNase digestion to obtain probes of the correct size. This is a 5-step procedure (Fig. 1):

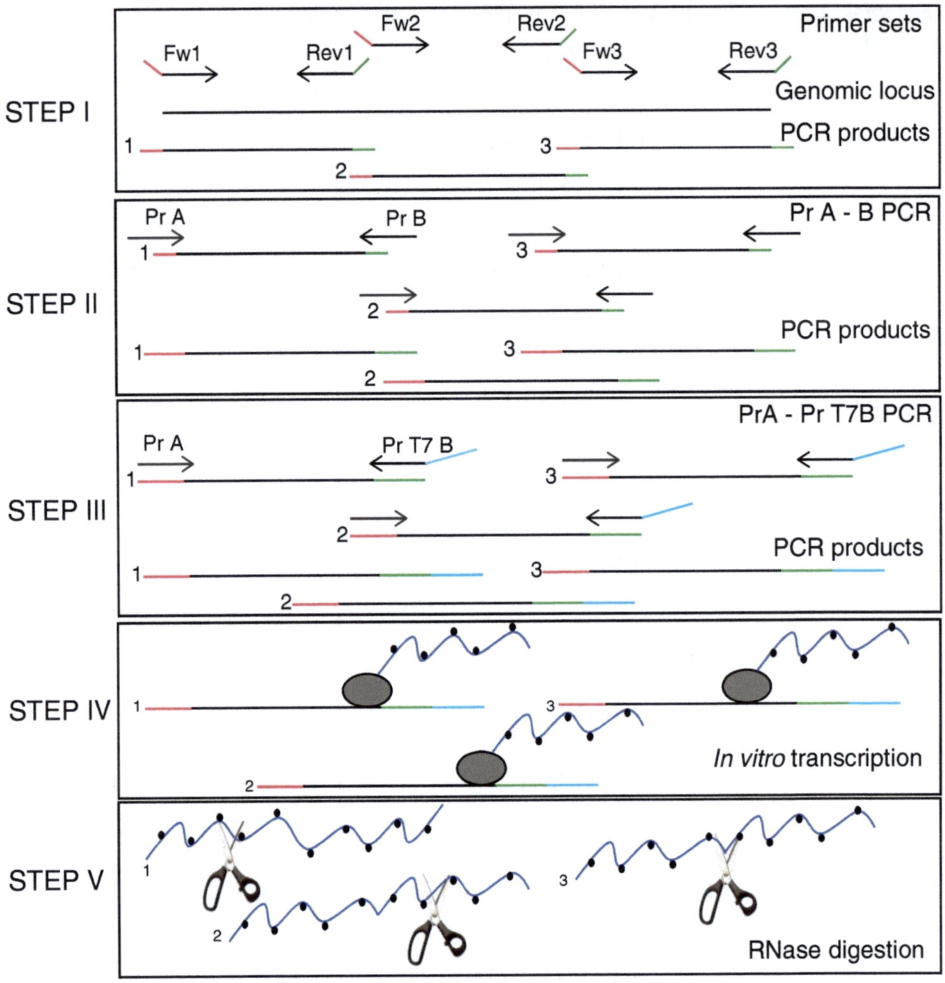

Fig. 1 Scheme of biotin-labelled riboprobe generation for RNA FISH. PCR products covering the region of interest will be produced using specific primers containing at the 5′-end sequences complementary to Primer A and B, in red and green, respectively (step I). PCR products will be further amplified with Primer A and Primer B (step II) and with primer A and Primer T7B in order to introduce a T7 promoter (step III). Antisense riboprobes will be transcribed in vitro and biotin labelled (step IV). Riboprobes will be further reduced in size with a mild RNase digestion (step V)

1. First PCR (*see* **Note 3**): The primers used in this PCR are as follows: Fw primer carrying the specific sequence of interest (e.g., a specific class of TEs or a lincRNA) plus the sequence in red and Rev. primer carrying the specific sequence of interest plus the sequence in green.

   Example of PCR primers:

Fw: 5-ATCGCACCAGCGTGTnnnnnnnnnnnnnnnnn-3

Rev: 5-TGAGGAGCCGCAGTGnnnnnnnnnnnnnnnnn-3

Reaction: 50 μL per well (primers 0.5 μM, dNTPs 0.2 mM, 1× HF buffer, 0.5 μL Phusion High-Fidelity DNA Polymerase).

Template: 100 ng of human gDNA.

Program:

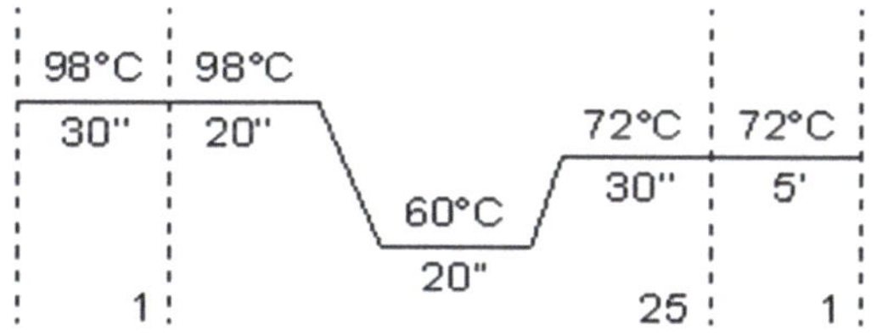

2. Second PCR: The primers used in this PCR are Primer A (Fw), containing the red sequence used before, and Primer B (Rev), containing the green sequence used before, both extended in cue:

Primer A: 5-CTCACTATAGGGATCGCACCAGCGTGT-3

Primer T7B: 5-GGATTCTAATACGACTCACTATAGGGATGAGGAGCCGCAGTG-3

Reaction: 50 μL per well (primers 0.5 μM, dNTPs 0.2 mM, 1× HF buffer, 0.5 μL Phusion High-Fidelity DNA Polymerase).

Template: 1–2 μL of first PCR.

Program:

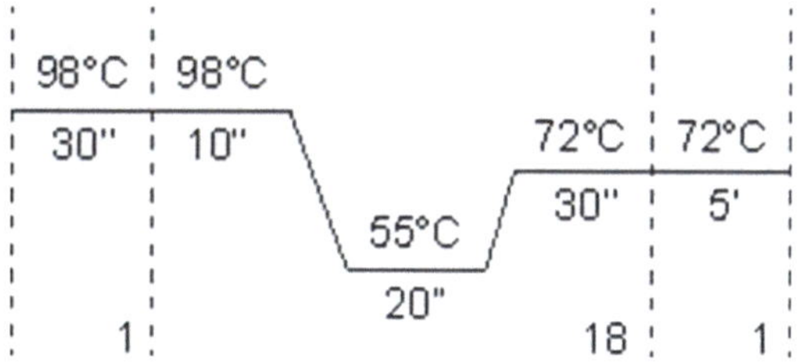

3. Third PCR: The primers used in this PCR are primer A (Fw) and primer T7B (Rev). This step will include the T7 promoter needed for in vitro transcription. Carefully study where the T7 promoter should be included to produce antisense probes.

Primer A: 5-CTCACTATAGGGATCGCACCAGCGTGT-3

Primer T7B: 5-GGATTCTAATACGACTCACTATAGGGATGAGGAGCCGCAGTG-3

Reaction: 50 μL per well (primers 0.5 μM, dNTPs 0.2 mM, 1× HF buffer, 0.5 μL Phusion High-Fidelity DNA Polymerase).

Template: 1 μL from second PCR.
Program:

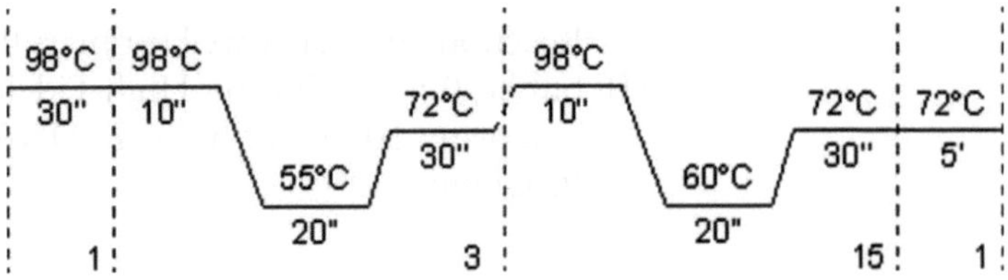

Load PCR products on an agarose preparative gel to check amplicon size. For the third PCRs, proceed with extraction/purification with the PureLink PCR Purification Kit. Elute in 40 μL and measure concentration with Qubit dsDNA HS Assay Kit.

4. Transcription of biotinylated riboprobes: Transcription is performed with MAXIscript T7 transcription kit and Biotin RNA labeling mix (*see* **Note 4**). Reaction volume is 50 μL.

   Template: 500 ng DNA (equimolar pool of third PCR samples).

Biotin RNA labeling mix:	5 μL
10× buffer:	5 μL
T7 enzyme:	5 μL
RNase inhibitors (Ambion):	0.5 μL ddH_2O up to 50 μL

Put the sample at 37 °C in a PCR thermoblock overnight (ON).

Stop reaction with 0.01 M EDTA. Add DEPC H_2O up to 150 μL. Precipitate with 1/10 volume of 3 M Na acetate pH 5.2, 3 volumes of 100% ethanol, and incubate the sample at −20 °C ON. Centrifuge at max speed at 4 °C for 60 min, wash the pellet twice with 70% ethanol, dry and suspend pellet in 40 μL DEPC H_2O, add 1 μL RNase Cocktail, and measure concentration with Qubit RNA HS Assay Kit.

5. RNase digestion of biotinylated riboprobes

   Dilute 1 μg of pooled biotinylated riboprobes in a final volume of 20 μL with DEPC H_2O and treat with 1 μL of RNase Cocktail diluted in DEPC H_2O (1:1000) for 2 min at 37 °C. Stop reaction with 0.01 M EDTA. Check the size of the riboprobes on a 2.2% agarose gel; optimal size is a smear around 100 pb (Fig. 2a).

6. Probe precipitation (for each slide).

   (a) 100–50 ng of riboprobes.

   (b) 10 μg salmon sperm DNA.

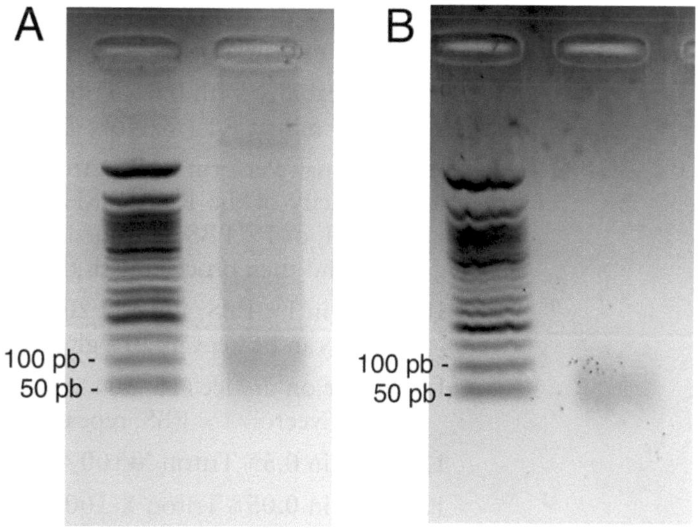

Fig. 2 Example of probe sizes for use in RNA and DNA FISH. (**a**) Riboprobes after RNase digestion run on an 2.2% agarose gel, 50-bp marker. (**b**) Nick-translated probes for DNA FISH run on an 2.2% agarose gel, 50-bp marker

(c) 10 μg yeast tRNA.

(d) Add DEPC H_2O up to 150 μL.

(e) Add 3 volumes of 100% ethanol.

(f) Add 1/10 volume of 3 M Na acetate pH 5.2.

Precipitate at −80 °C for 1 h, better at −20 °C ON. Centrifuge at maximum speed for 1 h; wash twice with 70% ethanol. Resuspend the pellet in 2 μL of 100% formamide pH 7 per slide.

3.1.2 Cell Seeding, Fixation, Pretreatment, and Permeabilization (See ***Note 5****)*

1. Use Glass Coverslips (10 mm).
2. Wash the glass with DEPC H_2O, then with 70% ethanol, and let them dry. Put one glass for each well of a 24-multiwell plate.
3. Add 200 μL of poly-L-lysine 0.1% (w/v) directly on the glass, making a drop (*see* **Note 6**).
4. Leave out the drop carefully, let the glass dry for 15 min, wash with DEPC H_2O, let the glass dry for 15 min, and repeat previous passages twice.
5. Put 150 μL of ex vivo sorted, primary human T cells (2×10^6/mL) directly on the glass making a drop (*see* **Note 6**), and let the drop stay at room temperature (RT) for 30 min.
6. Quickly replace the drop and add freshly made PFA 3% for 10 min at RT. During the last minute, add a few drops of 1× PBS/0.5% Triton X-100/10 mM ribonucleoside vanadyl complexes (RVC).

7. Wash with 0.05% Triton X-100/1× PBS/10 mM RVC 3 × 5 min at RT.
8. Permeabilize with 0.5% Triton X-100/1× PBS/10 mM RVC for 10 min at RT (*see* **Note 7**).
9. Optional: Perform RNAse treatment on few slides to verify the specificity of the RNA FISH. Use 2.5 μL RNase Cocktail in 250 μL of 1× PBS per slide in 24-multiwell plate, at 37 °C for 1 h, and then proceed with the following steps.
10. Rinse in 1× PBS, and add 20% glycerol/1× PBS, ON at 4 °C. Slides can be kept in 20% glycerol/1× PBS up to 7 days.
11. Freeze on dry ice (15–30 s), thaw gradually at RT, and soak in 20% glycerol/1× PBS; repeat 4 times.
12. Wash in 0.5% Triton X-100/1× PBS for 5 min at RT.
13. Wash in 0.05% Triton X-100/1× PBS 2 × 5 min at RT.
14. Incubate in 0.1N HCl DEPC H_2O for 10 min at RT (*see* **Note 7**).
15. Rinse in 2× SSC.

3.1.3 Hybridization

1. Put the formamide resuspended probe at 80 °C for 2 min for denaturation and then quickly on ice. Add an equal amount of 4× SSC/20% dextran sulfate.
2. Load the probes on a clean microscope slide.
3. Turn the coverslip with cells upside down on the drop of probe in hybridization mixture.
4. Seal the coverslip with rubber cement.
5. Place slides on a hot block and denature for 5 min at 50 °C.
6. Hybridize ON at 37 °C in a metallic box floating in a water bath.

3.1.4 Detection

1. Peel off rubber cement and strip off the coverslip and transfer it to 2× SSC.
2. Wash in 50% formamide/2× SSC 3 × 5 min at 39 °C.
3. Wash in 2× SSC 3 × 5 min at 39 °C.
4. Wash in 1× SSC 5 min on an orbital shaker at RT.
5. Wash in 4× SSC/0.2% Tween 20 2 × 5 min on an orbital shaker at RT.
6. Block in TBN/BSA for 30 min at RT on an orbital shaker.
7. Incubate with the appropriate concentration of streptavidin HRP (1:1000) (*see* **Note 4**) diluted in TNT/BSA for 30 min at RT on an orbital shaker.
8. Wash in TNT 4 × 5 min at RT on an orbital shaker.

9. For signal amplification, use TSA Plus Fluorescent kit Cy3.5: Incubate TSA working solution (1:150) in 1× amplification buffer for 3 min at RT on an orbital shaker (*see* **Note 8**).
10. Wash in TNT 4 × 5 min at RT on an orbital shaker.
11. Equilibrate cells in 1× PBS and post-fix preparation in 2% formaldehyde/1× PBS for 2 min at RT.
12. Wash 5 times briefly in 1× PBS at RT (if you are interested in COMBO chrRNA–DNA FISH, proceed with protocol step in Subheading 3.2).
13. Stain with 1 ng/μL DAPI (4,6-diamidino-2-phenylindole)/ 1× PBS for 5 min at RT.
14. Wash 5 times briefly in 1× PBS at RT.
15. Mount in antifade prolong mounting (you can proceed with COMBO chrRNA–DNA FISH steps also after visualization, detaching the glasses from the microscope slides and rinsing quickly with 2× SSC).

3.2 3D COMBO RNA–DNA FISH (See Note 9)

3.2.1 Probe Preparation and Labelling Procedures (See Note 10)

Probes for DNA FISH can be prepared from a PCR equimolar pool (of a size of 1000–2000 pb) covering the genomic region of interest, or rather from a BAC containing the genomic region of interest subjected to nick translation.

1. (a) Nick translation of PCR pool:

   Use 2–3 μg of PCR pool that are equivalent to 10 DNA FISH reactions:

Reagents	Initial conc.	Final conc.
dNTPs (C-G-A)	0.5 mM	0.05 mM
dTTP	0.1 mM	0.01 mM
Biotin/Dig/Cy3 dUTP	1 mM	0.02 mM
Tris–HCl pH 7.8	1 M	50 mM
$MgCl_2$	100 mM	5 mM
B-Mercaptoethanol	100 mM	10 mM
BSA	100 ng/μL	10 ng/μL
DNA pol I	10 U/μL	0.1 U/μL
DNAse I	1 U/μL	0.001 U/μL
DNA	x	x
ddH_2O		Up to 50 μL

Incubate at 16 °C for 45 min for small sequences. Check the size of the probes on 2.2% agarose gel. We prefer probes of size <200 pb (Fig. 2b). Stop the reaction with 0.5 mM EDTA.

(b) Nick translation of BAC:

Use 2–3 μg of BAC.

Reagents	Initial conc.	Final conc.
dNTPs (C-G-A)	0.5 mM	0.05 mM
dTTP	0.1 mM	0.01 mM
Biotin/Dig/Cy3 dUTP	1 mM	0.02 mM
Tris–HCl pH 7.8	1 M	50 mM
$MgCl_2$	100 mM	5 mM
B-Mercaptoethanol	100 mM	10 mM
BSA	100 ng/μL	10 ng/μL
DNA pol I	10 U/μL	0.1 U/μL
DNAse I	1 U/μL	0.002 U/μL
DNA	x	x
ddH_2O		Up to 50 μL

Incubate at 16 °C for up to 4 h for BAC. Check the size of the probes on 2.2% agarose gel. We prefer probes of size <200 pb (Fig. 2b). Stop the reaction with 0.5 mM EDTA.

2. Probe precipitation (for each slide):
   (a) 5 μL of nick-translated PCR pool or BAC.
   (b) 20 μg unlabelled salmon sperm DNA.
   (c) Add water up to 150 μL.
   (d) Add 3 volumes of 100% ethanol.
   (e) Add 1/10 volume of 3 M Na Acetate pH 5.2.
   (f) Add 3.5 μg of species-specific *Cot-1* DNA (if you are not studying repetitive elements).

   Precipitate at −80 °C for 1 h. Centrifuge at maximum speed for 1 h, and wash twice with 70% ethanol. Resuspend the pellet in 2 μL of 100% formamide pH 7 per slide (shake at 40 °C, can take up to a few hours) and then add the equal volume of 4× SSC/20% dextran sulfate. Hybridization probes can be stored at −20 °C for a long period of time.

3.2.2 Fixing and Denaturation

1. Fix in 4% PFA for 3 min at RT.
2. Incubate in 50% formamide pH 7/2× SSC (at least 30 min at RT, better ON [*see* **Note 11**]).

3.2.3 Hybridization

1. Denature the probes at 80 °C for 5 min and then put quickly on ice.
2. Load hybridization solution on a clean microscope slide.
3. Turn the coverslip with cells upside down on the drop of hybridization mixture.
4. Seal the coverslip with rubber cement. Let the rubber cement dry up completely.
5. After rubber cement is dried up, place slides on a hot block and denature at 75 °C for 4 min.
6. Hybridize at 37 °C ON (*see* **Note 12**) in a metallic box floating in a water bath.

3.2.4 Detection

1. Peel off rubber cement and strip off the coverslip and transfer it to 2× SSC.
2. Wash in 2× SSC 3 × 5 min at 37 °C, shaking 90 rpm in 6-multiwell plate.
3. Wash in 0.1× SSC 3 × 5 min at 60 °C, shaking 90 rpm in 6-multiwell plate.
4. Rinse briefly in 4× SSC/0.2% Tween 20.
5. Block in 4× SSC/0.2% Tween 20/4% BSA for 20 min at 37 °C in 24-multiwell plate.
6. Incubate with the appropriate concentration of anti-digoxigenin 488 (1:150) (*see* **Note 13**) diluted in 4× SSC/0.2% Tween 20/4% BSA in a dark and wet chamber for 35 min at 37 °C, shaking 90 rpm in 24-multiwell plate (*see* **Note 14**).
7. Wash in 4× SSC/0.2% Tween 20 3 × 5 min at 37 °C, shaking 90 rpm in 6-multiwell plate.
8. Equilibrate cells in 1× PBS and post-fix preparation in 2% formaldehyde/1× PBS for 2 min at RT in 24-multiwell plate.
9. Wash 5 times briefly in 1× PBS (if you are interested in performing Immunofluorescence proceed directly with Subheading 3.3).
10. Stain with 1 ng/μL DAPI/1× PBS for 5 min at RT.
11. Wash 5 times briefly in 1× PBS.
12. Mount in antifade prolong mounting (you can proceed with immunofluorescence steps also after visualization, rinsing the glasses quickly with 2× SSC and then equilibrating in 1× PBS).

3.3 3D COMBO chrRNA–DNA–ImmunoFISH

1. Incubate primary antibody dilution in 2% goat serum/1% BSA/1× PBS-Tween 0.1%, (Ex. H3K4me3, 1:250 Millipore 07-473) at 4 °C ON, mild shaking.
2. Wash in PBS-Tween 0.1% 3 × 5 min at RT, mild shaking.

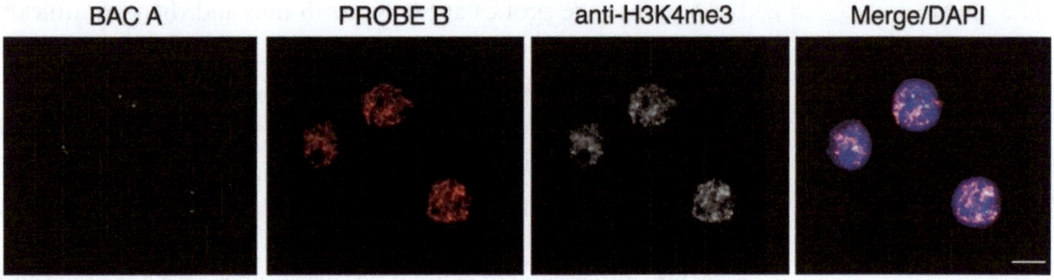

Fig. 3 Representative 3D COMBO chrRNA–DNA–ImmunoFISH. Representative nuclei of 3D multicolor COMBO chrRNA–DNA–ImmunoFISH using probe mapping on a genomic locus (BAC A, green), antisense riboprobes for TE RNAs (probe B), and anti-H3K4me3 (gray). Nuclei are counterstained in DAPI (blue). Original magnification ×63. Scale bar 5 μm

3. Incubate Alexa Fluor 647 fluorescent secondary antibody dilution (1:1000) in 2% goat serum/1% BSA/PBS-Tween 0.1% for 1 h at RT, mild shaking.
4. Wash in PBS-Tween 0.1% 3 × 5 min at RT, mild shaking.
5. Stain with 1 ng/μL DAPI/1× PBS for 5 min at RT.
6. Wash 5 times briefly in 1× PBS.
7. Mount in antifade prolong mounting.
8. Visualize at confocal microscopy (Fig. 3).

4 Notes

1. For an optimal protocol, we suggest to freshly prepare all the solutions at use. Solutions should be prepared in DEPC water. Prepare all solutions following appropriate safety rules (i.e., fume hoods, gloves).
2. 20× SSC should be autoclaved, filtered, and aliquoted to avoid contamination.
3. Set up all amplification and PCR steps using typical PCR precautions.
4. UTP labelling can vary depending on the experimental plan, on the cell type autofluorescence, and on RNA abundance/availability for hybridization. Instead of biotin, digoxigenin can be use. Directly fluorescent-labeled UTP can be used.
5. After PFA fixing, place the multiwell on ice for permeabilization and deproteinization passages.
6. Note that the drop has to remain on the glass surface without touching the well.
7. Permeabilization and deproteinization passages are tricky points and time of reaction and concentration of the reagents

strongly depend on cell type and on cytoplasm abundance and degree of nuclei compaction.

8. UTP biomarker and fluorophore choice can vary depending on the cell type autofluorescence and on the experimental plan. TSA amplification is often not needed for very abundant RNAs or RNAs that are not complexed with proteins and chromatin; in those cases you can do the detection using directly labelled streptavidin.
9. This protocol can be adapted for COMBO RNA–DNA FISH, RNA–ImmunoFISH, and DNA FISH or DNA–ImmunoFISH purposes. If you are interested in DNA FISH-based protocols uncoupled from the RNA FISH procedure, please follow Subheading 3.1.2, use PFA 4% instead of 3% do the optional **step 9**, and proceed with Subheading 3.2.2 **step 2**.
10. Choose the labelling of DNA FISH probes accordingly to RNA FISH probes (e.g., use digoxigenin-UTP for DNA nick-translated probes accordingly to biotin-UTP riboprobes).
11. Timing of formamide denaturation is a crucial point and can vary depending on cell type. For lymphocytes, we recommend to prolong the timing to ON.
12. You can prolong time of hybridization up to 2–3 days if you are performing DNA FISH uncoupled from RNA FISH. For COMBO 3D chrRNA–DNA FISH, we suggest to respect the indicated timing.
13. In case you choose biotin to label DNA FISH probes, we suggest streptavidin 1:1000.
14. For directly labelled probes, leave the slides in 4× SSC/0.2% Tween 20/4% BSA for another 35 min in a dark and wet chamber at 37 °C.

References

1. Deplancke B, Alpern D, Gardeux V (2016) The genetics of transcription factor DNA binding variation. Cell 166(3):538–554. https://doi.org/10.1016/j.cell.2016.07.012
2. Kouzarides T (2007) Chromatin modifications and their function. Cell 128(4):693–705. https://doi.org/10.1016/j.cell.2007.02.005
3. Allis CD, Jenuwein T (2016) The molecular hallmarks of epigenetic control. Nat Rev Genet 17(8):487–500. https://doi.org/10.1038/nrg.2016.59
4. Pombo A, Dillon N (2015) Three-dimensional genome architecture: players and mechanisms. Nat Rev Mol Cell Biol 16(4):245–257. https://doi.org/10.1038/nrm3965
5. Bonev B, Cavalli G (2016) Organization and function of the 3D genome. Nat Rev Genet 17(12):772. https://doi.org/10.1038/nrg.2016.147
6. Rowley MJ, Corces VG (2018) Organizational principles of 3D genome architecture. Nat Rev Genet 19(12):789–800. https://doi.org/10.1038/s41576-018-0060-8
7. Malecova B, Morris KV (2010) Transcriptional gene silencing through epigenetic changes mediated by non-coding RNAs. Curr Opin Mol Ther 12(2):214–222
8. Bohmdorfer G, Wierzbicki AT (2015) Control of chromatin structure by long noncoding RNA. Trends Cell Biol 25(10):623–632. https://doi.org/10.1016/j.tcb.2015.07.002

9. Engreitz JM, Ollikainen N, Guttman M (2016) Long non-coding RNAs: spatial amplifiers that control nuclear structure and gene expression. Nat Rev Mol Cell Biol 17 (12):756–770. https://doi.org/10.1038/nrm.2016.126
10. Guttman M, Donaghey J, Carey BW, Garber M, Grenier JK, Munson G, Young G, Lucas AB, Ach R, Bruhn L, Yang X, Amit I, Meissner A, Regev A, Rinn JL, Root DE, Lander ES (2011) lincRNAs act in the circuitry controlling pluripotency and differentiation. Nature 477(7364):295–300. https://doi.org/10.1038/nature10398
11. Rinn JL, Chang HY (2012) Genome regulation by long noncoding RNAs. Annu Rev Biochem 81:145–166. https://doi.org/10.1146/annurev-biochem-051410-092902
12. Kelley D, Rinn J (2012) Transposable elements reveal a stem cell-specific class of long noncoding RNAs. Genome Biol 13(11):R107. https://doi.org/10.1186/gb-2012-13-11-r107
13. Hall LL, Carone DM, Gomez AV, Kolpa HJ, Byron M, Mehta N, Fackelmayer FO, Lawrence JB (2014) Stable C0T-1 repeat RNA is abundant and is associated with euchromatic interphase chromosomes. Cell 156 (5):907–919. https://doi.org/10.1016/j.cell.2014.01.042
14. Jachowicz JW, Bing X, Pontabry J, Boskovic A, Rando OJ, Torres-Padilla ME (2017) LINE-1 activation after fertilization regulates global chromatin accessibility in the early mouse embryo. Nat Genet 49(10):1502–1510. https://doi.org/10.1038/ng.3945
15. Percharde M, Lin CJ, Yin Y, Guan J, Peixoto GA, Bulut-Karslioglu A, Biechele S, Huang B, Shen X, Ramalho-Santos M (2018) A LINE1-nucleolin partnership regulates early development and ESC identity. Cell 174(2):391–405.e319. https://doi.org/10.1016/j.cell.2018.05.043
16. Nelles DA, Fang MY, O'Connell MR, Xu JL, Markmiller SJ, Doudna JA, Yeo GW (2016) Programmable RNA tracking in live cells with CRISPR/Cas9. Cell 165(2):488–496. https://doi.org/10.1016/j.cell.2016.02.054
17. Fort A, Hashimoto K, Yamada D, Salimullah M, Keya CA, Saxena A, Bonetti A, Voineagu I, Bertin N, Kratz A, Noro Y, Wong CH, de Hoon M, Andersson R, Sandelin A, Suzuki H, Wei CL, Koseki H, Consortium F, Hasegawa Y, Forrest AR, Carninci P (2014) Deep transcriptome profiling of mammalian stem cells supports a regulatory role for retrotransposons in pluripotency maintenance. Nat Genet 46(6):558–566. https://doi.org/10.1038/ng.2965
18. Cui C, Shu W, Li P (2016) Fluorescence in situ hybridization: cell-based genetic diagnostic and research applications. Front Cell Dev Biol 4:89. https://doi.org/10.3389/fcell.2016.00089
19. Gall JG (2016) The origin of in situ hybridization—a personal history. Methods 98:4–9. https://doi.org/10.1016/j.ymeth.2015.11.026
20. Levsky JM, Singer RH (2003) Fluorescence in situ hybridization: past, present and future. J Cell Sci 116(Pt 14):2833–2838. https://doi.org/10.1242/jcs.00633
21. Rudkin GT, Stollar BD (1977) High resolution detection of DNA-RNA hybrids in situ by indirect immunofluorescence. Nature 265 (5593):472–473
22. Mastenbroek S, Twisk M, van der Veen F, Repping S (2011) Preimplantation genetic screening: a systematic review and meta-analysis of RCTs. Hum Reprod Update 17 (4):454–466. https://doi.org/10.1093/humupd/dmr003
23. Ross FM, Avet-Loiseau H, Ameye G, Gutierrez NC, Liebisch P, O'Connor S, Dalva K, Fabris S, Testi AM, Jarosova M, Hodkinson C, Collin A, Kerndrup G, Kuglik P, Ladon D, Bernasconi P, Maes B, Zemanova Z, Michalova K, Michau L, Neben K, Hermansen NE, Rack K, Rocci A, Protheroe R, Chiecchio L, Poirel HA, Sonneveld P, Nyegaard M, Johnsen HE, European Myeloma N (2012) Report from the European Myeloma Network on interphase FISH in multiple myeloma and related disorders. Haematologica 97(8):1272–1277. https://doi.org/10.3324/haematol.2011.056176
24. Fudenberg G, Imakaev M (2017) FISH-ing for captured contacts: towards reconciling FISH and 3C. Nat Methods 14(7):673–678. https://doi.org/10.1038/nmeth.4329
25. Giorgetti L, Heard E (2016) Closing the loop: 3C versus DNA FISH. Genome Biol 17 (1):215. https://doi.org/10.1186/s13059-016-1081-2
26. Buxbaum AR, Haimovich G, Singer RH (2015) In the right place at the right time: visualizing and understanding mRNA localization. Nat Rev Mol Cell Biol 16(2):95–109. https://doi.org/10.1038/nrm3918
27. Wu B, Buxbaum AR, Katz ZB, Yoon YJ, Singer RH (2015) Quantifying protein-mRNA interactions in single live cells. Cell 162

(1):211–220. https://doi.org/10.1016/j.cell.2015.05.054

28. Cortesi A, Pesant M, Sinha S, Marasca F, Sala E, Gregoretti F, Antonelli L, Oliva G, Chiereghin C, Soldà G, Bodega B (2019) 4q-D4Z4 Chromatin architecture regulates the transcription of muscle atrophic genes in facioscapulohumeral muscular dystrophy. Genome Res. 29 (6):883–895. https://doi.org/10.1101/gr.233288.117. Epub 2019 May 16
29. Femino AM, Fay FS, Fogarty K, Singer RH (1998) Visualization of single RNA transcripts in situ. Science 280(5363):585–590
30. Cabili MN, Dunagin MC, McClanahan PD, Biaesch A, Padovan-Merhar O, Regev A, Rinn JL, Raj A (2015) Localization and abundance analysis of human lncRNAs at single-cell and single-molecule resolution. Genome Biol 16:20. https://doi.org/10.1186/s13059-015-0586-4
31. Dunagin M, Cabili MN, Rinn J, Raj A (2015) Visualization of lncRNA by single-molecule fluorescence in situ hybridization. Methods Mol Biol 1262:3–19. https://doi.org/10.1007/978-1-4939-2253-6_1
32. Raj A, van den Bogaard P, Rifkin SA, van Oudenaarden A, Tyagi S (2008) Imaging individual mRNA molecules using multiple singly labeled probes. Nat Methods 5(10):877–879. https://doi.org/10.1038/nmeth.1253
33. La Manno G, Gyllborg D, Codeluppi S, Nishimura K, Salto C, Zeisel A, Borm LE, Stott SRW, Toledo EM, Villaescusa JC, Lonnerberg P, Ryge J, Barker RA, Arenas E, Linnarsson S (2016) Molecular diversity of midbrain development in mouse, human, and stem cells. Cell 167(2):566–580.e519. https://doi.org/10.1016/j.cell.2016.09.027
34. Samacoits A, Chouaib R, Safieddine A, Traboulsi AM, Ouyang W, Zimmer C, Peter M, Bertrand E, Walter T, Mueller F (2018) A computational framework to study sub-cellular RNA localization. Nat Commun 9(1):4584. https://doi.org/10.1038/s41467-018-06868-w
35. Boisset JC, Vivie J, Grun D, Muraro MJ, Lyubimova A, van Oudenaarden A (2018) Mapping the physical network of cellular interactions. Nat Methods 15(7):547–553. https://doi.org/10.1038/s41592-018-0009-z
36. Jungmann R, Avendano MS, Dai M, Woehrstein JB, Agasti SS, Feiger Z, Rodal A, Yin P (2016) Quantitative super-resolution imaging with qPAINT. Nat Methods 13(5):439–442. https://doi.org/10.1038/nmeth.3804
37. Legant WR, Shao L, Grimm JB, Brown TA, Milkie DE, Avants BB, Lavis LD, Betzig E (2016) High-density three-dimensional localization microscopy across large volumes. Nat Methods 13(4):359–365. https://doi.org/10.1038/nmeth.3797
38. Ricci MA, Manzo C, Garcia-Parajo MF, Lakadamyali M, Cosma MP (2015) Chromatin fibers are formed by heterogeneous groups of nucleosomes in vivo. Cell 160(6):1145–1158. https://doi.org/10.1016/j.cell.2015.01.054
39. Lubeck E, Cai L (2012) Single-cell systems biology by super-resolution imaging and combinatorial labeling. Nat Methods 9 (7):743–748. https://doi.org/10.1038/nmeth.2069
40. Chen B, Gilbert LA, Cimini BA, Schnitzbauer J, Zhang W, Li GW, Park J, Blackburn EH, Weissman JS, Qi LS, Huang B (2013) Dynamic imaging of genomic loci in living human cells by an optimized CRISPR/Cas system. Cell 155(7):1479–1491. https://doi.org/10.1016/j.cell.2013.12.001
41. Coons AH (1960) Immunofluorescence. Public Health Rep 75:937–943
42. Shimomura O, Johnson FH, Saiga Y (1962) Extraction, purification and properties of aequorin, a bioluminescent protein from the luminous hydromedusan, Aequorea. J Cell Comp Physiol 59:223–239
43. Sigal YM, Zhou R, Zhuang X (2018) Visualizing and discovering cellular structures with super-resolution microscopy. Science 361 (6405):880–887. https://doi.org/10.1126/science.aau1044
44. Cremer M, Grasser F, Lanctot C, Muller S, Neusser M, Zinner R, Solovei I, Cremer T (2008) Multicolor 3D fluorescence in situ hybridization for imaging interphase chromosomes. Methods Mol Biol 463:205–239. https://doi.org/10.1007/978-1-59745-406-3_15
45. Marasca F, Cortesi A, Manganaro L, Bodega B (2020, Jan) 3D Multicolor DNA FISH tool to study nuclear architecture in human primary cells. J Vis Exp. 25(155). https://doi.org/10.3791/60712

Chapter 17

An Algorithm for the Analysis of the 3D Spatial Organization of the Genome

Francesco Gregoretti, Alice Cortesi, Gennaro Oliva, Beatrice Bodega, and Laura Antonelli

Abstract

We present an algorithm, and its MATLAB implementation, based on mathematical methods to detect and localize 3D multicolor DNA FISH spots in fluorescence cell image z-stacks. This algorithm provides a method to measure the relative positioning of spots in the nucleus and inter-spot distances with the aim to enrich our understanding of the 3D spatial organization of the genome within the cell nucleus.

Key words Fluorescence microscopy, Multicolor 3D-FISH, Nuclear organization, Image processing and analysis, Cellular and subcellular segmentation, Variational segmentation model, Mathematical morphology method

1 Introduction

We have developed an algorithm based on mathematical methods to analyze 3D multicolor DNA FISH in fluorescence cell image z-stacks and we will show how it can be carried out in MATLAB environment. The algorithm performs the 2D segmentation of cell nuclei and the detection of spots for each slice of the stack followed by the 3D reconstruction and identification of nuclei and spots. It then measures the relative positioning of spots in the nucleus and inter-spot distances. The algorithm has been implemented in a tool called NuCLεD (Nuclear Contacts Locator in 3D) and used to successfully analyze several experimental sets of 3D multicolor DNA FISH in three or four colors: "blue" (corresponding to the staining with DAPI, always present, used for the identification of nuclei), "green," "red," and "violet" (for the different genomic regions assessed) [1]. In particular, 1257 three-color stacks and 1110 four-color stacks were analyzed, each consisting of 15 slices (images), for a total of more than 120,000 images. The total number of nuclei analyzed was 6314. For each of the three

Beatrice Bodega and Chiara Lanzuolo (eds.), *Capturing Chromosome Conformation: Methods and Protocols*, Methods in Molecular Biology, vol. 2157, https://doi.org/10.1007/978-1-0716-0664-3_17,

fluorescent pairs ("green-red," "green-violet," and "red-violet"), the algorithm had been able to identify exactly four spots (two for each fluorescent channel) in about 30% of the nuclei.

The algorithm relies on a region-based segmentation model [2] for the segmentation of cell nuclei and on the h-dome transformation [3] for the detection of spots.

We decided to not use a global operator after having observed that a spot in one part of the image could be lighter or darker than the background in another part. This is due to the facts that the spots have inhomogeneous intensity distribution over the image and that the image may have an uneven background.

2 Materials

1. *Hardware test bed*: We carried out our tests on a Linux workstation with Intel Core i5 CPU, 2.67 GHz processor speed, and 4 GB RAM.
2. *MATLAB*: The algorithm has been implemented and tested with MATLAB versions 8.4 (2014a) and 9.1 (2016b) on Linux.
3. *3D multicolorDNAFISHimages*: An 8-bit 2D fluorescence cell image z-stack acquired counterstaining nuclei and one or more different genomic loci using specific probes (two or more channels). Example image stacks, used in this Chapter to test the code, show the DAPI (blue) and the genomic regions on chr4 (green), chr10 (violet), and chr8 (red), and have 15 slices with an interslice distance of 0.21 μm and a pixel size of 0.065 μm in x and y. You can download the example image stacks from this repository: https://github.com/talisman4/methods-in-molecular-biology

   Create a directory "images" in your current working directory and put all images there.
4. *Nuclei segmentation MATLAB mex function*: For the segmentation of cell nuclei, we use an algorithm that implements one of the most efficient minimization methods for a nonconvex segmentation problem [2, 4, 5]. To this purpose, we use a C code to be turned in MEX file so that it can be called directly from MATLAB:

   Download the file *ac_mex.c* from https://github.com/talisman4/methods-in-molecular-biology to your current working directory.

   Start MATLAB and run.

```
mex ac_mex.c
```

5. *H-dome transform MATLAB function*: A method to detect the local regional maxima (as the bright objects in biological images) is based on a mathematical morphology method: the h-dome transformation [3]. We use the h-dome transform function (*hdTransform2*) that is part of the code package released as a supplementary material for the article [6]. Download the file *hdTransform2.m* from https://github.com/talisman4/methods-in-molecular-biology to your current working directory.
6. *Plot a 3D array using patch function*: Download the file *PATCH_3Darray.m* from https://www.mathworks.com/matlabcentral/fileexchange/28497-plot-a-3d-array-using-patch to your current working directory.

3 Methods

We show how our algorithm can be carried out in MATLAB environment. The algorithm aims to detect exactly two 3D spots for each fluorescent channel. This detection is performed through two main stages (Subheadings 3.1 and 3.2): 2D segmentation of DNA FISH spots for each slice of the stack and 3D reconstruction. Then for each nucleus, the algorithm checks the number of identified spots for each fluorescent channel. If there are exactly two spots for each fluorescent channel, that nucleus is deemed suitable for the analysis and spatial measurements are performed (Subheading 3.3). The presented algorithm concerns the case of four channels but can easily be adapted to the case of less or more than four channels. Figure 1 represents a flowchart with steps of the algorithm.

3.1 2D Segmentation of DNA FISH Spots

In the first stage, we first need to detect nucleus regions that enclose DNA FISH signals. To this purpose, we use the algorithm described in Subheading 2 to segment nuclei. Then, subsequently, DNA FISH spot detection is performed by an h-dome transformation-based algorithm followed by two threshold operations on the dome-like structures. The h-dome algorithm has four parameters, related to object size, that need to be tuned based on the data in order to obtain effective results.

Suppose we want to analyze the image stacks stored in the repository https://github.com/talisman4/methods-in-molecular-biology, let us show how this first stage can be easily accomplished for each slice of each stack.

The algorithm can be sketched as follows:

1. Read the images and convert them to grayscale (*see* Fig. 2a, b).
2. Make the partition of DAPI cell image in nucleus regions and background: initialize the minimizer (*see* **Note 1**), set the vector

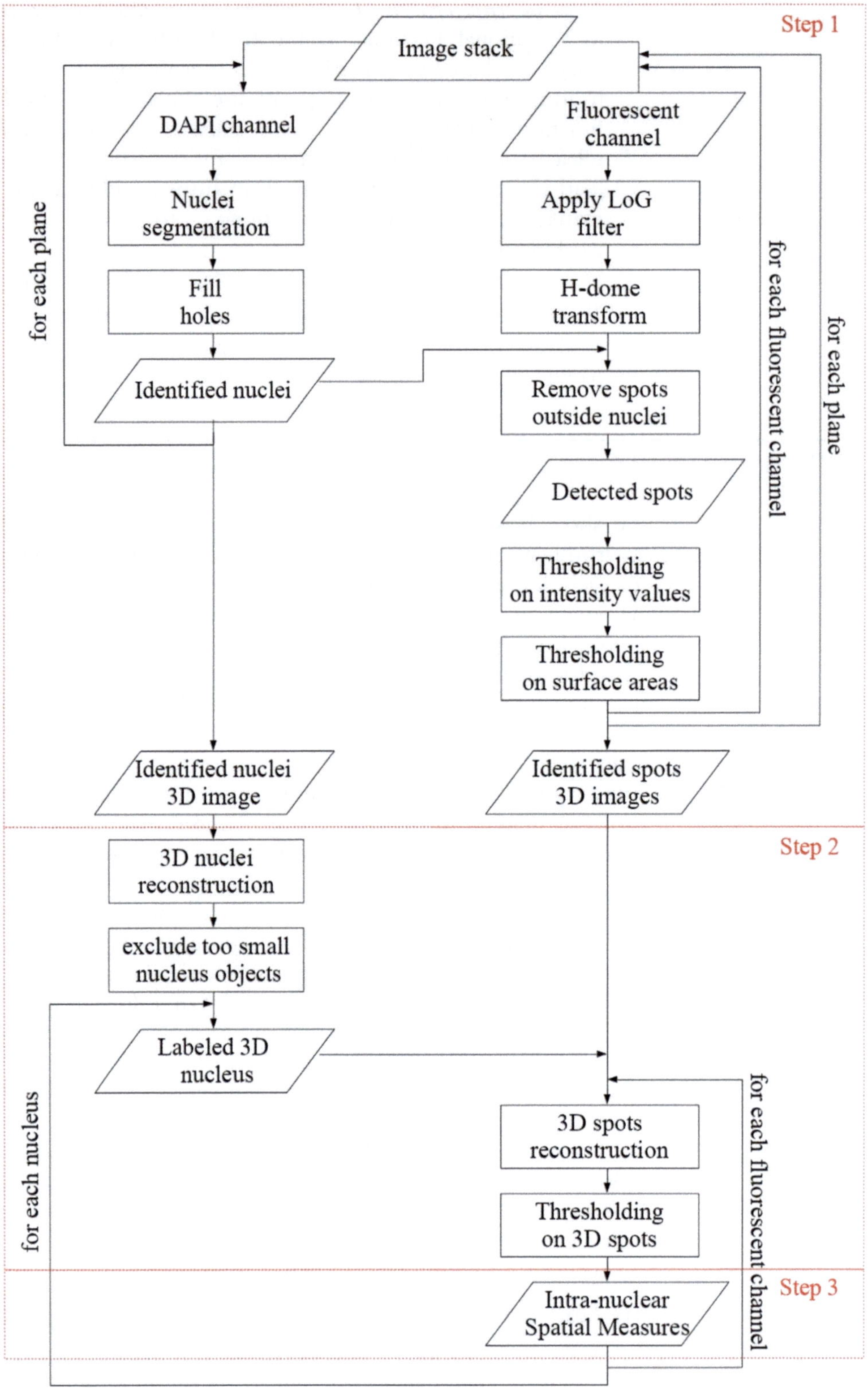

Fig. 1 Flowchart with steps of the MATLAB algorithm for the analysis of 3D multicolor FISH

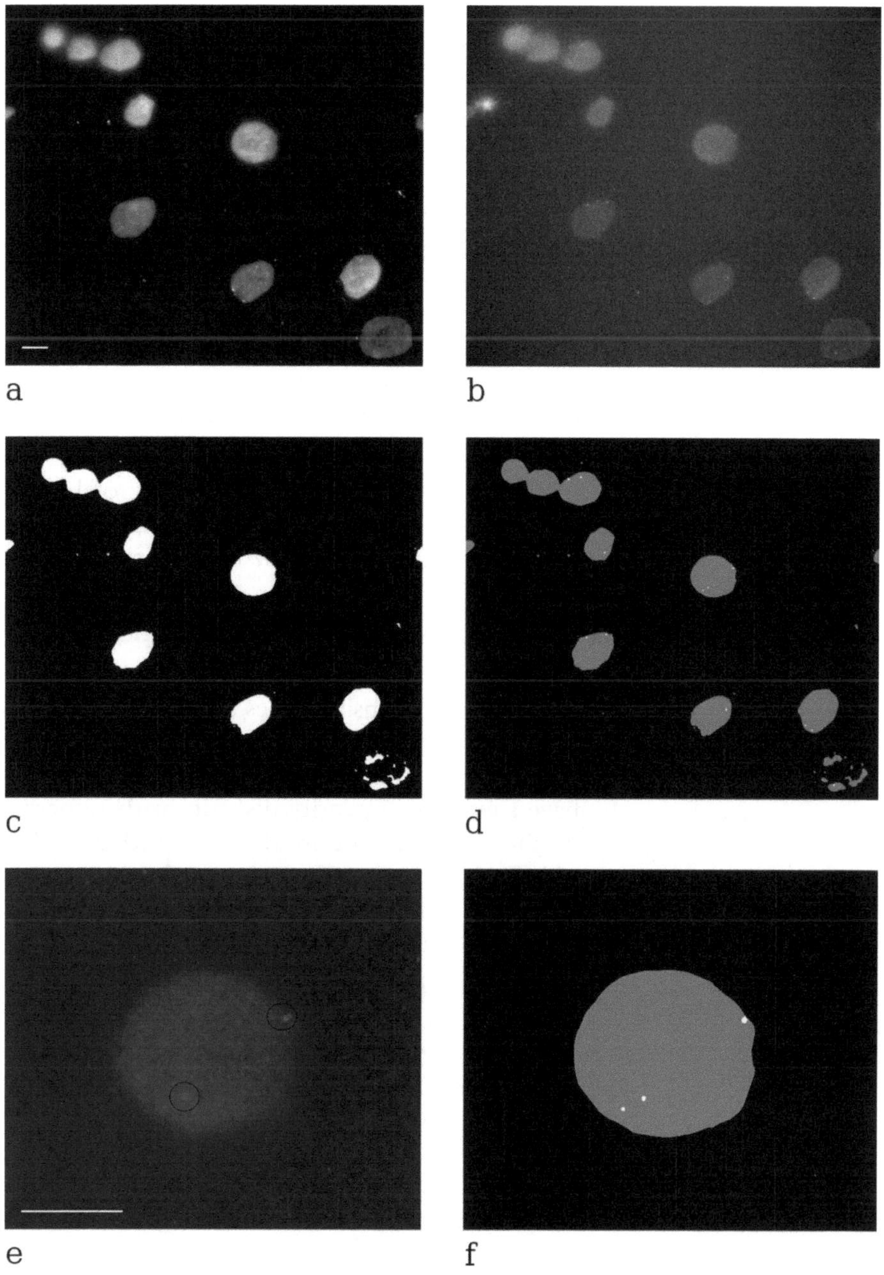

Fig. 2 Representative images of 2D fish spot segmentation stage (middle slices): grayscale DAPI (blue) image (**a**); grayscale chr4 (green) image (**b**); partition of DAPI image in nucleus regions and background (**c**); "trinary" segmented images (regions are segmented as background [black colored], nuclei [gray colored], and fish spots [white colored]) (*see* **Note 7**) (**d**); detail of a single cell with manual annotation (**e**); detail of a single cell with the detected spots (**f**). Bar, 10 μm

of parameters for the active contour function (*see* **Note 2**), and call the *ac_mex* function. Output image highlights foreground objects (nucleus regions, *see* Fig. 2c).

3. Invert colors in nucleus region image (*see* **Note 3**).
4. Fill holes in nucleus regions.
5. For each fluorescent channel:
   (a) Filter the channel image applying the Laplacian of Gaussian (LoG) operator (*fspecial* MATLAB function) of size (*win_size*) 13 and standard deviation 7 (*see* **Note 4**). This enhances the signal in the areas where objects (spots) are present. First, convert the input image to type "double" before applying *imfilter* (*see* **Note 5**).
   (b) Apply the h-dome transformation that extracts bright structures by cutting off the intensity of height h from the top, around local intensity maxima. We use $h = 0.5$ with a neighborhood size (*nb_size*) of 19×19 (*see* **Note 6**).
   (c) Remove spots outside the nucleus regions.
   (d) Perform a thresholding on h-dome image that excludes pixels whose intensity values are below a threshold. The threshold is 1.96 standard deviations above the mean of dome intensity values. We therefore assume that spot areas have significant intensity disparity with respect to other bright areas present in the cell nucleus. We use the MATLAB function *find* to get the nonzero intensity values, and then we compute their mean and standard deviation by using *mean* and *std*.
   (e) Apply a threshold classification based on the surface areas of the spots, in order to discard too small/big objects which are probably just noise. First, compute the label matrix[1] that contains labels for the 8-connected[2] objects (spots) found in the previously thresholded image. Then retrieve area values for all spots and filter out spots smaller than a surface area of 25 or bigger than a surface area of 225 (*see* **Note 6**), obtaining an image that highlights 2D spots (*see* Fig. 2d–f).
6. Update the 3D arrays used for the 3D reconstruction of nuclei and spots.

[1] The pixels labeled 0 are the background. The pixels labeled 1 make up one object; the pixels labeled 2 make up a second object; and so on.

[2] Pixels are connected if their edges or corners touch.

Resulting MATLAB code follows:

```
% initialize the variables used to set the filename of the image to be read
filenameformat = ['%s' '%03d_z%03d_c%03d_8bitrgb.tif'];
population = 'images/CN6_chr4-chr10-rp11';
FluoC=struct;

% start a for loop on the number of slices
for plane = 1:15
  % read the images and convert them to grayscale
  ImageFIlename=sprintf( filenameformat, population, 1, plane, 4 );
  ImTemp=imread(ImageFIlename);
  ImDapi = double(rgb2gray(ImTemp));
  ImageFIlename=sprintf( filenameformat, population, 1, plane, 1 );
  ImTemp=imread(ImageFIlename);

  % FluocC(1) refers to green channel
  FluoC(1).Im = double(rgb2gray(ImTemp));
  ImageFIlename=sprintf( filenameformat, population, 1, plane, 2 );
  ImTemp=imread(ImageFIlename);

  % FluocC(2) refers to red channel
  FluoC(2).Im = double(rgb2gray(ImTemp));
  ImageFIlename=sprintf( filenameformat, population, 1, plane, 3 );
  ImTemp=imread(ImageFIlename);

  % FluocC(3) refers to violet channel
  FluoC(3).Im = double(rgb2gray(ImTemp));

  % initialization of minimizer
  pfu0 = InitFunU(ImDapi, size(ImDapi));

  % setting vector of parameters for active contour function
  SegParameters = [plane; size(ImDapi,1); size(ImDapi,2); 0.01; 1; 30; 30];

  % detect nucleus regions
  % ac_mex output list:
  % regIm highlights foreground objects
```

```
% cntIm shows object contours
% pfu is the segmented image
% fMeanIn is the mean fluorescence intensity value of objects
% fMeanOut is the mean fluorescence intensity value of background
% den1 is the number of object pixels
% den2 is the number of background pixels
[regIm, cntIm, pfu, fMeanIn, fMeanOut, den1, den2] = ...
    ac_mex(ImDapi, FluoC(1).Im, FluoC(2).Im, FluoC(3).Im, ...
           pfu0, double(SegParameters));
% invert colors in regIm
regIm2 = 255 - regIm;
% fill holes in nucleus regions
regIm2 = imfill(regIm2,'holes');

% start a for loop on the number of fluorescent channels
for fc=1:3
  % convert the input image to type double
  ImPrime = double(FluoC(fc).Im);
  ImPrime = ImPrime/max(ImPrime(:));

  % apply the log operator
  H = -1*fspecial('log', 13, 7);

 % filter fluorescent channel image
  filtImage = imfilter(ImPrime,H,'replicate');

  % scale filtered image values to interval [0,1]
  filtImage = filtImage - min(filtImage(:));
  filtImage = filtImage/max(filtImage(:));

  % apply h-dome transformation
  hdImage = hdTransform2(filtImage, 0.5, 19);

  % remove spots outside nucleus regions
  hdImage(regIm2 == 0) = 0;

  % compute mean and standard deviation of nonzero intensity values
  [rows,cols,vals] = find(hdImage);
  mean_spot = mean(vals);
  std_spot = std(vals);

  % compute the threshold
  thresh =mean_spot + 1.96*std_spot;

  % perform a thresholding on h-domes image
  clusterImage = hdImage;
  clusterImage(hdImage < thresh) = 0;
```

```
    % compute the label matrix
    LabelMatrix = bwlabel(clusterImage);

    % retrieve area values for all spots
    stats = regionprops(LabelMatrix, 'Area');
    area_values = [stats.Area];

    % filter out spots
    idx = find((area_values >=25) & (area_values <= 225));
    bw = ismember(LabelMatrix, idx);
    FluoC(fc).imSpt = clusterImage;

    % FluoC(fc).imSpt highlights 2D spots
    FluoC(fc).imSpt(bw == 0) = 0;
  end

  % update the 3D arrays used for the 3D reconstruction of nuclei and spots
  nuclei_vol(:,:,plane) = regIm2;
  nuclei_vol(:,:,plane) = imclearborder(nuclei_vol(:,:,plane),8);
  for fc=1:3
    FluoC(fc).Spt_vol(:,:,plane)=FluoC(fc).imSpt>0;
  end
end
```

At the end of the first stage, we have obtained the 3D images *nuclei_vol*, *FluoC(1).Spt_vol*, *FluoC(2).Spt_vol*, *FluoC(3).Spt_vol* by stacking the 2D segmented slices representing the nucleus regions and spots (*see* Fig. 3a). Therefore, they contain the positions of the detected spots and nuclei from all slices.

3.2 3D Reconstruction

In the second stage, we carry out the 3D reconstruction and identification of nuclei and spots.

The algorithm can be sketched as follows:

1. Remove small objects from the 3D nuclei image: removes all objects (connected components) that have fewer than 650 pixels (*see***Note 8**) with a pixel connectivity of 6.[3]
2. Create the conversion of the cleaned 3D nuclei image in binary form: black values become 0, white values become 1.

[3] Pixels are connected if their faces touch.

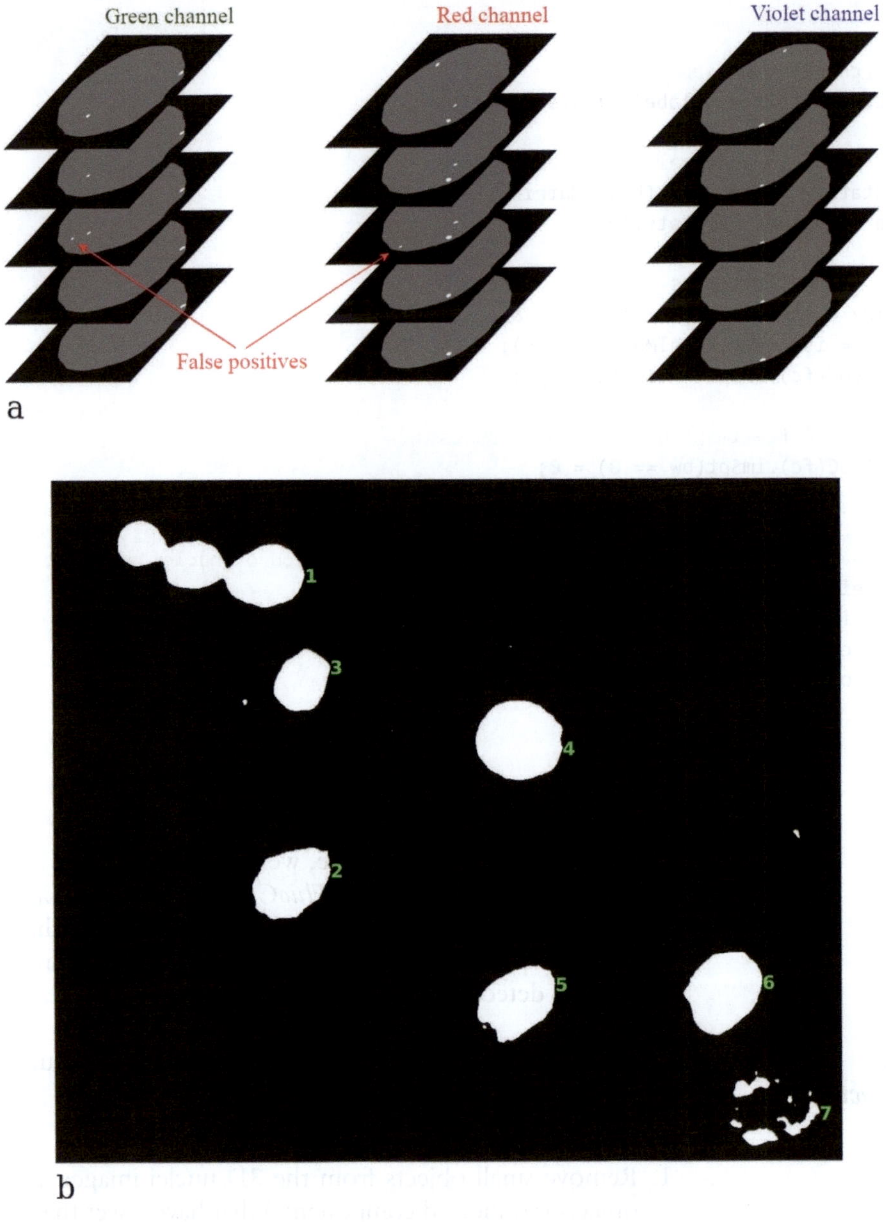

Fig. 3 Segment from slice 5 to slice 9 in the 3D images of identified spots (false-positives on single slices that are not connected to anything above and below do not give rise to any 3D spot) (**a**); image of cell nuclei along with the corresponding labels (*see* **Note 9**) (**b**)

3. Obtain the 3D reconstructions of nuclei through the connected component algorithm (*bwconncomp* MATLAB function, using a connectivity of 26[4]). The 3D nuclei are then labeled by applying the *labelmatrix* MATLAB function so they can be easily separated each from the others.
4. Exclude nuclei whose volume is less than 10% of mean volumes and that are present in only one slice, which are just noise (*see* Fig. 3b for an image of cell nuclei along with the corresponding labels).
5. For each detected nucleus:
   (a) Crop 3D spot images belonging to the nucleus.
   (b) Compute the nuclear centroid, nuclear isosurface, and nuclear isosurface end-caps from reconstructed 3D nucleus data.
   (c) For each fluorescent channel:
   - Obtain the 3D reconstructions of spots through the connected component algorithm (*bwconncomp* MATLAB function, using a connectivity of 6).
   - Perform a threshold operation on the 3D spots to obtain the more significant ones: first, select spots that have more than 75 pixels and that are present in at least three slices (*see***Note 10**), and then keep all the selected spots whose volume is above the threshold of the standard deviation of the volumes.

[4] Pixels are connected if their faces, edges, or corners touch.

Resulting MATLAB code follows:

```
% remove small objects from 3D nuclei image
nuclei_vol = bwareaopen(nuclei_vol, 650, 6);

% convert 3D nuclei image in binary form
PSI = boolean(nuclei_vol);

% 3D reconstructions of nuclei
nuclei_CC = bwconncomp(PSI);

% label 3D nuclei
nuclei_L = labelmatrix(nuclei_CC);

% compute volume for each 3D nucleus object
nuclei_dim = [];
nuclei_h = [];
for k = 1:nuclei_CC.NumObjects
  nuclei_dim(end+1) = size(nuclei_CC.PixelIdxList{k},1);
  [x,y,z] = ind2sub(size(PSI),nuclei_CC.PixelIdxList{k});
  nuclei_h(end+1) = max(z)-min(z);
end

% exclude too small nucleus objects
nuclei_mean_dim = mean(nuclei_dim);
idx_PSI = find(nuclei_dim > .1*nuclei_mean_dim & nuclei_h > 1);
mask_PSI = ismember(nuclei_L, idx_PSI);
nuclei_L(mask_PSI == 0) = 0;

% M is the number of identified nuclei
M = size(idx_PSI,2);

% PSI2(:,:,:,k), with k from 1 to M, is the nucleus
% corresponding to the k-th label
PSI2 = false([size(nuclei_vol),M]);
for k=1:M
  PSI2(:,:,:,k) = (nuclei_L==idx_PSI(k));
end
```

```
fieldsize = size(PSI2);

% NCL holds pixel data regarding each nucleus
for n=1:M
  [i] = find(PSI2(:,:,:,n));
  [x,y,z] = ind2sub(size(PSI2),i);
  NCL{n} = struct;

  % n-th nucleus
  NCL{n}.Nucleus = PSI2( min(x):max(x), min(y):max(y), min(z):max(z), n );

  % Spt_data holds pixel data regarding detected spots within nucleus n
  for fc=1:3
    FluoC(fc).Spt_data{n} = cell(0);
    % crop 3D spot images belonging to the n-th nucleus;
    FluoC(fc).Spt_data{n} = ...
      FluoC(fc).Spt_vol( min(x):max(x), min(y):max(y), min(z):max(z)) .* ...
      NCL{n}.Nucleus;
    FluoC(fc).Spt_data{n} = bwareaopen(FluoC(fc).Spt_data{n},17,6);
  end
end

% if there is more than one nucleus object remove all but the biggest
for n=1:M

  temp = bwconncomp(NCL{n}.Nucleus,6);

  if temp.NumObjects>1
    thamax = 0; thamaxk = 0;

    for k=1:temp.NumObjects
      if thamax<size(temp.PixelIdxList{k},1)
        thamax = size(temp.PixelIdxList{k});
        thamaxk = k;
      end
    end

    for k=1:temp.NumObjects
      if k~=thamaxk
        NCL{n}.Nucleus(temp.PixelIdxList{k}) = 0;
      end
    end

    for fc=1:3
      % remove spot outside of remaining obj
      FluoC(fc).Spt_data{n}=FluoC(fc).Spt_data{n}.*NCL{n}.Nucleus;
```

```
      end
    end
  end

  % initialize some cell arrays used to store the positions of the nuclear
  % centroids and of the detected spots and nuclear isosurface data
  C = cell(0); Nucleus_surface = cell(0); Nucleus_caps = cell(0);
  for fc=1:3
    FluoC(fc).Spt = cell(0);
    FluoC(fc).Spt_selected = cell(0);
  end
  % start a for loop on the number of detected nuclei
  for n=1:size(NCL,2)
    % compute nuclear isosurface
    Nucleus_surface{n} = isosurface(permute(NCL{n}.Nucleus,[2,1,3]),.5,'noshare');
    % compute nuclear isosurface end-caps
    Nucleus_caps{n} = isocaps(permute(NCL{n}.Nucleus,[2,1,3]));
    % compute the nuclear centroid
    [Cx,Cy,Cz] = ind2sub(size(NCL{n}.Nucleus),find(NCL{n}.Nucleus));
    C{n} = [mean(Cx),mean(Cy),mean(Cz)];
    for fc=1:3
      % 3D reconstructions of spots
      FluoC(fc).Spt{n} = bwconncomp(FluoC(fc).Spt_data{n},6);
      % threshold operation on the 3D spots
      if FluoC(fc).Spt{n}.NumObjects > 0
        % the field pick of the struct Spt_selected keeps track
        % of the final selected spots
        FluoC(fc).Spt_selected{n}=struct( ...
          'volume', zeros(FluoC(fc).Spt{n}.NumObjects,1), ...
          'height', zeros(FluoC(fc).Spt{n}.NumObjects,1), ...
          'max_height', zeros(FluoC(fc).Spt{n}.NumObjects,1), ...
          'min_height', zeros(FluoC(fc).Spt{n}.NumObjects,1), ...
          'pick', zeros(FluoC(fc).Spt{n}.NumObjects,1), ...
          'sortIndexes', zeros(FluoC(fc).Spt{n}.NumObjects,1) ...
          );
        for k=1:FluoC(fc).Spt{n}.NumObjects
          [x,y,z] = ind2sub(size(NCL{n}.Nucleus), ...
                    FluoC(fc).Spt{n}.PixelIdxList{k});

          FluoC(fc).Spt{n}.Centroid{k} = [mean(x),mean(y),mean(z)];

          if ( size(FluoC(fc).Spt{n}.PixelIdxList{k},1) >= 75 && ...
               (max(z)-min(z)) >= 2 )
            FluoC(fc).Spt_selected{n}.volume(k) = ...
              size(FluoC(fc).Spt{n}.PixelIdxList{k},1);
            FluoC(fc).Spt_selected{n}.height(k) = max(z)-min(z);
            FluoC(fc).Spt_selected{n}.max_height(k) = max(z);
```

```
          FluoC(fc).Spt_selected{n}.min_height(k) = min(z);
          FluoC(fc).Spt_selected{n}.pick(k)=1;
        end
      end
      if (size(nonzeros(FluoC(fc).Spt_selected{n}.pick),1) > 2)
        [ ...
          sortVolumes ...
          FluoC(fc).Spt_selected{n}.sortIndexes ...
        ] = sort(FluoC(fc).Spt_selected{n}.volume,'descend');

        meanVolumes = mean(nonzeros(( ...
                      FluoC(fc).Spt_selected{n}.pick .* ...
                      FluoC(fc).Spt_selected{n}.volume)));

        stdVolumes = std(nonzeros(( ...
                      FluoC(fc).Spt_selected{n}.pick .* ...
                      FluoC(fc).Spt_selected{n}.volume)));

        for k=1:FluoC(fc).Spt{n}.NumObjects
          if (FluoC(fc).Spt_selected{n}.pick(k) == 1)
            if (size(FluoC(fc).Spt{n}.PixelIdxList{k},1) <= stdVolumes)
              FluoC(fc).Spt_selected{n}.pick(k)=0;
            end
          end
        end
        if (size(nonzeros(FluoC(fc).Spt_selected{n}.pick),1) == 1)
          FluoC(fc).Spt_selected{n}.pick( ...
            FluoC(fc).Spt_selected{n}.sortIndexes(2)) = 1;
        end
      end
    end
  end
end
```

At the end of the second stage, *C{n}* contains the nuclear centroid for the nucleus*n*, and *Nucleus_surface{n}.vertices* and *Nucleus_caps{n}.vertices* contain the vertices of the isosurface extracted from the reconstructed 3D data of the nucleus*n*, while *FluoC(1).Spt{n}*, *FluoC(2).Spt{n}*, *FluoC(3).Spt{n}* contain the positions of the points corresponding to the 3D reconstruction of the detected spots within the nucleus*n*. *FluoC(1).Spt_selected{n}*, *FluoC(2).Spt_selected{n}*, *FluoC(3).Spt_selected{n}* contain relevant data for the more significant detected spots within the nucleus*n* (*see* Fig. 4).

3.3 Intranuclear Spatial Measurements

For each nucleus, the algorithm checks the number of identified spots for each fluorescent channel. If there are exactly two spots for each fluorescent channel, that nucleus is deemed suitable for the

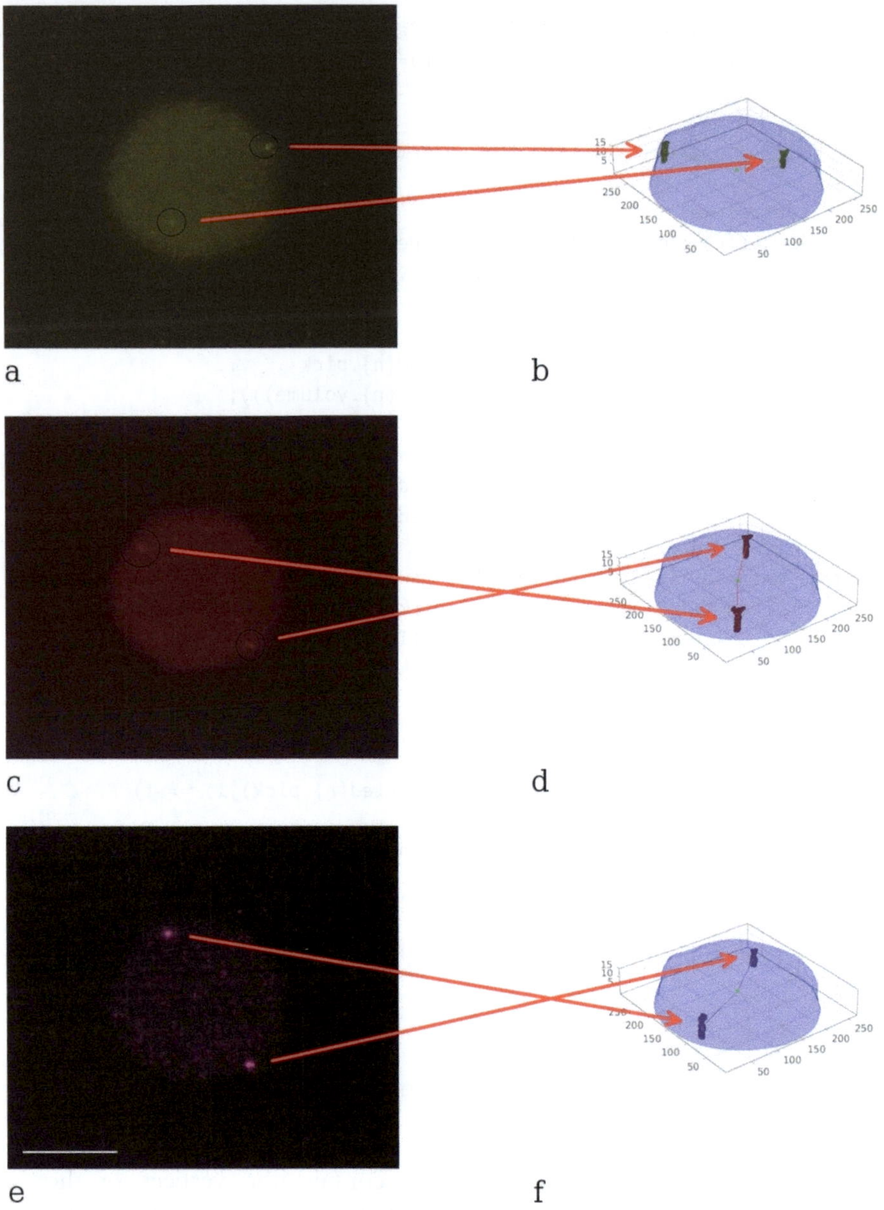

Fig. 4 Representative images of 3D reconstruction stage: detail of the cell labeled “4” with “green” spots manually annotated (middle plane) (**a**); 3D reconstruction of the nucleus “4” with “green” spots (*see***Note 11**) carried out by our procedure (**b**); detail of the cell labeled “4” and “red” spots manually annotated (middle plane) (**c**); 3D reconstruction of the nucleus “4” and “red” spots (**d**); detail of the cell labeled “4” with “violet” spots manually annotated (middle plane) (**e**); 3D reconstruction of the nucleus “4” and “violent” spots (**f**). Bar, 10 μm

analysis: it computes the distances from the centroid of any spot from the centroid of any other spot, the distances of the centroid of any spot from the nuclear periphery and the nuclear centroid, and the distance from the nuclear periphery to the nuclear centroid.

1. Compute the distances from the centroid of any spot from the centroid of any other spot: we build the 3×6 matrix of spot centroids (*CentroidMatrix*) and use the MATLAB function *dist* to get the 6×6 matrix of distances for each nucleus suitable for the analysis.

```
xyscale = 0.065; zscale = 0.2;
DistTable=cell(0);
CentroidMatrix=cell(0);
for n=1:size(NCL,2)
  if (FluoC(1).Spt{n}.NumObjects > 0 && FluoC(2).Spt{n}.NumObjects > 0 && ...
      FluoC(3).Spt{n}.NumObjects > 0)

    if (size(nonzeros(FluoC(1).Spt_selected{n}.pick),1) == 2 && ...
        size(nonzeros(FluoC(2).Spt_selected{n}.pick),1) == 2 && ...
        size(nonzeros(FluoC(3).Spt_selected{n}.pick),1) == 2)

      CentroidMatrix{n}=[];
      ColDistTable_idx=0;
      for fc=1:3
        for k=1:FluoC(fc).Spt{n}.NumObjects
          if (FluoC(fc).Spt_selected{n}.pick(k) == 1)
            ColDistTable_idx=ColDistTable_idx+1;
            CentroidMatrix{n}(:,ColDistTable_idx) = ...
              FluoC(fc).Spt{n}.Centroid{k} .* [xyscale xyscale zscale];
          end


        end
      end
      DistTable{n} = dist(CentroidMatrix{n});
    end
  end
end
```

For each suitable nucleus*n*, *DistTablet{n}* contains the 6×6 matrix of distances. The inter-spot distance matrix for the nucleus 4 is shown in Table 1.

2. Compute the distances of the centroid of any spot from the nuclear periphery: for each spot centroid (column of *Centroid-Matrix*), by using the MATLAB function *pdist*, we compute the minimum of distances from any vertex of nuclear periphery (row of *Nucleus_all_vertices*).

Table 1
Inter-spot distances for the cell nucleus labeled "4"

	Green spot 1	Green spot 2	Red spot 1	Red spot 2	Violet spot 1	Violet spot 2
Green spot 1	0	12.7589	13.9086	9.2127	11.5947	10.1526
Green spot 2	12.7589	0	9.6396	7.92	11.3493	7.9867
Red spot 1	13.9086	9.6396	0	15.3005	3.7247	15.7823
Red spot 2	9.2127	7.92	15.3005	0	15.2085	1.0856
Violet spot 1	11.5947	11.3493	3.7247	15.2085	0	15.8577
Violet spot 2	10.1526	7.9867	15.7823	1.0856	15.8577	0

```
Nucleus_all_vertices = [Nucleus_surface{n}.vertices;
Nucleus_caps{n}.vertices];
min_dist=cell(0);
for n=1:size(NCL,2)
  if (FluoC(1).Spt{n}.NumObjects > 0 && FluoC(2).Spt{n}.NumObjects > 0 && ...
      FluoC(3).Spt{n}.NumObjects > 0)

    if (size(nonzeros(FluoC(1).Spt_selected{n}.pick),1) == 2 && ...
        size(nonzeros(FluoC(2).Spt_selected{n}.pick),1) == 2 && ...
        size(nonzeros(FluoC(3).Spt_selected{n}.pick),1) == 2)

      dist_from_nucleus_verts=[];

      for col = 1:size(CentroidMatrix{n},2)
        for row = 1:size(Nucleus_all_vertices,1)
          dist_from_nucleus_verts(row) = pdist([CentroidMatrix{n}(:,col)';
          Nucleus_all_vertices(row,:).*[xyscale xyscale zscale]]);
        end
        min_dist{n}(col) = min(dist_from_nucleus_verts);
      end

      min_dist{n};
    end
  end

end
```

For each suitable nucleus*n*, *min_dist{n}* contains the distances of the centroid of any spot from the nuclear periphery.

3. Compute the distances of the centroid of any spot from the nuclear centroid.

```
dist_from_centroid=cell(0);
for n=1:size(NCL,2)
  if (FluoC(1).Spt{n}.NumObjects > 0 && FluoC(2).Spt{n}.NumObjects > 0 && ...
      FluoC(3).Spt{n}.NumObjects > 0)

    if (size(nonzeros(FluoC(1).Spt_selected{n}.pick),1) == 2 && ...
        size(nonzeros(FluoC(2).Spt_selected{n}.pick),1) == 2 && ...
        size(nonzeros(FluoC(3).Spt_selected{n}.pick),1) == 2)

      for col = 1:size(CentroidMatrix{n},2)
        dist_from_centroid{n}(col) = pdist([CentroidMatrix{n}(:,col)';
        [C{n}(1),C{n}(2),C{n}(3)].*[xyscale xyscale zscale]]);;
      end

      dist_from_centroid{n};
    end
  end
end
```

For each suitable nucleus*n*, *dist_from_centroiid{n}* contains the distances of the centroid of any spot from the nuclear centroid.

4. Compute the minimum and maximum distances from the nuclear centroid to the nuclear periphery: we compute the minimum and the maximum of distances from any vertex of nuclear periphery (row of *Nucleus_all_vertices*) to the nuclear centroid.

```
min_nucleous_dist=cell(0);
max_nucleous_dist=cell(0);

for n=1:size(NCL,2)
  if (FluoC(1).Spt{n}.NumObjects > 0 && FluoC(2).Spt{n}.NumObjects > 0 && ...
      FluoC(3).Spt{n}.NumObjects > 0)

    if (size(nonzeros(FluoC(1).Spt_selected{n}.pick),1) == 2 && ...
        size(nonzeros(FluoC(2).Spt_selected{n}.pick),1) == 2 && ...
        size(nonzeros(FluoC(3).Spt_selected{n}.pick),1) == 2)
      dist_from_centroid_verts=[];
      for row = 1:size(Nucleus_all_vertices,1)
        dist_from_centroid_verts(row) = ...
          pdist([[C{n}(1),C{n}(2),C{n}(3)] .* [xyscale xyscale zscale];
          Nucleus_all_vertices(row,:) .* [xyscale xyscale zscale]]);
      end
      min_nucleous_dist{n} = min(dist_from_centroid_verts);
      max_nucleous_dist{n} = max(dist_from_centroid_verts);
    end
  end
end
```

For each suitable nucleus$n$, *min_nucleus_dist{n}* and *max_nucleus_dist{n}* contain the minimum and maximum distances from the nuclear periphery to the nuclear centroid.

4 Notes

1. Following [7], we choose as initial minimizer function u, the image to be segmented with intensity values scaled between 0 and 1. A listing of our *InitFunU.m* MATLAB function is given below.

```
function [pfu] = InitFunU(img, iNy, iNx)
  pfu = abs(img);
  maxI = double(max(img(:)));
  minI = double(min(img(:)));
  pfu = (pfu-minI)/(maxI - minI); % pfu has values in [0,1]
```

2. The parameters 0.01, 1, 30, 30 are, respectively, the regularization parameter λ, the Bregman distance parameter μ, maximum Bregman iterations, and maximum Gauss Seidel iterations. Please refer to [5] for the meaning of these parameters.
3. *regIm* as returned by *ac_mex* has black nuclei regions and white background, while we want nucleus regions to be white.
4. The standard deviation can be related to the size of the objects to be detected and in our experiments was equal to 7 pixels. The size has been chosen to be 13 in the tuning parameter phase (*see* **Note 6**).
5. The *imfilter* output image has the same data type as the input image. The *imfilter* function computes the value of each output pixel using double-precision, floating-point arithmetic. If the result exceeds the range of the input data type, the *imfilter* function truncates the result to the allowed range of the data type. To avoid any sort of truncation, we convert the image to type "double" before calling *imfilter*.
6. The parameter values have a significant effect on the detection accuracy and need to be tuned specifically for your data. In order to find the optimal parameters for the h-dome transformation, as well as the optimal thresholds for the thresholding operations, we performed our algorithm on a real image dataset, for which we obtained expert manual spot annotations (used as ground truth) to compare with. We suggest to refer to these parameter value ranges: h = [0.05:0.0.05:1], *win_size* = [3:2:13], *nb_size* = [3:2:19] (*see* also ref. 6). Optimal parameters and thresholds were chosen in order to maximize

the number of true positives (spots that were both manually annotated and detected by the algorithm) and minimize the true negatives (spots manually annotated but not detected by the algorithm).

7. Use the following code lines to produce a "trinary" segmented image of the middle plane:

```
regImG=zeros(size(nuclei_vol(:,:,7)));
regImG( nuclei_vol(:,:,7) == 255 & FluoC(1).Spt_vol(:,:,7) > 0 ) = double(255);
regImG( nuclei_vol(:,:,7) == 255 & FluoC(1).Spt_vol(:,:,7) == 0 ) = double(100);
figure; imshow(regImG,[])
```

8. This number can vary according to the resolution of the images being analyzed. In all the datasets we analyzed, 3D nuclei were usually made of a number of pixels far greater than 650. Therefore, removing objects with fewer than 650 pixels serves as a means to remove noise.

9. Use the following code lines to produce an image of cell nuclei along with the corresponding labels.

```
figure; imshow((nuclei_vol(:,:,floor(size(nuclei_vol,3)/2))));
for h=1:M
  [ti,tj] = find(PSI2(:,:,floor(size(PSI2,3)/2),h));
  if (size(ti,1) == 0) | (size(tj,1) == 0)
    for plane=1:size(PSI2,3)
      [ti,tj] = find(PSI2(:,:,plane,h));
      if (size(ti,1) ~= 0) && (size(tj,1) ~= 0)
        break
      end
    end
  end
  text(tj(floor(size(tj,1))), ti(floor(size(ti,1))), int2str(h), ...
    'FontWeight', 'bold','Color','g','FontSize',13);
end
```

10. By selecting only connected components that are spread out across more slices, false-positives on single slices, which are not connected to anything above and below, do not give rise to any 3D spot.

11. Suppose we want to produce an image of the 3D reconstruction of the cell nucleus labeled "4" and of green/red/violet spots. Use the following code lines:

```
ch_color(1,:)=[102,204,0]./255;
ch_color(2,:)=[255,51,51]./255;
ch_color(3,:)=[178,102,255]./255;

for fc=1:3
  figure(fc); plot3(C{4}(1),C{4}(2),C{4}(3),'go','MarkerSize',2,'LineWidth',2);
  set(gcf,'DoubleBuffer','on');
  for k=1:FluoC(fc).Spt{4}.NumObjects
    if (FluoC(fc).Spt_selected{4}.pick(k) == 1)
      temp = zeros(size(NCL{4}.Nucleus));
      temp(FluoC(fc).Spt{4}.PixelIdxList{k}) = 1;
      PATCH_3Darray(temp,ch_color(fc,:));
      line([C{4}(1),FluoC(fc).Spt{4}.Centroid{k}(1)], ...
           [C{4}(2),FluoC(fc).Spt{4}.Centroid{k}(2)], ...
           [C{4}(3),FluoC(fc).Spt{4}.Centroid{k}(3)], ...
           'Color', [102,204,0]./255);
      p_s = patch(Nucleus_surface{4});
      set(p_s,'FaceColor',[0,0,175]./255,'EdgeColor','none','FaceAlpha',.2);
      p_c = patch(Nucleus_caps{4});
      set(p_c,'FaceColor',[0,0,175]./255,'EdgeColor','none','FaceAlpha',.2);
      set(gca,'projection','perspective');
      daspect([1/.065 1/.065 1/.21]), view(3); box('on');
      axis('tight'); axis('vis3d'); grid('on');
    end
  end
end
```

References

1. Cortesi A, Pesant M, Sinha S, Marasca F, Sala E, Gregoretti F, Antonelli L, Oliva G, Chiereghin C, Soldà G, Bodega B (2019) 4q-D4Z4 chromatin architecture regulates the transcription of muscle atrophic genes in facioscapulohumeral muscular dystrophy. Genome Res 29(6):883–895
2. Chan TF, Esedoglu S, Nikolova M (2006) Algorithms for finding global minimizers of image segmentation and denoising models. SIAM J Appl Math 66(5):1632–1648
3. Vincent L (1993) Morphological grayscale reconstruction in image analysis: applications and efficient algorithms. IEEE Trans Image Process 2(1):176–201
4. Goldstein T, Bresson X, Osher S (2010) Geometric applications of the split Bregman method: segmentation and surface reconstruction. J Sci Comput 45(1–3):272–293
5. Antonelli L, De Simone V (2018) Comparison of minimization methods for nonsmooth image segmentation. Commun Appl Indust Math 9 (1):68–86
6. Ruusuvuori P, Äijö T, Chowdhury S, Garmendia-Torres C, Selinummi J, Birbaumer M, Dudley AM, Pelkmans L, Yli-Harja O (2010) Evaluation of methods for detection of fluorescence labeled subcellular objects in microscope images. BMC Bioinformat 11(248):1–17
7. Antonelli L, De Simone V, Di Serafino D (2016) On the application of the spectral projected gradient method in image segmentation. J Math Imaging Vis 54(1):106–116

INDEX

Beatrice Bodega and Chiara Lanzuolo (eds.), *Capturing Chromosome Conformation: Methods and Protocols*, Methods in Molecular Biology, vol. 2157, https://doi.org/10.1007/978-1-0716-0664-3,

I

L

M

N

P

Q

R

S

T

MIX
Papier aus verantwortungsvollen Quellen
Paper from responsible sources
FSC® C105338

If you have any concerns about our products,
you can contact us on
ProductSafety@springernature.com

In case Publisher is established outside the EU,
the EU authorized representative is:
Springer Nature Customer Service Center GmbH
Europaplatz 3, 69115 Heidelberg, Germany

Printed by Libri Plureos GmbH
in Hamburg, Germany